D1519589

MICROTECHNOLOGY AND MEMS

Springer
Berlin
Heidelberg
New York
Barcelona
Hong Kong
London
Milan
Paris
Singapore
Tokyo

MICROTECHNOLOGY AND MEMS

The series Microtechnology and MEMS comprises text books, monographs, and state-of-the-art reports in the very active field of microsystems and microtechnology. Written by leading physicists and engineers, the books describe the basic science, device design, and applications. They will appeal to researchers, engineers, and advanced students.

Mechanical Microsensors
By M. Elwenspoek and R. Wiegerink

M. Elwenspoek R. Wiegerink

Mechanical Microsensors

With 235 Figures

Professor Dr. Miko Elwenspoek
Dr. Remco Wiegerink

MESA Research Institute
Twente University of Technology
P.O. Box 217
7500 AE Enschede
The Netherlands
e-mail: m.elwenspoek@el.utwente.nl
r.j.wiegerink@el.utwente.nl

ISSN 1615-8326
ISBN 3-540-67582-5 Springer-Verlag Berlin Heidelberg New York

Library of Congress Cataloging-in-Publication Data

Elwenspoek, M. (Miko), 1948-
Mechanical microsensors / M. Elwenspoek, R. J. Wiegerink. p. cm. -- (Microtechnology and MEMS)
Includes bibliographical references and index.
ISBN 3540675825 (alk. paper)
1. Detectors -- Design and construction. 2. Microelectromechanical systems -- Design and construction. 3. Transducers. 4. Silicon. 5. Micromachining. I. Wiegerink, Remco J., 1964-
II. Title. III. Series.
TK 7875 .E49 2001
681'.2--dc21 00-057363

Springer-Verlag Berlin Heidelberg New York
a member of BertelsmannSpringer Science+Business Media GmbH

Printed in Germany

Typesetting: Camera-ready copy from author
Cover-concept: eStudio Calamar Steinen
Cover production: *design & production* GmbH, Heidelberg
Computer-to-plate and Printing: Saladruck, Berlin
Binding: Stürtz AG Universitätsdruckerei, Würzburg

Printed on acid-free paper SPIN: 10733906 57/3020 CU - 5 4 3 2 1 0 -

Preface

This book on mechanical microsensors is based on a course organized by the Swiss Foundation for Research in Microtechnology (FSRM) in Neuchâtel, Switzerland, and developed and taught by the authors. Support by FSRM is herewith gratefully acknowledged.

This book attempts to serve two purposes. First it gives an overview on mechanical microsensors (sensors for pressure, force, acceleration, angular rate and fluid flow, realized by silicon micromachining). Second, it serves as a textbook for engineers to give them a comprehensive introduction on the basic design issues of these sensors. Engineers active in sensor design are usually educated either in electrical engineering or mechanical engineering. These classical educational programs do not prepare the engineer for the challenging task of sensor design since sensors are instruments typically bridging the disciplines: one needs a rather deep understanding of both mechanics and electronics. Accordingly, the book contains discussion of the basic engineering sciences relevant to mechanical sensors, hopefully in a way that it is accessible for all colours of engineers. Engineering students in their 3rd or 4th year should have enough knowledge to be able to follow the arguments presented in this book. In this sense, this book should be useful as textbook for students in courses on mechanical microsensors (as is currently being done at the University of Twente).

At this place we wish to acknowledge colleagues who contributed in one way or the other to the text. This is in first place the whole micromechanics group at the University of Twente: Han Gardeniers, Theo Lammerink, Gijs Krijnen, Erwin Berenschot, Meint de Boer, Dick Ekkelkamp, Hans Hassink, Cees van Rijn, Wietse Nijdam, Remco Sanders, Henk van Wolferen, Gui Chengun, Peter Leussink, Johannes Burger, Niels Tas, Edwin Oosterbroek, Willem Tjerkstra, Stein Kuijper, Robert Zwijze, Theo Veenstra, Erik van Veenendaal, Jasper Nijdam, Henk Wensink, John van Baar, Joost van Honschoten, Marko Blom, Han van Egmond, Albert van den Berg, Jaap van Suchtelen, Henri Jansen, Hans-Elias de Bree, Cristina Neagu, Frans Blom, Frans van de Pol, Siebe Bouwstra, Wim Hendriks, Harrie Tilmans, Albert Prak, Cees van Mullem, Vincent Spiering, Gert-Jan Burger, Rob Legtenberg, Joost van Kuijk, Jan van Vlerken, Edwin Smulders, Bob Haverkort, Thijs van Thor, Jan-Cees te Riet-Scholten,

Hans Ijntema, Michiel Hamberg, Job Elders. Many of them have assisted in proof reading, however the responsibility for any errors rests with the authors. Jan Fluitman as the former professor in the micromechanics group and former director of the MESA+ institute deserves special thanks.

Enschede,
July 2000

Miko Elwenspoek
Remco Wiegerink

Contents

1. Introduction

The use of silicon microsensors for pressure, acceleration, angular rate and fluid flow is increasing at high rates since micromachining has become a more or less mature technology. These sensors are used in great numbers especially in automobiles, process control, in the medical field and for scientific instrumentation. Market studies in the past years (mid nineties) have predicted an enormous increase in the need of these sensors. Recent predictions on market volumes of microcomponents (besides mechanical sensors ink jet printer heads and hard disk heads) are in the range of US$100 billion annually in Europe alone (Micromachine 1998).

The production price for pressure sensors has dropped well below one dollar per piece. Similar developments are expected from sensors for acceleration and angular rate. This dramatic development is due in first instance to the way microsensors are fabricated. The technology derives from integrated circuit fabrication technology where the production price per piece is roughly reciprocal to the number of fabricated units. This production method is called "batch processing", where a large number of components are made at the same time. Basically, silicon is machined using etching techniques, thin film deposition and waferbonding. This fabrication method is now known as "silicon micromachining". Silicon micromachining has become reliable, which is a second important reason for the commercial success of microsensors.

The working principle of mechanical sensors (except a certain class of flow sensors) relies on the mechanical deformation of a construction (deflection of a membrane or a mass suspended by a beam). This deformation is translated into an electrical signal.

Silicon happens to be an optimal material for mechanical sensors because of quite extraordinary mechanical properties. For sensors one needs a reproducible signal which means for the case of mechanical sensors that the structure must deform under an equal load in the same way.

Thus one needs a material free from mechanical hysteresis and free of creep. Hysteresis is due to yield of the material, or in other words due to plastic deformation. Silicon fails before it is deformed plastically, at least at room temperature. In fact, the stress at which silicon fails is considerably larger than the yield stress of stainless steel: in this sense, silicon is much stronger than all metals. Note however, that this does not mean that silicon is preferable over steel for all types of construction. If steel is stressed at some point of the construction above the yield strength, it deforms until nowhere the yield stress is exceeded. This way the structure as a whole will not fail. This is quite different in silicon constructions: if the yield strength is surpassed somewhere, silicon fails and the structure breaks down. This property of silicon – being brittle – is advantageous for sensors: If the sensor is overloaded, it breaks and will not function at all, in place of giving a false signal.

The other important property for sensing reproducible mechanical deformation is creep. This is a phenomenon, shown by all materials, in which a construction continues to deform at constant external load. For example, if a load from an elastically deformed spring is released the spring does not resemble its initial length immediately, only approximately, and from this length it "creeps" slowly back to its original length. This phenomenon occurs even if the load is small enough so that the yield strength is nowhere exceeded. This is a process, which can take minutes or hours. Single crystalline silicon belongs to the best materials with respect to creep; the effects are of the order of ten parts per million. The amount of creep depends on the geometry of the samples. Bethe's study revealed that creep of silicon and other materials becomes much more serious for thin cantilever beams (Bethe 1989).

The mechanical deformation due to mechanical forces can be measured in a number of ways. Use can be made of piezoelectricity, changes of the electric resistance due to geometric changes of resistors or due to strain in the resistors, changes of electric capacitance, changes of the resonant frequency of vibrating elements in the structure, or changes of optical resonance.

Silicon is not piezoelectric. Therefore, in order to translate a deformation to an electrical signal using piezoelectric effects one needs other materials, usually in the form of thin films. To date there are no reliable thin film materials available for this purpose. On the other hand, quartz is piezoelectric accordingly there are indeed mechanical microsensors machined from quartz. This book concentrates on silicon sensors.

Very common is the use of strain gauges in conventional sensors. The resistance of a conductor depends on its geometry, so conductors assembled on deforming bodies will give information about the deformation. Semiconductors have an additional materials effect: the conductivity depends on the strain in the material. This effect is called piezoresistivity. The relative change of the resistance per strain is called the gauge factor. For conventional metal strain gauges the gauge factor is of the order of one, while the piezoresistive effect in silicon increases the gauge factor of silicon strain gages to one hundred. The disadvantage of the piezoresistive effect is its temperature dependence.

Vibrating elements in the sensor construction also play the role of strain gauges: resonant strain gauges. The interest in resonant sensors has its root in the following attractive features: The output signal is a frequency. A frequency is much easier transferred into a digital signal; no AD-converter is required to feed the signal into a computer. A frequency as a signal is much more robust to disturbances than an amplitude (e.g. a voltage). Vibrating microbridges can replace strain gauges. They have much greater resolution than metal or piezoresistive strain gauges. We know two silicon resonant sensors on the market. One is fabricated by Druck, based on a design by John Greenwood (Greenwood 1984), and the other one is sold by Yokogawa Electric, designed by K. Ikeda (1990). The disadvantages of resonant silicon sensors are that they are not easy to make (which makes them quite expensive) and the technology is yet not well established. The high costs might be compensated by new simpler mechanical constructions for the load supply and by the simpler electronics.

A third important way to measure the mechanical deformation of a body is the capacity of a charged distribution of conductors. The capacitance depends on the geometric distribution, so any measurements of a capacitance of conductors assembled to the deforming body will give information about its form. Capacitive sensors have some advantages over piezoresistive sensors: they are less sensitive to temperature variations, and the sensitivity is larger, which is due to the mechanical construction, as we shall see in Chap. 5. The electronics however is more complex.

A category of microsensors, which do not always rely on a mechanical deformation, belongs to mechanical sensors: flow sensors. Many of these sensors rely on thermal effects, where a heater is used and the temperature is measured, either of the heater itself or of the medium in its surrounding. The flow changes the temperature distribution. There are also types of flow sensors where a deformation is induced by forces exerted on a sensor element by the streaming fluid: the fluid flow causes shear forces, drag forces and pressure gradients.

Designing a microsensor is a formidable task, which involves the whole range from the basic physics of the device and its interaction which the ambient, its fabrication to systems issues like electronic interfacing, electronic circuitry and packaging. This book has a double function: it is written as a review, and as a textbook to teach students the subject of mechanical microsensors. We make the attempt to describe all ingredients necessary to perform this task. We expect that the reader has basic knowledge of mathematics, in particular vectors and calculus. The book is directed to engineers and scientists of any of the fundamental engineering disciplines (mechanical and electrical engineering, physics and chemistry). This means that in deriving the basic models we start from a fairly basic fundament, however we make use of rather advanced mathematics of a level a student in engineering should have mastered after his second year.

We tried to give a comprehensive overview of the mechanical microsensors described so far, with the emphasis on the basic ideas. We describe the functioning of the sensors both, on a qualitative level and on a quantitative level.

We start the discussion with the description of MEMS. MEMS is the modern acronym for microelectromechanical systems, the sensors are a subset of this field. More important, sensors are quite often *part* of MEMS. This chapter also contains a summary on scaling, to give the reader some feeling about changes of our world when dimensions shrink. An overview on silicon micromachining follows. This chapter is a very condensed version of "Silicon Micromachining", by one of the authors and Henri Jansen (Elwenspoek and Jansen 1998). The fabrication technology of course is one of the essentials for the design of microsensors. The next two chapters are devoted to theory of mechanical deformation and on the two most important transduction mechanisms for mechanical microsensors, namely piezoresistive and capacitive. Chapters on the sensors themselves follow: sensors for pressure, force, acceleration, angular rate and fluid flow. The latter contains some basic theory of flow and heat transfer. A most advanced (from a technological point of view) and most complex (regarding systems design) type of sensors, resonant sensors, will be described next. The last two chapters are devoted to sensor interfacing: electronics and packaging.

2. MEMS

In many technical systems there is a strong trend for miniaturisation. This trend results on one hand from the fact that small components and systems perform differently: small systems can perform actions large systems cannot (example: minimal invasive surgery). In many cases a miniaturisation makes the systems more convenient (example: GSM telephone). On the other hand technology derived from IC-fabrication processes allows the production of miniature components in large volumes for low prices (examples: pressure sensors for automobile applications, ink jet printers).

The technology for miniaturisation develops from a number of fabrication methods: We mentioned IC-fabrication methods, but there are many groups at universities and companies that aim at the fabrication of small systems using technologies that derive from more conventional machining, such as cutting, drilling, sand blasting, spark erosion, embossing, casting, mould injection etc. In this book we concentrate on silicon micromachining and on the mechanical microsensors that can be fabricated by this method.

2.1 Miniaturisation and Systems

There are two basic notions used to indicate the science of miniaturised mechanical components and systems: Micro systems technology (MST, this notion was coined originally in Germany) and Microelectromechanical Systems (MEMS, invented in the USA). While MEMS is more specific to mechanical components, MST includes also microoptical systems, chemical sensors, analysis systems etc. However, in all microsystems mechanical and electromechanical functions must be realised, therefore in practice, both notions have a very large overlap. The important word in both notions is *system*. Here, system is used in contrast to *component*. Microthings are necessarily systems because of the small details in the systems. One cannot assemble a microsystem from components from the shelf. Example: if a factory wants to build a car, which is a pretty complex system, the whole car is designed using (in principle) available components (motor, transmission, doors, windows, wheels, etc.). These are purchased or fabricated following the specification of the designers and assembled. Such a procedure is impossible with microthings. First of all, the components are too small to be assembled at reasonable costs. Further, there are no components that fit together. In designing a microsystem, the components have to be designed during the design of the whole system. There is no design phase of the system separate from the design phases of components: the design of a microsystem is an integral process. Even more: the fabrication of the microsystem is an integral process. This process is being provided by silicon micromachining. We shall illustrate the design- and fabrication process a few times.

Referring to microsensors, we have leaned by a painful process that the sensing element, the electronic interfacing and the package (if you wish, the mechanical interface) must be designed as a whole. Designing one part of the system after the other in most cases will lead to sensors which cannot be packaged, or of which the package is too expensive.

A second systems aspect plays a major role in the fabrication process. Micromachining is very complex and not developed to a standard technology. The process space is still largely unexplored. When designing a microsystem, the fabrication process of the system must be designed, too. Thus system design and process design are integrated in a single design process. Microsystems designers therefore must have a large number of skills and a rather broad experience. This is quite different from IC-design and processing. For IC-processes there are strict and clear design rules which guide the circuit designer, and he does not have to worry how the circuit is actually made. Microsystems designer must be able to design both: the system and the fabrication process.

A third systems aspect is the following: Usually microsystems have complex functions. The functions have roots in different physical domains. For a sensor this is clear: a sensor must transmit a signal from a particular domain (chemical, mechanical, thermal) to the electrical domain. A sensor designer must of course know the domains to which the sensor has ports. As an example, in a pressure sensor a membrane is deformed and the deformation is the measure for the pressure. The sensor designer must know the mechanical properties of membranes. The deflection of the membrane can be measured by optical interferometry. In this case, the designer must know optics as well. The designer must know how to treat his (usually) small electrical signal, how to amplify it, how to realise the interfaces to a computer, so he must have knowledge of electronics, too. Sensor designers therefore are Jacks of all trades. In practice, they have to work in groups of engineers of different colours, and understand enough of other disciplines in order to be able to communicate.

2.2 Examples of MEMS

Since this book is devoted to microsensors, there will be no extensive discussion of other microsystems such as actuators, microstructures, microrobots etc. We give a few examples here. We find this important because the technology – silicon micromachining – enables us to fabricate complex systems and simple systems for more or less the same price. So there is the chance (and the challenge) to integrate many functions in a single piece in an integrated production process.

Using silicon it is obvious to think of the combination of mechanical and electronic functions. Much work has been done to develop integrated sensors, or "smart sensors". A few are on the market (e.g. accelerometers from Bosch and Analog Devices, a pressure sensor from Bosch), however it seems that the development of the production process for the integration of electronic and mechanical functions is extremely expensive. The hybrid integration of several mechanical and electromechanical functions seems to be much simpler and economically more feasible.

2.2.1 Bubble Jet

The ink jet printer is one of the successful microsystems now. One of the ways to dose drops of ink on a piece of paper is to squeeze the liquid ink close to an opening – the nozzle – in a way that a single drop with a reproducible volume and a certain speed will leave the nozzle. This can be accomplished by a piezoelectric element; however, the first successful technology uses thermal expansion of ink vapour. Schematically the ink jet printer head looks like as shown in Fig. 2.1.

The thermal inkjet printer head consists of a reservoir, a channel, a nozzle and a heating element. When the heating element heats the ink, its temperature is raised above the critical point, so a bubble is formed without a nucleation barrier (which would result in a too slow and not reproducible formation of a bubble). The pressure in the bubble is used to displace the ink, and because the flow resistance to the nozzle is smaller than to the ink reservoir ink is ejected. When switching off the heating power the ink cools and the vapour bubble collapses and an ink bubble (in fact, often many bubbles) breaks off. This sequence is shown in Fig. 2.2.

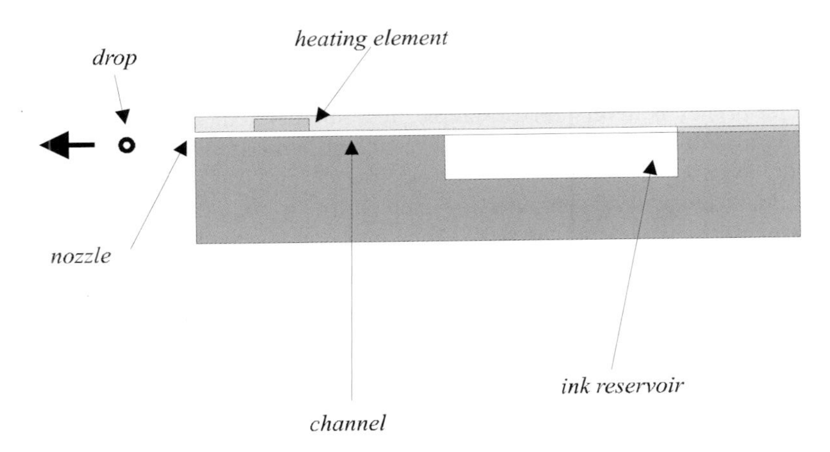

Fig. 2.1. Schematic of an ink jet printer head

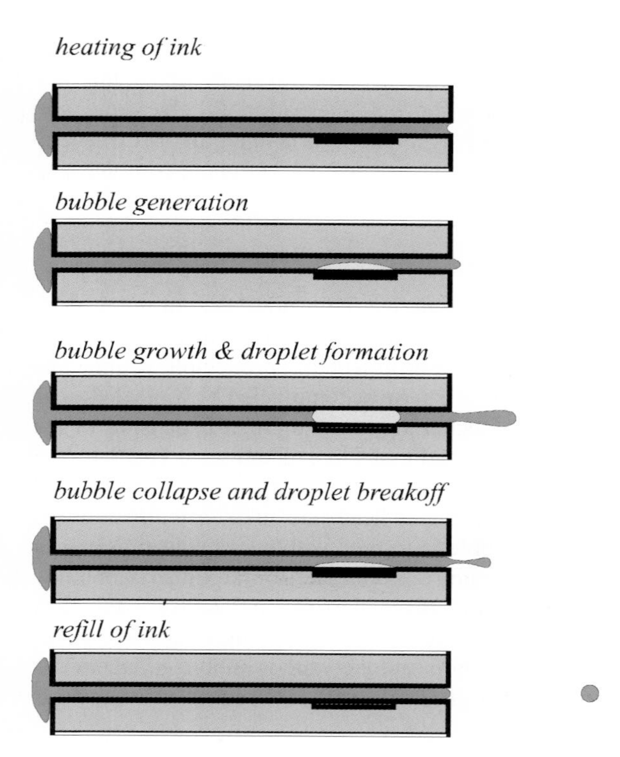

Fig. 2.2. Cycle of drop formation

The ink jet printer is a very nice example of a microsystem in a number of respects. It is a system. In fact, what we have drawn here is only a very small portion of the system. There must be some electric wiring. The printer must be able to communicate with a computer, so there must be an electric interface. When designing the channel and the nozzle, one must know what type of ink will be used: the viscosity is a very important parameter and is strongly temperature dependent. When choosing the ink, one must know on what type of paper will be printed. One needs to know the size of the droplets. The ink jet printer as an example par excellence for multidisciplinary work: thermodynamics (critical points and phase transitions), hydrodynamics (very complex because of the drop formation), and electronic engineering for the head, mechanical and electric engineering. The ink jet printer head has a heater, so micromachining allows to integrate more heating elements into the channel, some can be used as actual heaters, and some as temperature sensing element e.g. by exploiting the temperature dependence of the electrical resistance of the materials used. The distribution is changed by convection and measured.

2.2.2 Actuators

Microactuators can be used for positioning of small masses. An example of such an actuator is shown in Fig. 2.3. This is a so-called comb drive actuator, in which interdigiting teeth contract under a voltage difference. The maximum force these actuators can deliver is quite small (in the order of 10 pN!), therefore one needs many of them in parallel. Future applications of comb drive actuators are in positioning mirrors or tips in read-write heads in data storage devices. Currently, much effort is taken to develop microresonators for telecommunication purposes. Connecting a comb drive to a spring gives a resonator, which can be used in an electromechanical high quality factor filter.

Currently there is a search for actuators, which are fast and strong. Electrostatic actuators are fast but the maximum force, even using clever designs, seems to be limited to about 1 mN. Therefore considerable effort is undertaken to fabricate piezoelectric actuators. In general, thermal actuators tend to be strong but slow.

Comb drives have important applications in mechanical microsensors. Since they represent a capacity the value of which depends on the overlap of the fingers, they can be used as position sensitive devices. Deflection of seismic masses in inertia sensors (see Chap. 7) and shear stress in fluid flow sensors (see Chap. 8) is measured using comb drive actuators. In both types of devices comb drives are also applied in servo systems where the force is the sensor output, needed to keep the fingers in a fixed position.

Fig. 2.3. Micromechanical comb drive actuator. The gap between the teeth is approx. 2 μm. In the upper half there is a shuttle (Legtenberg 1996)

2.2.3 Micropumps

A great promise in MEMS comes from micro fluid handling systems (MFS), after the first components for MFS have been demonstrated. Among them a silicon micropump. Historically the first silicon micropump was a peristaltic pump, designed by Jan Smits during the first half of the eighties. Smits' work was published not before Transducers '89 (Smits 1990), two years after the follow up of this work was presented at Eurosensors 87 by van Lintel, van de Pol and Bouwstra (van Lintel 1988), see Fig. 2.4, which in fact was the first working silicon micropump ever published.

Van Lintel used a piezoelectric disc glued to a glass membrane; under a voltage difference the piezoelectric disc changes its lateral dimension which results in a bending moment in the dimorph. The original prototype was able to produce a maximum pressure of 100 cm water and had a maximum yield of ca. 10 μl/min at 1 Hz block wave actuation.
Mainly from aesthetic reasons van de Pol suggested an alternative for the piezoelectric actuation which he called thermopneumatic actuation. By heating a gas or a liquid under its vapour pressure the pressure is increased and a membrane can be deflected. This pump, when using ambient air as expanding medium in the actuator, was able to produce a maximum pressure of 40 cm water and had a maximum yield of ca. 30 μl/min. Maximum yield occurred at 1 Hz block wave actuation.

There are many groups now active in the design of micropumps (examples: Forschungszentrum Karlsruhe, Universtät Freiburg, Institut für Mikrotechnik Mainz (IMM), Hahn-Schickart Gesellschaft, Institute for Micromachining and Information Technology, HSG-IMIT, in Villingen - Schwenningen), and the performances has been increased considerably during the past years. A recent overview can be found in van Kuijk (1996). The latest pumps are self-priming and resistant against the formation of gas bubbles in the system (H. Linnemann 1998, K.-P. Kämper 1998).

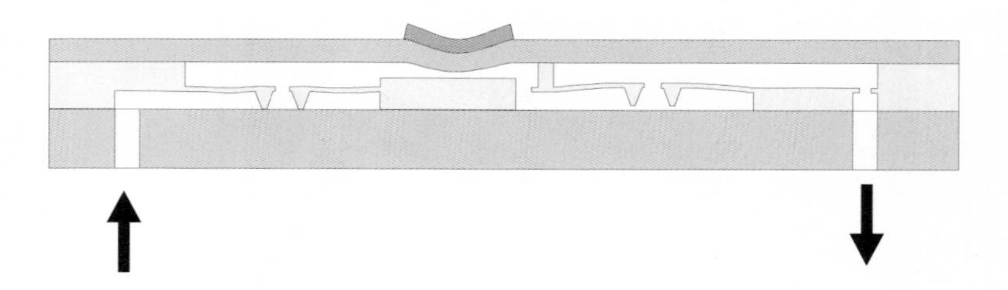

Fig. 2.4. Schematic of van Lintels membrane pump (van Lintel 1988). In the centre the pump chamber and the pump membrane are drawn. If the membrane bends down, the right valve opens and the left valve is forced to close; this state is shown. If the membrane is released or bends upwards, the left valve opens. Note that this pump is asymmetric. Reproduced from 'Silicon Micromachining', M. Elwenspoek and H. Jansen, 1999, by permission of Cambridge University Press

The integration of the thermopneumatic pump and the flow sensor is rather straightforward. Figure 2.5 shows an optical photograph of the wafer with four dosage systems, and in Fig. 2.6. we show the cross section of a dosing system. It has been described first at MEMS '93 in Ft. Lauderdale, USA (Lammerink 1993). Once having demonstrated the feasibility of integrating several functions of a MFS on one wafer the progress is obvious: tedious and expensive assembly of micro components can be avoided. Flow sensors are important components in microliquid handling systems and will be described in detail in Chap. 8.

Micro fluid handling systems will possibly replace chemical sensors. In chemical sensors the sensing material reacts directly with the chemical whether or not in the ambient. MFS can be used as chemical laboratory on a chip, which includes functions such as sample pre-treatment, calibration and possibly a chemical reaction if the detection of a reaction product is easier than that of the chemical itself.

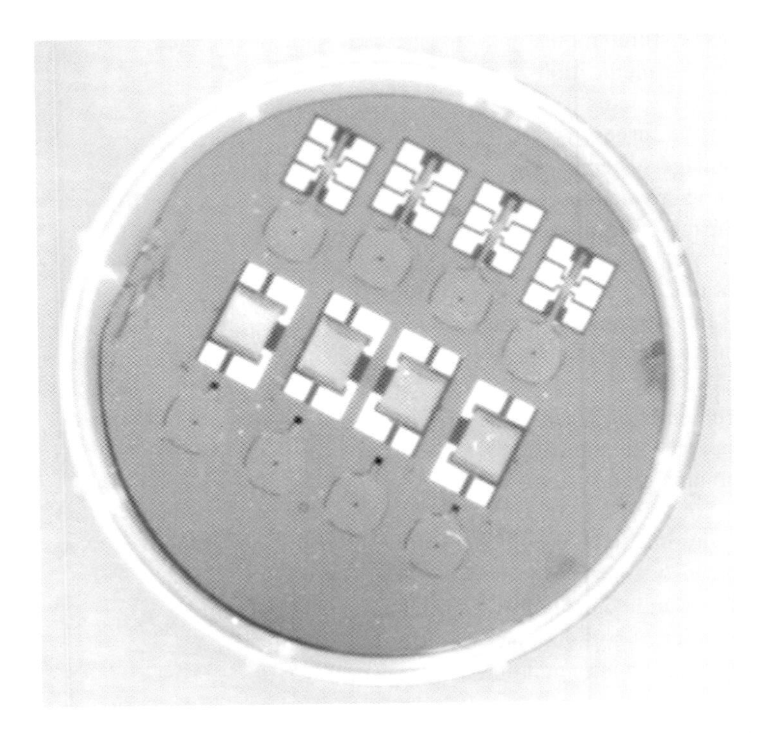

Fig. 2.5. A dosage system with a micropump and a flow sensor integrated on one chip. Photograph of the wafer containing four dosage systems (Lammerink et al. 1993). Reproduced from 'Silicon Micromachining', M. Elwenspoek and H. Jansen, 1999, by permission of Cambridge University Press

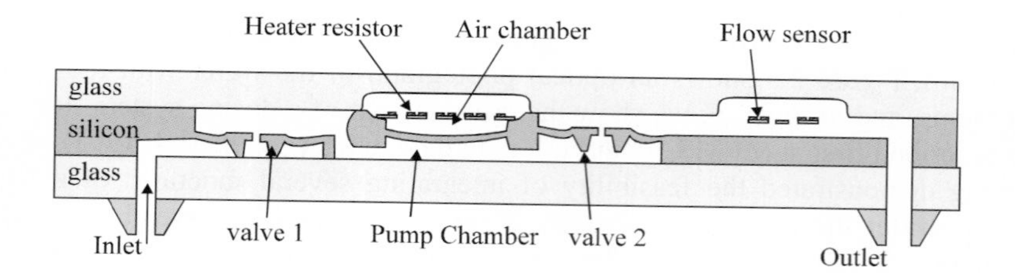

Fig. 2.6. Cross-section of the dosage system. Reproduced from 'Silicon Micromachining', M. Elwenspoek and H. Jansen, 1999, by permission of Cambridge University Press

Below we give a list (far from complete) of applications of MEMS. A recent overview can be found in Fluitman (1996).

Medical applications:

- hearing and seeing aids, prosthesis for eye and ear
- systems for nerve stimulation
- dosage systems for medication
- implanted dosage systems
- minimum invasive surgery
- implanted drainage systems
- endoscopy with active endoscopes
- surface structuring of implanted systems

Consumer applications:

- microsystems for information storage
- CD-players in the size of match boxes, CD's in the size of coins
- ink jet printers
- control instrumentation (sensors) for products such as cars, houses, copymachines, telecommunication
- (optical) communication systems

Applications in environmental control, chemistry, pharmacy, agriculture:

- (bio-)chemical sensors, analysis systems, robots
- on-line monitoring and control of waste production and gas emission
- fight vermin by microrobots (in place of hazardous chemicals)
- noise control (microphones, soundintensity sensors)

Applications in fabrication:

- production facilities for chips in the size of an average living room
- control instrumentation in industry
- industrial robots
- microrobots for product control

Applications in science:

- micro instrumentation for:
  - * wind tunnels (pressure flow patters drag forces)
  - * aeronautics (sensors and actuators)
  - * space travel, "micro" and "nano" satellites
  - * cell micro manipulation
  - * distributed sensing for the study of e.g. turbulence
  - * sensors for friction and wear
- materials science
- scanning microscopy

2.3 Small and large: Scaling

There are a few matters one needs to keep in mind when miniature technical systems are designed. Many children imagine worlds on the top of a pin. There is an impressive movie (The fantastic voyage, written by Isaak Asimov) in which a submarine is shrunken to the size of a few hundred micrometers and injected into the human body for microsurgery. From a scientific point of view, these thoughts are ridiculous. Shrinking the linear dimensions does not shrink forces in the same way. A few things are quite obvious. Linear dimensions shrink slower than surfaces and surfaces shrink slower than volumes. So, friction associated with surfaces becomes relatively more important than masses: weight and inertia tend to be negligible in the microworld. But also stiffness, electric forces, magnetic forces, adhesive forces, the character of convection, the importance of boundary layers, surface finish, characteristic times associated with diffusion and resonance all scale in their own way.

It is instructive to analyse scaling by comparing forces and effects by studying the exponent of the length scale of linear dimensions l. We then have:

Linear dimension: $l \sim l^1$

Surfaces: $A \sim l^2$

Volume: $V \sim l^3$

2.3.1 Electromagnetic Forces

The electrostatic force between capacitor plates is given by the derivative of its electric energy W with the gap distance x:

$$F = -\left(\frac{\partial W}{\partial x}\right)_Q \tag{2.1}$$

where the charge Q must be held constant. W can be found in the easiest manner by the energy density w:

$$w = \frac{1}{2} ED = \frac{\varepsilon_0}{2} E^2 \tag{2.2}$$

E and D are electric field strength and electric displacement, respectively. We assumed that the dielectric between the capacitor plates is air (ε_r=1). It follows that

$$W = \int w dV = wAx \tag{2.3}$$

with A the surface of the electrodes and x their distance. For w = constant, meaning E = constant we obviously have

$$F_{\text{el}} = -wA = -\frac{\varepsilon_0 E^2 A}{2} = -\frac{Q^2}{2\varepsilon_0 A} \tag{2.4}$$

We have used that the electric field of a plane is proportional to the charge density Q/A. Note that taking the derivative in (2.4) at constant Q corresponds to constant E or constant w. If we are able to shrink the whole capacitor at constant energy density, meaning at constant electric field or constant charge density (this sounds reasonable), the energy of the capacitor scales with the volume and the force between the capacitor plates like the surfaces:

Electric Energy: $W_{\text{el}} \sim l^3$.

Electric force: $F_{\text{el}} \sim l^2$.

Consequently, electrostatic forces *increase* relative to the volume when the systems shrink.

The maximum field in air is not independent of the distance of capacitor plates if they approach closely. This has been observed first by Paschen. In Fig. 2.7 we show the breakdown voltage in air and Ar as a function of the product of plate separation and pressure. If the separation between the plates is roughly of the order of the mean free path of molecules in air, the maximum field increases.

Magnetic forces can be calculated in a analogous way. The magnetic field energy density is given by (with B, H and μ the magnetic flux density, magnetic induction and the permeability, respectively)

$$w_{\text{mag}} = \frac{1}{2} BH = \frac{B^2}{2\mu} \tag{2.5}$$

so one is tempted to assume that the energy in a gap of a magnetic circuit scales with the volume and the forces with the surface, but this conclusion is too quick. The reason is that the magnetic field in the gap is given by

$$B = \frac{1}{\mu_0} \frac{n}{L} I \tag{2.6}$$

with n the number of turns of wire wound around a magnetic jock, L the total length of the coil and I the current through the coil. When shrinking the coil and the magnet, it may be possible to keep n constant, but the cross-section S of the wires must shrink, so that it will be impossible to keep I constant. It is more reasonable to keep the current density constant, but this means that $B\sim S$: B scales like l, see Eq. (2.6)! The energy density then scales like l^2. We have:

Magnetic energy: $W_{\mathrm{mag}} \sim l^5$

Magnetic force: $F_{\mathrm{mag}}\sim \sim l^4$

Some care is necessary here: The maximum current density is limited by the dissipation, accordingly by the maximum temperature the system can stand. Small systems loose their heat faster than larger systems: Heat flow is proportional to the temperature gradient, which is proportional to l^{-1}. Taking this into account, one would end up with $F_{mag}\sim l^2$. However, it depends very much on the details of the design if one really can use the large temperature gradients. Furthermore, the assumption that the number of wires is independent on the system size becomes dubious if the coils shrink below 1 mm^3. In practice, magnetic actuators smaller than 1 mm^3 are hardly feasible.

Any comparison between magnetic motors and electrostatic motors will end up that at the small end the electrostatic motors are stronger, while at the large end the magnetic motors are stronger.

If one is able to use materials with a high dielectric constant (e.g. PZT – lead circonate titanate has ε_r=1300) the energy density and the total outputforce increases by the amount of the dielectric constant. This increases the maximum force of electrostatic micromotors to about 1 N.

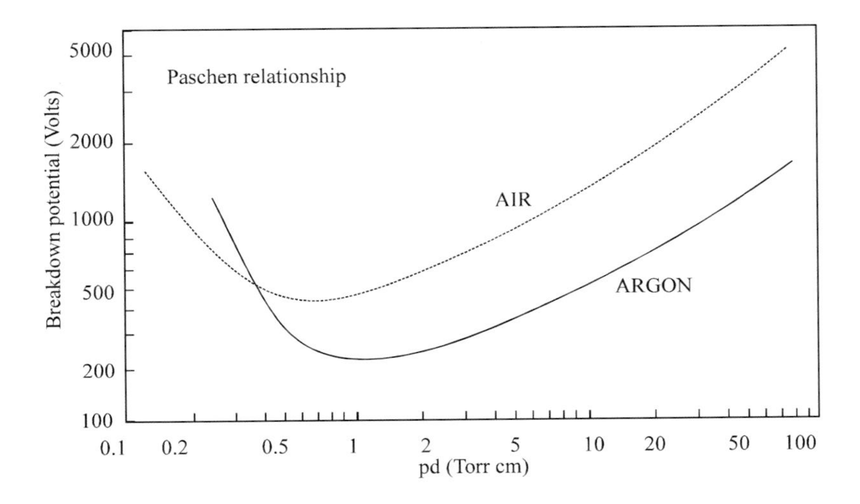

Fig. 2.7. The DC breakdown voltage as a function of the gas pressure p and the electrode spacing d for plane parallel electrodes in air and Argon. Such curves are determined experimentally and are known as Paschen Curves

2.3.2 Coulomb Friction

In the macroscopic world friction forces are independent of the contact area. The physical reason is that two rough bodies without load touch each other only at three points. In silicon micromachining, this seems to be different in some cases because silicon is extremely smooth (the micro surface roughness of silicon wafers is of the order of 1 nm, rather less). The adhesive forces can be very large, which is being used in wafer bonding and it leads to process problems in surface micromachining, as we will show in the next chapter. These surface forces scale with l^2, and become large in small systems.

2.3.3 Mechanical strength

Small mechanical constructions are stiff. One can see this easily in models of sailing ships: the form of the cables does not follow what gravity tries to impose, but it is the internal stress, which results in their arbitrary shape. The differential equation for a cantilever beam loaded by its own weight per length P is

$$\frac{\partial^4 \zeta}{\partial x^4} = \frac{P}{EI} \tag{2.7}$$

with E Young's modulus, I the moment of inertia, $I = bh^3/12$, and $\zeta = \zeta(x)$ the deflection. b and h are width and thickness respectively. See Fig. 2.8. This equation will be derived and discussed in Chap. 4. Obviously with $P = g\rho bL$ (ρ = density, L = length of the beam, g = gravitational acceleration)

$$\zeta \sim L^4 bh / bh^3 = L^4 / h^2 \sim l^2 \tag{2.8}$$

Deflection under own weight $\zeta \sim l^2$

A ten times smaller (thinner and shorter) beam bends 100 times less due to its own weight.

A beam under a lateral compressive stress buckles if a critical load is exceeded. This load is given by

$$T_{\mathrm{cr}} \sim \frac{EI}{L^2} \tag{2.9}$$

Again if we assume that the beam is loaded by its own weight, T = $\rho gLbh$, (g = acceleration) and we find that the critical length is given by

$$L_{\mathrm{cr}}{}^3 \sim \frac{EI}{\rho g} = \frac{E}{\rho g}\frac{bh^3}{12} \tag{2.10}$$

from which we may conclude that

Stable length $L_{\mathrm{cr}} \sim l^{4/3}$.

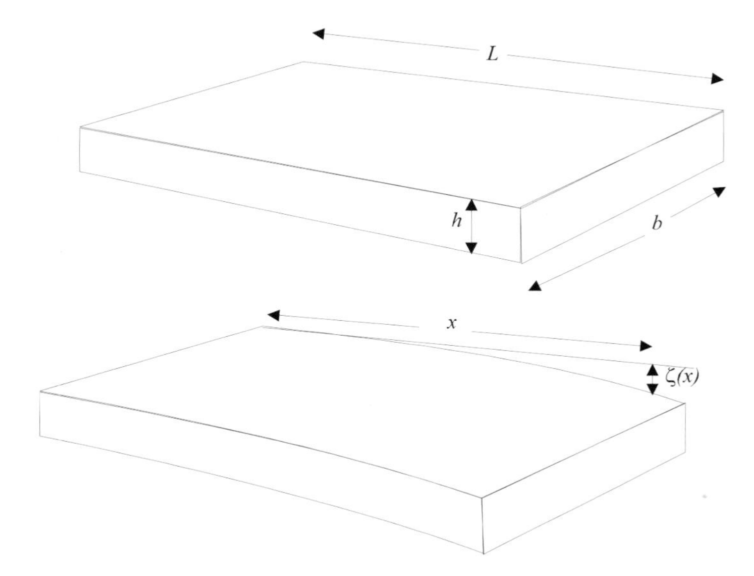

Fig. 2.8. Geometry of a bent cantilever beam

2.3.4 Dynamic Properties

The time a particle needs to travel a distance L by diffusion is given by:

$$\tau = \frac{L^2}{\alpha D} \tag{2.11}$$

where D is the diffusion constant and α is a numerical value two times the dimension of the diffusion process (e.g. $\alpha = 6$ for diffusion in three dimensions). Self diffusion coefficients in normal liquids at room temperature (water, mercury) is $D = 10^{-5}$ cm^2/s. In gases D is 1 cm^2/s. We see that diffusion tends to be very fast in small systems:

Diffusion times $\tau \sim l^2$

This equation is also valid for thermal diffusion. However, in liquids heat diffusion coefficients are much larger than particle diffusion coefficients.

A mass subject to viscous friction moves after a certain time with constant velocity, which depends on the force acting on the mass. Taking again the mass's weight as the force, the velocity of a spherical mass (radius r) is given by

$$v_\infty = 18\eta r / 4\rho g r^3 \tag{2.12}$$

with η the dynamic viscosity. We find that viscous forces heavily damp any motion. Shrinking a ball falling though water by a factor of 10 leads to a steady state velocity 100 times smaller. The transient time also varies like l^2.

Drift velocity $v_\infty \sim l^2$

Transient time $\tau \sim l^2$

The electrical resistance of conductor increases when scaling down. We have for the resistance R_{el} of a wire with specific conductivity σ, length L and cross-section S

$$R_{el} = \frac{1}{\sigma}\frac{L}{S} \tag{2.13}$$

Electrical resistance $R_{el} \sim l^{-1}$

The hydraulic resistance of a fluid streaming through a pipe with length L and cross-section S increases much faster. The physical reason is that at the walls of the pipe the velocity of the fluid vanishes. The hydrodynamic resistance is proportional to L/S^2,

$$R_{hy} = a\eta L / S^2 \tag{2.14}$$

Hydraulic resistance $R_{hy} \sim l^{-3}$

With high velocities the hydrodynamics become unstable and the convection becomes turbulent. The Reynolds number Re is the criterion: If in pipes Re > 70, there is a chance for vortices, and the transition to turbulent flow is around Re = 2300. The Reynolds number is given by

$$\mathrm{Re} = \frac{\rho U r}{\eta} \tag{2.15}$$

If the velocity U scales linearly with the size, we have

Reynolds number $\mathrm{Re} \sim l^2$

If we take the drift velocity, the Reynolds number scales even like l^3. There is no turbulence in microsystems in which liquids flow. There can be turbulence in microsystems with gas flow.

In Fig. 2.9 we show the Reynolds number of animals swimming or flying as observed in nature. The empirical exponent is very close to 2, indeed (solid line in the figure). Note that also man-made "animals" (the submarine) follow the trend.

Resonance frequencies become large for small systems. The easiest way to see this is by studying the eigen frequencies of a string or a tube. The lowest eigen frequency corresponds to a state where the length of the string equals half the wavelength λ. From the well-known relation $c = \lambda\nu$ (with c the phase velocity of the wave, ν the frequency) it is immediately seen that eigen frequencies scale like l^{-1}.

Resonant frequency $\nu \sim l^{-1}$

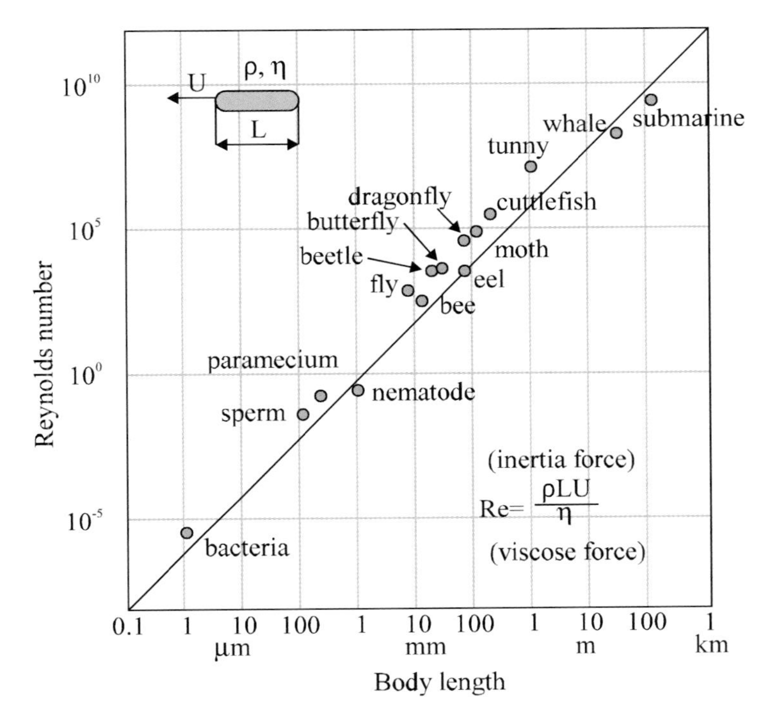

Fig. 2.9. Reynolds number as a function of size (Hayashi 1994) Reproduced from 'Silicon Micromachining', M. Elwenspoek and H. Jansen, 1999, by permission of Cambridge University Press

This relation is also valid for the eigen frequency of a beam where the restoring force is due to stiffness. The resonant frequency is given by

$$\nu_{\mathrm{r}} = \frac{\alpha}{L^2}\sqrt{\frac{EI}{\rho S}} \tag{2.16}$$

α is a number, the value of which depends on the mode and the boundary conditions. A beam with a rectangular cross-section ($S = bh$) has $I = bh^3/12$, so

$$\nu_{\mathrm{r}} \sim h/L^2 \sim l^{-1}$$

We conclude this section with commenting on the price of machined mechanical parts and systems. In Fig. 2.10 the price of system or component per gram is plotted as a function of the mass. It can be seen that prices generally increase steeply if the systems become small. This indicates that it is difficult to machine small components.

In Table 2.1 a summary of the scaling law is given.

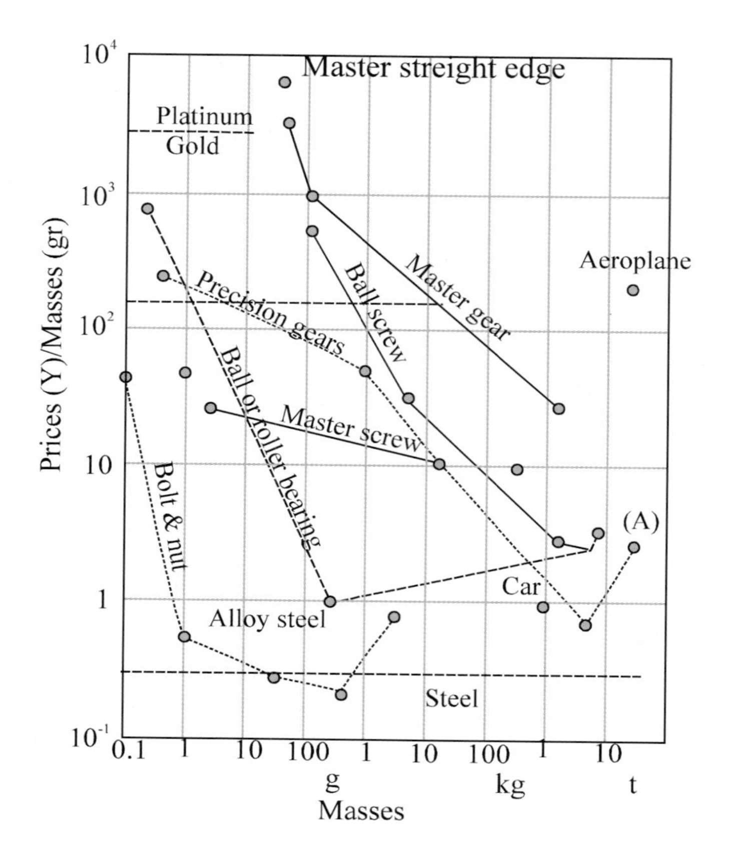

Fig. 2.10. Price per mass of machines as a function of the size (data from Hayashi (1994)) Reproduced from 'Silicon Micromachining', M. Elwenspoek and H. Jansen, 1999, by permission of Cambridge University Press

2.4 Available Fabrication Technology

2.4.1 Technologies Based on Lithography

In photolithography light impinges a transparent plate with opaque regions, called a mask. Below the mask a photosensitive material is placed, which can be "developed" after illumination. In this process, the illuminated regions (or the non-illuminated ones) are dissolved, and the underlying material is accessible to further processing. Figure 2.11 shows an example.

Table 2.1. Scaling of physical properties

Linear Dimension	$l \sim l^1$
Surfaces	$A \sim l^2$
Volume	$V \sim l^3$
Electric Energy	$W_{el} \sim l^3$
Electric Force	$F_{el} \sim l^2$
Magnetic Energy	$W_{mag} \sim l^5$
Magnetic Force	$F_{mag} \sim l^5$
Deflection Under Own Weight	$\zeta \sim l^2$
Stable Length	$L_{cr} \sim l^{4/3}$
Diffusion Times	$\tau \sim l^2$
Drift Velocity	$v_\infty \sim l^2$
Transient Time	$\tau \sim l^2$
Electric Resistance	$R_{el} \sim l^{-1}$
Hydraulic Resistance	$R_{hy} \sim l^{-3}$
Reynolds number	$\mathrm{Re} \sim l^2$
Resonant Frequency	$\nu \sim l^{-1}$

2.4.1.1 Silicon Micromachining

The basis of silicon micromachining is photolithography. Photolithography defines regions on a silicon wafer where the machining is done. The machining includes etching (in a liquid, plasma, gas or by high energy ions or particles), doping and deposition of thin films. It is clear then that micromachining allows the fabrication of three-dimensional structures with complex forms only in two dimensions, everything in the third dimension is in some way extended from two dimensions. Some extension to really three-dimensional structures can be accomplished by means of wafer bonding. These technologies will be described in greater detail later in Chap. 3.

The main difference of silicon micromachining with most other machining techniques is that the first technology is suitable for batch processing, similar to the fabrication techniques of integrated circuits. The technology is quite expensive, mainly because a clean room is required, but the price of components is not proportional to the number of fabricated pieces. In the limit of very large production volumes, the production costs of the whole production is independent from the number of components fabricated. The other technologies described below (with the exception of the LIGA process, which is in some respects similar to silicon micromachining) do not have this property. Here each component must be made piece by piece. Low prices for large production volumes are the result of mechanisation.

2.4.1.2 LIGA

LIGA is an acronym based on the German words Lithography, Galvanik (electroplating) and Abformung (moulding). The technology originated at the Kernforschungszentrum Karlsruhe in Germany, where attempts were made to machine microfluidic elements for isotope separation. The process involves illumination of a thick X-ray sensitive resist by high-intensity radiation from an X-ray source. Since X-rays are practically not scattered or diffracted at the dimensions of interest (μm), this first illumination step results in structures with tremendously high aspect ratios and very flat sidewalls. Illumination is done through an X-ray absorbing mask; to prepare this mask, often the same X-ray source – a synchrotron – must be used. The resist is deposited on top of a metal to allow electroplating into the opened features. The resulting metal structure then is used as a mould insert. Polymer microcomponents fabricated this way are on the market now.

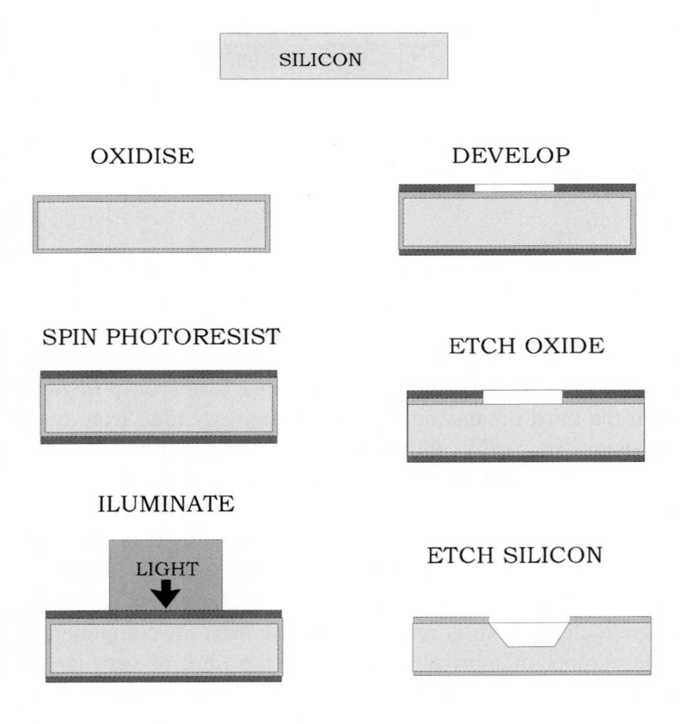

Fig. 2.11. Basic photolithographic process. Reproduced from 'Silicon Micromachining', M. Elwenspoek and H. Jansen, 1999, by permission of Cambridge University Press

Obviously the illumination step is the most expensive one in the process. The development of microsystems using LIGA is very expensive, because many masks must be made before a process is stable and the system has the desired properties. LIGA is economically only feasible for very large production volumes. Therefore alternatives have been looked for, with some interesting results.

One of the alternatives is to use conventional photolithography to illuminate thick mask layers (10–30 μm), of course with all the trouble caused by light scattering. A new photoresist (SU-8) performs much better in this respect (Eyre 1998).

A second alternative is based on silicon micromachining by first fabricating the high aspect ration structures with the help of reactive ion etching (RIE). This step replaces the illumination step in the LIGA process.

2.4.2 Miniaturisation of Conventional Technologies

The most important difference with photolithography based fabrication technologies is that using conventional technology the components must be made piece by piece. Finest details that can be machined are in the range of 10 to 100 μm, one to two orders larger than what photolithography makes possible. One of the greatest advantages is that one is not restricted to flat substrates.

Cutting, turning, drilling, milling and *shaving* can be done using highly specialised precision machines. An example and a first guide to the literature can be found in (Yamagata 1996).

Ultrasonic machining is a well known technique to fabricate quartz tuning forks, but the technology is being extended to other materials and the precision increases (example: Xi-Qing Sun 1996).

Also techniques such as *sandblasting*, *laser ablation* and *spark erosion* are miniaturised. Powderblasting has been used to fabricate features in glass in the size range of 100 μm and larger for flat panel displays at Philips Research (Slikkerveer 1996). The aspect ratio seems to be limited by 1:1 and the minimum feature size by fracture mechanics (10–20 μm) and by the particle size (powders with particles below 10 μm are very dangerous). Laser ablation works by heating the substrate locally so that the material evaporates. The technique suffers from heat production. If the heat is not carried away by the gas but diffuses into the substrate it deforms plastically due to large temperature gradients and thermal expansion. Light can also be used to enhance etch rates locally and to enhance the nucleation and growth of materials. Mark Madou gives a rather recent overview in his book (Madou 1997).

3. Silicon Micromachining

The arsenal of technologies for silicon micromachining comprises photolithography, thin film deposition, doping, etching by wet chemicals and plasmas and waferbonding. Mainly for the sake of clarity and to describe the notions necessary to understand the limits and possibilities of micromachining we give a brief account of these fabrication methods. More information on silicon micromachining can be found in the literature mentioned in the text and in a few recent textbooks (Büttgenbach 1991, Elwenspoek 1999, Heuberger 1989, Kovacs 1998, Madou 1997, Menz 1993, Muller 1991, Muller 1999, Ristic 1994, Sze 1994, Tabib-Azar 1998).

3.1 Photolithography

The basic photolithographic process includes a drawing, which defines transparent and opaque areas on a mask. The mask can be produced by direct writing on a glass plate in the desired size, but it can also be drawn much larger and subsequently reduced by photographic means.

Since for pattern transfer light is used, the minimum feature size feasible is defined by the wavelength of the light.

The pattern transfer from the mask to the substrate requires the following steps:

- resist spinning
- prebake (depending on the resist, typically 10 minutes at 90°C)
- illumination in a mask aligner. The mask aligner allows precise alignment of the mask with features on the substrate and the crystallographic orientation of the silicon wafer
- postbake (depending on the resist, normally 20 minutes at 120°C)
- development

Usually, a mask is expensive. Typical prices are in the range of 1000$ per mask. Emulsion masks are considerably less expensive, but subject to wear during use.

There are three types of mask aligners: for contact printing, proximity printing and projection printing. Proximity printing uses the shadow of the opaque regions of the mask, and suffers from diffraction of light at the mask edges. The result is a resolution inferior to contact printing where diffraction effects are minimised. Using light of a wavelength of 400 nm, the minimum linewidth is in the range of 1 μm. Contact printing has the obvious disadvantage that the mask touches the substrate, which can result in damage of the mask (particularly for emulsion masks) or the resist film on the substrate.

Projection printing is most expensive but leads to the best results (minimum linewidth close to the wavelength of light). Working at the limit of minimum feature size is quite difficult and must be avoided.

In the transparent resist films there are standing light waves during illumination. They cause the wavy geometry of the sidewalls of the resist after development.

Basically there are two types of photoresist: negative and positive. Positive resist becomes soluble after illumination as a result of depolymerisation. Negative resist becomes insoluble after illumination. It is much harder to work with negative resist, due to competing chemical reaction of the material with ambient air and a poorer adhesion. Furthermore, the achievable minimum linewidth is greater (2-3 μm) than the properties of light in principle allows. The resolution of positive resist is much better (0.5μm).

Practically all processes need a wafer cleaning before the process actually is carried out. Standard in micromachining is RCA-cleaning. The RCA1 and RCA2 procedure is as follows:

- RCA1: 1 part NH_3 (25% aqueous solution) in 5 parts water, heat up to the boiling point, add 1 part H_2O_2 and then immerse the wafer for 10 minutes.
- RCA2: 1 part HCl in 6 parts water, heat up to the boiling point, add 1 part H_2O_2 and then immerse the wafer for 10 minutes.

NH_3 etches silicon! The etching is prevented if peroxide is added to the solution. RCA1 removes organic dirt (resist), and RCA2 removes metal ions. The second cleaning process is required to keep the oven tubes for thermal oxidation and indiffusion free of metals. Just for an etch step, RCA2 is not necessary but RCA1 is. Both cleaning processes leave the wafers with a thin oxide film (thickness in the range of nm).

In order to be able to fabricate small features a dust free environment is required. The maintenance of a clean room is very costly.

3.2 Thin Film Deposition and Doping

It is impossible to make a sensor only from silicon. One has to be able to etch silicon, therefore materials are required which are inert against the etching agents used to machine silicon. Sensors have some kind of electric function, so resistors and conductors are needed. Finally one needs special materials for transduction.

These materials typically are used in the form of thin films deposited on top of silicon wafers. By doping (electrical) properties of silicon are altered. Here we shall give a very brief account for the most important techniques

3.2.1 Silicon Dioxide

Silicon dioxide (SiO_2, in jargon "oxide") usually is used as a mask material for wet chemical etching. SiO_2 etches slowly in KOH solutions and in practically not at all in EDP (see below).

The growth rate of oxide is strongly dependent on the temperature. The rate controlling process seems to be the diffusion of oxygen atoms or molecules through the growing layer of silicon dioxide, once the films are thicker than a few nm. The growth rate of wet and dry oxide is given in Fig. 3.1. In the "wet" process, steam is added to the oxygen that flows though the tube. A standard way to add steam, is to direct the oxygen through (nearly) boiling demineralized water. As seen from Fig. 3.1 water vapour increases the growth rate quite considerably.

The thickness of the oxide film required depends on a number of things. For a following etch step, one has to think of the etch rate of the oxide, which is not insignificant in KOH. This etch rate will be discussed later, but to give a figure: etching through a 300 μm wafer requires an oxide layer of 2 μm.

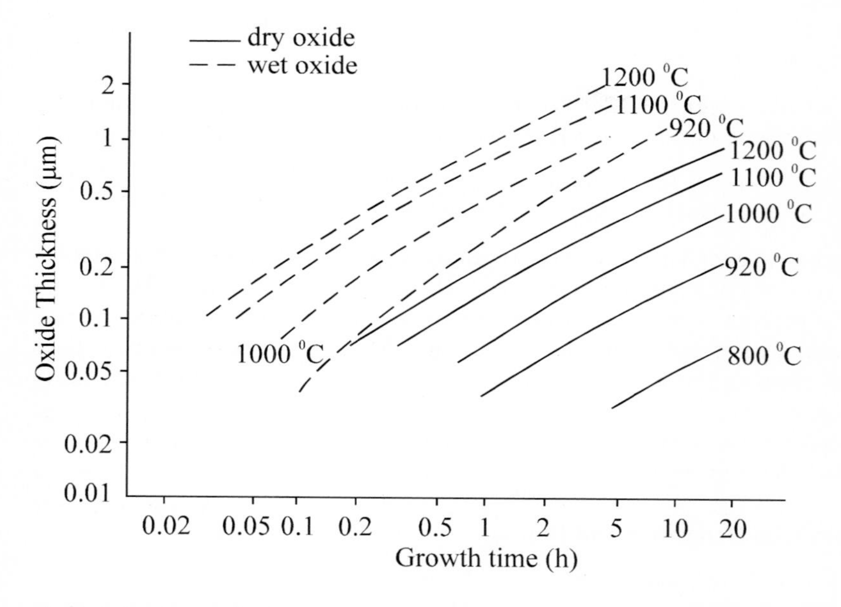

Fig. 3.1. Thickness of a thermally grown oxide layer as a function of the growth time for several temperatures. solid line: dry oxide, broken line: wet oxide (redrawn from Büttgenbach 1991)

The density of silicon dioxide differs of course from that of Si. The oxidised silicon has a volume larger than the silicon itself. Upon oxidation a oxidised layer occupies a volume larger by 46%. The result is a compressive stress[1] in the layer with respect to the underlying bulk silicon. Thermal oxide therefore is useless for most applications. It is under a compressive stress in the order of $2–4\times10^{10}$ Pa. Membranes and microbridges buckle; this material is only useful, when buckling is aimed at.

3.2.2 Chemical Vapour Deposition

Chemical vapour deposition (CVD) is a very important deposition method in micromachining. Prominent materials are silicon in the polycrystalline form (”polysilicon” or simply “poly”), silicon nitride and phosphor silicate glass (PSG). The latter is mainly used as a ”sacrificial layer” in a process called “surface micromachining” (see below), while the former two are often used as structural materials. Poly has properties comparable to single crystalline silicon, and silicon nitride is a very hard, chemically inert and strong – though brittle – material with very small thermal conductivity.

In the CVD process basically a gas is fed through a pipe which contains wafers at elevated temperature. The gas is cracked close to the hot substrate, and the reaction products are deposited. We distinguish between APCVD (atmospheric pressure CVD), LPCVD (low pressure CVD) and PECVD (Plasma enhanced CVD). The main difference between APCVD and LPCVD is that in the latter usually the step that determines the growth rate of the film is the integration of atoms in the surface, while in the former diffusion of gas to the wafers is dominant. On says that the former process is controlled by surface kinetics, the latter is controlled by volume transport kinetics. This has important consequences for micromachining. If surface integration governs the kinetic processes one gets a conformal coverage of the surface by the thin film. In particular, steps are covered (c.f. Fig.3.2) using LPCVD much better than using APCVD. Also the mechanical and chemical quality (in terms of impurities, pinholes and density) of LPCVD films usually is much batter that APCVD films.

Polysilicon is generally deposited by Low Pressure Chemical Vapour Deposition (LPCVD) from pyrolysis of silane (SiH_4) at temperatures around 600 °C and pressures of several hundred mTorr. Both compressive and tensile stresses of large magnitude are known to exist in polysilicon. Deposition temperatures below the silicon crystallisation temperature (ca. 605 °C) are generally found to produce tensile films, while a higher deposition temperature generally results in compressive films. Polysilicon films, deposited near 600 °C have been shown to grow initially in the amorphous state, and subsequently recrystallise during

---

[1] The stress of a film grown on top of substrate is indicated with respect to the underlying material. An expanding layer is then said to be under compressive stress, a contracting layer under tensile stress.

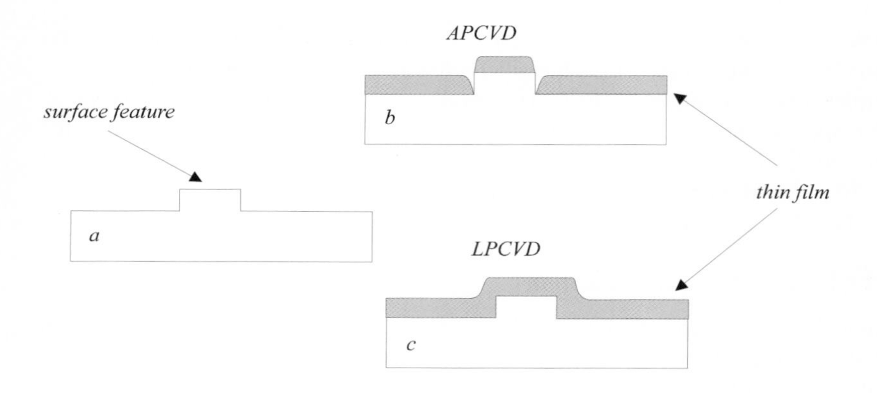

Fig. 3.2. Good and bad step coverage. (a) shows a step on the substrate surface. This step is covered poorly by a thin film deposited in a process governed by volume transport in is the case for APCVD (b). If surface kinetics control the process, a good step coverage results (LPCVD, c)

further deposition. As the density of amorphous silicon is smaller than the density of polysilicon, it contracts against boundary constraints, inducing a tensile stress. Tension occurs only in films that are deposited in the amorphous state and subsequently recrystallise. Annealing at high temperature induces recrystallisation, which allows intrinsic stresses in the silicon films to relax. In Fig. 3.3 the strain in polysilicon as a function of the annealing temperature and time is shown for undoped and boron doped samples (Guckel 1988).

Silicon nitride is deposited from a mixture of dichlorosilane (DCS, SiH_2Cl_2) and ammonia (NH_3) at elevated temperature (800 – 850 °C). The resulting film has properties that depend on the composition of the gas fed into the reactor. Stochiometric silicon nitride (Si_3N_4) deposited on silicon is under tensile stress in the range of 10^9 – 10^{10} Pa. Resulting structures can be very fragile because the strain in the film is close to the fracture strain. Deposited nitride at larger DCS flow compared to ammonia flow result in films having much smaller stress, and it is even possible to deposit films under compressive stress. Low stress nitride films can be deposited in an LPCVD reactor at 835 °C, 0.15 mbar pressure and a flow ratio DCS/ammonia of 40:8 (Bouwstra 1990).

Nitride can also be deposited in a PECVD reactor (PE = plasma enhanced) at much lower temperature (400 – 500 °C). These films however suffer from pinholes and a high H concentration in the film. They are less suitable as structural material.

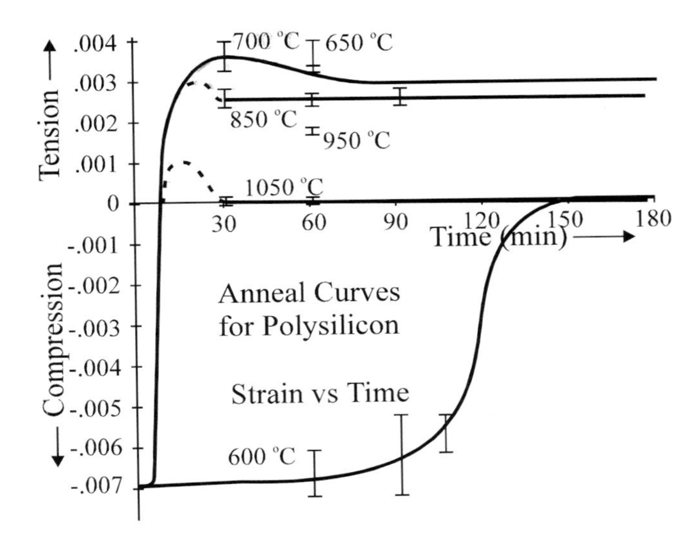

Fig. 3.3. Internal strain for LPCVD polysilicon films deposited on oxidised silicon substrates as a function of post-deposition anneal time and temperature. These films were deposited using the pyrolysis of 100 % silane at a temperature of 580 °C and a pressure of 300 mTorr (Guckel 1988). Reproduced from 'Silicon Micromachining', M. Elwenspoek and H. Jansen, (1999), by permission of Cambridge University Press

3.2.3. Evaporation

Materials with sufficiently high vapour pressure can be evaporated by resistive heating. In practice the target (this is the material to be evaporated) can be heated up to 1200 °C at most so this defines the range of materials that can be evaporated by resistive heating. Evaporation is largely restricted to deposition of elements since the composition of deposited compounds is difficult to control. If evaporation of a compound is done it is advisable to use separate targets for each element. Evaporation can also be done by RF heating, laser ablation or electron beam heating. E-beam heating is more common than resistive heating due to a number of advantages, among which less contamination, a better process control and a more efficient heat transfer to the target, so that practically everything can be evaporated using E-beam heating.

Evaporation is a fast process if the vapour pressure is large enough. Growth rates up to 1 μm/min are common. If the pressure is small enough so that the mean free path (λ) of atoms in the bell jar is larger than the distance d between the target and the substrate, the atoms move in a straight line. We have for the mean free path (Madou 1997)

$$\lambda = \sqrt{\frac{\pi RT}{2M}}\frac{\eta}{p}$$

Here, R is the gas constant, T the temperature, M the mass of the atoms, p the pressure and η the dynamic viscosity. An estimate for practical purposes is

$$\lambda p \approx 50\,\mathrm{mm} \times \mathrm{mTorr}\,.$$

The growth rate depends on the distance and respective orientation of target and substrate (see Fig. 3.4. for the definition of the symbols). The growth rate of the film is proportional to (Madou 1997)

$$A \propto \frac{\cos\beta \, \cos\Theta}{d^2}$$

This has some consequences for evaporation on substrates with surface features such as steps and trenches. Locally the angle β is changing, with the consequence that the film thickness will change accordingly. This effect implies a shadow effect which can be used in so called lift off processes as shown in Fig.3.5, quite in contrast to step coverage we have in LPCVD processes.

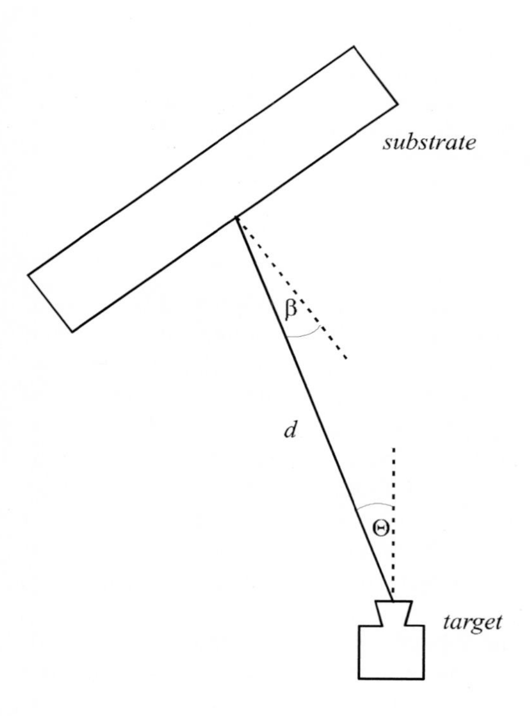

Fig. 3.4. Geometry of target and substrate during evaporation

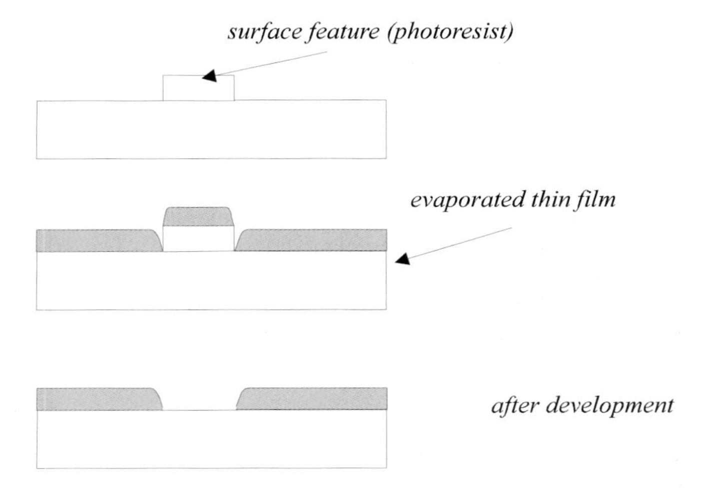

Fig. 3.5. Lift off process

3.2.4 Sputter Deposition

In deposition processes using sputtering high-energy ions hit the substrate and sputter material from the target. This material is deposited on the substrate.

Usually one uses Ar ions generated in a plasma between the target and the substrate. The target is negatively charged, so the Ar ions impinge of the target and not on the substrate which is grounded or floating.

There are some important advantages of sputter deposition compared to evaporation: practically all materials can be sputter deposited, there are more parameters for process control, the deposition may be more homogeneous over the wafer, one can use the sputtering for cleaning the target before deposition starts and the technology is suitable for upscaling. An important advantage is that the adhesion of the sputter deposited films is generally much better. This is due to the high kinetic energy of the sputtered atoms when they reach the substrate. The atoms penetrate a bit into the substrate.

3.2.5 Doping

The change of electric properties of semiconductors lies at the basis of integrated circuits. But not only electrical properties of silicon but also electrochemical, chemical and mechanical properties can be altered. Therefore doping is an important process in micromachining and micromechanics. Wet etch rates of silicon depend on the voltage difference between the silicon substrate and the etching solution, and the dependency is different for p-doped and n-doped silicon. We shall see later how this fact can be exploited. When etching electroless, the etch rate in certain anisotropically etching solutions can be reduced drastically by doping silicon with boron. Here we briefly describe the processes used for doping.

In general there are two possibilities for incorporating impurities: Ion implantation and indiffusion.

With *ion implantation* high energy ions from an accelerator hit the substrate and penetrate to a certain depth into the substrate. The penetration depth depends strongly on the ion energy. One needs keV for implantation to reach a few 100 nm. After implantation annealing helps to increase the thickness of the implanted layer, of course at the expense of reducing the concentration. Annealing is also required for restore the upper layer, which is damaged by high energy, impacts. Finally, annealing changes the electrical properties of the doped film by allowing the impurities to occupy lattice sites.

Indiffusion can be done in two ways. Solid source indiffusion is done in tube ovens in which one places silicon wafers next to boron or phosphorous wafers. The boron or phosphorous sublimates and the atoms diffuse into the silicon. During this process also an oxide layer grows, which must be removed before one continues the process. Boron silicate glass dissolves very slowly in HF, but the dissolution process can be speeded up considerably after placing the wafers in an wet oxidation oven. Indiffusion can also be done by depositing a boron or phosphor glass and anneal.

3.3 Wet Chemical Etching

Wet chemical etching is the oldest silicon micromachining technology. We distinguish between isotropic and anisotropic wet chemical etching. The chemistry of the etching is – in spite of great effort – still debated and poorly understood, and we shall not dwell into these discussions here but refer to (Elwenspoek 1999). In Fig. 3.6. we show the result of etching silicon through a mask using anisotropically and isotropically etching solutions.

The wet etching process in total is quite complex and requires considerable skill from the operators. We start here with brief descriptions of the isotropic processes, and pay more attention to anisotropic etching because it is much more important. The anisotropy of the etch rate is due to crystallographic properties of silicon, and a rather precise control of the geometry of etched structures is possible - within the limits of the crystallography. The shapes resulting from isotropic etching on the other hand result from concentration fields in the etching solution, so generally control is difficult.

3.3.1 Isotropic Etching

Etching solutions are mixtures of concentrated HF and HNO_3. These are very corrosive liquids; one of the few materials not attacked by these mixtures is Teflon, from which etching vessels and holders are made. Also noble metals are resistant and fortunately silicon nitride, which can be used best as a mask material.

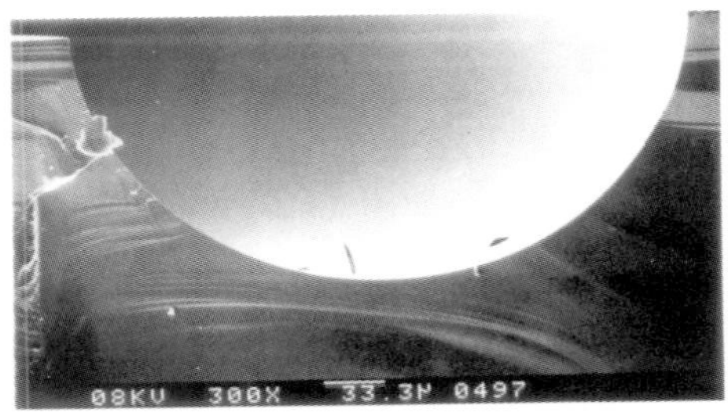

Fig. 3.6. Etchpits formed by etching silicon through a mask opening using (top) EDP (ethylenediamine, pyrocathecol, water mixture) and (bottom) HNO_3:HF:CH_3OOH. Reproduced from 'Silicon Micromachining', M. Elwenspoek and H. Jansen, 1999, by permission of Cambridge University Press

Maximum etch rates are close to 1 mm/min (! this is not a typing error), a 300 µm thick silicon wafer can be dissolved within 20 - 30 seconds. At this etch speed machining is uncontrollable, therefore the mixture must be diluted. HF and HNO_3 are mixed with water - or better - formic acid and used at room temperature. The etch rate depends strongly on the concentration, also the etched surfaces at high dilution with water or formic acid surfaces become rough. See (Elwenspoek 1999) for more details.

Tjerkstra et al. machined a chromatic column by isotropic etching (Tjerkstra 1997). They used a solution containing 5 vol% of a 50% HF solution, 15 vol% H_2O and 80 vol% of a 69% HNO_3 solution. This etches silicon isotropically, leaving half-circular channels with very smooth surfaces, see Fig. 3.7. The surface roughness, as estimated from SEM pictures, is about 3 nm. The etch rate at 20 °C is about 0.8 - 1 µm/min, depending on the structure density and mask opening size. This spread of etch rates is due to concentration gradients that vary over the wafer surface and which depend on the distribution of the open regions on the wafer.

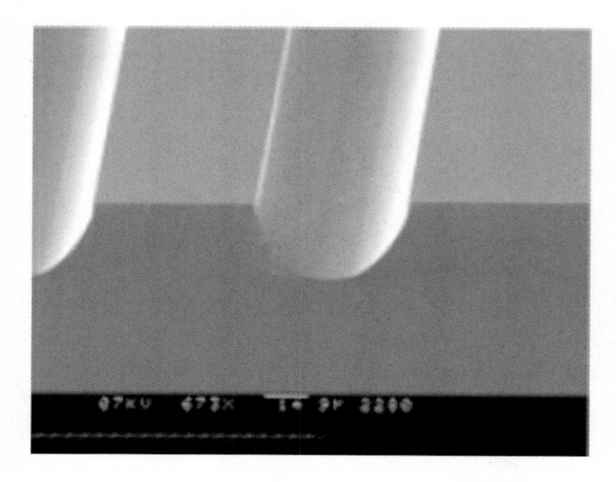

Fig. 3.7. Isotropically etched channel (Tjerkstra 1997)

3.3.2 Anisotropic Etching

The etch rate of silicon in OH-containing solutions is anisotropic. There are deep etch rate minima in the crystallographic $\langle 111 \rangle$ directions, and - depending on the solution - shallow minima in $\langle 001 \rangle$ (KOH) and $\langle 110 \rangle$ (EDP, KOH with isopropyl alcohol). The process temperature is at 60 °C or higher, and maximum etch rates are a few μm/min. The minimum etch rate in $\langle 111 \rangle$ direction is smaller by a factor typically of 100 or more, the etch rate in the direction of the shallow minima is smaller by a factor of 2 – 5 compared to the maximum etch rate. In certain cases, the shallow minima have been used for micromachining (Rosengren 1994, Bäcklund 1992), but generally, the $\langle 111 \rangle$ minima play the dominant role in micromachining using wet anisotropic etching.

In Fig. 3.8 we show the definition of the three crystal planes, which are important here.

Silicon wafers can be purchased with various orientations. Here we only consider $\langle 001 \rangle$ orientation. On the $\langle 001 \rangle$ wafers the slowly etching $\langle 111 \rangle$ planes are inclined by an angel of ca. 54°, and the edges run along the $\langle 110 \rangle$ direction on the wafer. This direction is indicated by a flat portion of the side of a wafer, the so-called wafer flat. Because of the deep minimum in the $\langle 111 \rangle$ direction, a pit etched through a mask opening on a $\langle 001 \rangle$ oriented wafer will ultimately be bounded by $\langle 111 \rangle$ planes, see Fig. 3.9.

Mask materials are silicon dioxide thermally grown in a wet oxidation oven or silicon nitride grown in an LPCVD reactor. While Si_3N_4 is practically not attacked by KOH, SiO_2 is. The etch rate ratio of Si $\langle 111 \rangle$ to SiO_2 is shown in Fig. 3.10 as a function of the KOH concentration.

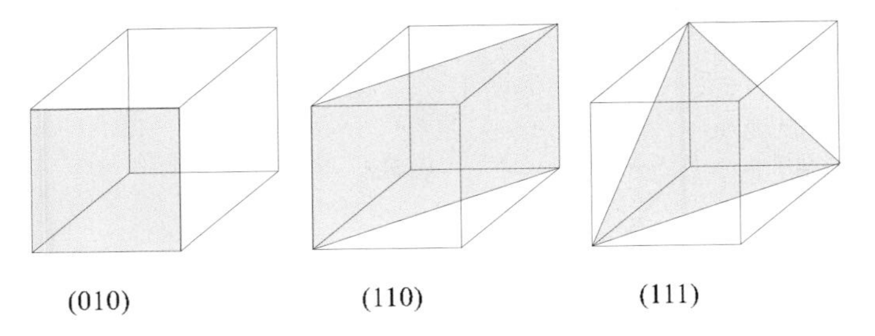

Fig. 3.8. The family of planes (from left to right) (010), (110) and (111) in cubic lattices

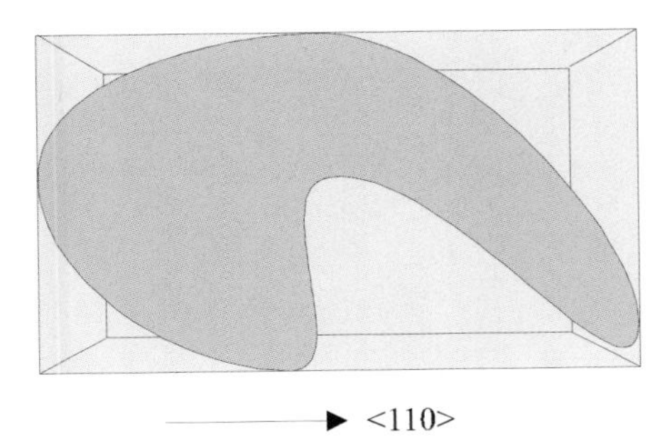

Fig. 3.9. Etch pit under a mask opening of arbitrary shape

Silicon etch rates depend on the temperature and the concentration. There is a maximum of the etch rate around 20 wt%, and this concentration is used normally because small changes in concentration do not affect the etch rate. Temperature and concentration dependency is shown in Figs. 3.11 and 3.12.

Anisotropic etching as such allows a good control of the lateral dimensions of features[2], but without additional measures the etch depth can be controlled only by the etch time. Normally thin structures must be machined which are made by etching right through wafers, leaving only a thin membrane or beam, much thinner than the wafer. There are some means to optimise the control of etch depth by optically monitoring the etching (Minami 1994), however in the research laboratory the fabrication of thinner structures than 10 μm is not practical, *a fortiori* for production.

[2] Although sometimes we hear that the underetching depends on the batch KOH: sometimes the reproducibility is bad [Moldovan 1998, Steckenborn 1998]

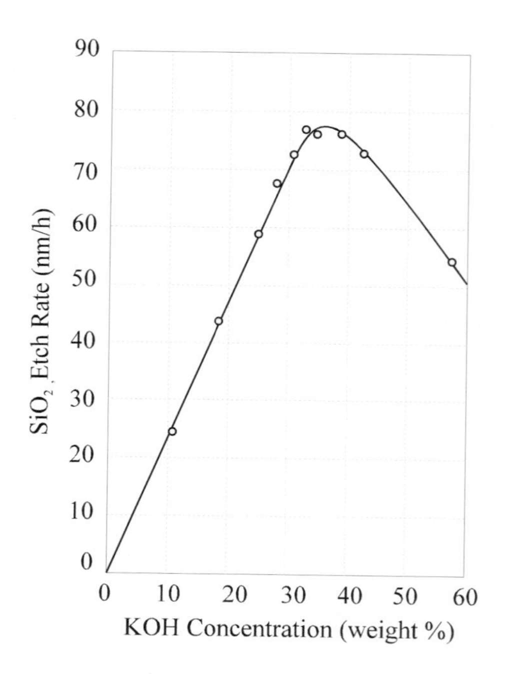

Fig. 3.10. SiO_2 etch rate in aqueous KOH solutions at 60 °C as a function of concentration. Reproduced from 'Silicon Micromachining', M. Elwenspoek and H. Jansen, 1999, by permission of Cambridge University Press. (Seidel 1990)

The last point we have to comment on is compensation of convex corners. Anisotropic etching stops at concave corners of two (111)-faces. Convex corners however are etched. During etching higher index planes appear such as (411). Often it is required to etch mesas, structures bound by (111) planes but convex, and it is necessary to prevent the etching of convex corners. This can be achieved by giving the etch mask an appropriate form. An example is given in Fig. 3.13. The width of the bars must be approximately equal to twice the height of the mesa. Programs that simulate the structure one gets by etching with a given mask are commercially available. For an overview and more details and references, we refer to Elwenspoek (1999) and Madou (1997).

3.3.3 Etch Stop

There rest three basic ways to stop the etching at a well controllable thickness. In the section on doping we referred to boron (p-doping) and phosphor (n-doping) for micromachining purposes. This is explained here.

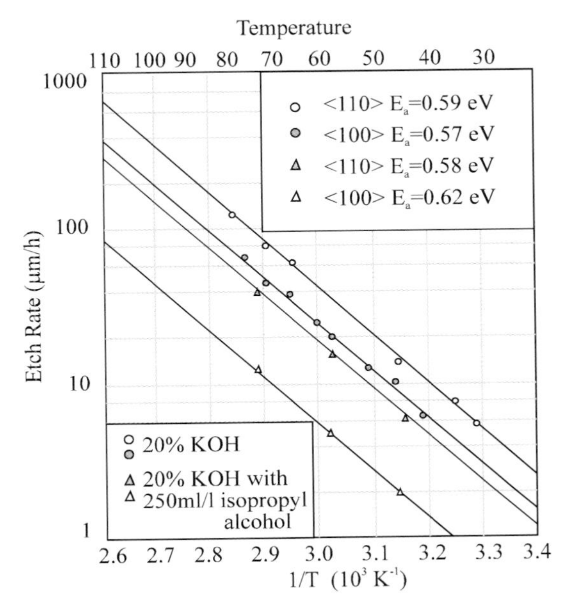

Fig. 3.11. Temperature dependence of etch rates in 20% KOH solutions in ⟨100⟩ and ⟨110⟩ directions, with and without the addition of isopropyl alcohol. Reproduced from 'Silicon Micromachining', M. Elwenspoek and H. Jansen, 1999, by permission of Cambridge University Press (Seidel 1990)

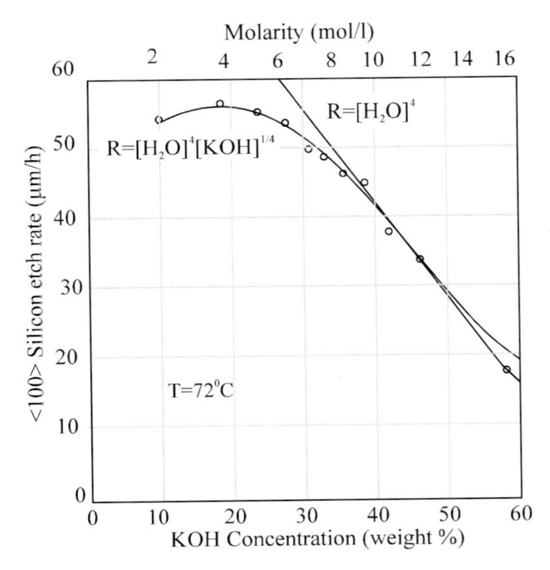

Fig. 3.12. Concentration dependence of the etch rate. Reproduced from 'Silicon Micromachining', M. Elwenspoek and H. Jansen, 1999, by permission of Cambridge University Press (Seidel 1990)

Fig. 3.13. Etch mask for a corner compensation

In Fig. 3.14 we show how the etch rate of silicon in EDP (ethylendiamide, pyrocathecol pyridine and water) type S^3 depends on the boron concentration. It is seen that around 2-3×10^{19} atoms/cm^3 the etch rate starts to decrease, and close to 10^{20} atoms/cm^3 the etch rate is smaller by a factor of 100. By indiffusion or ion implantation (large dose!) it is possible to reach this high concentration (close to the solubility of B in Si). Membrane thickness between a few 10's of nm to 20 μm have been achieved using this etch stop technique (van Huffelen 1991). The problems associated with B+-etch stop are high tensile stress and high conductivity of the material. Further, after indiffusion the wafer is too rough to bond (see below) a wafer on the B+ layer by silicon fusion bonding. It is also important to keep in mind that areas of indiffusion are a bit oxidised, after removal of the oxide there is a shallow recess compared with the undoped silicon.

The B+ etch stop does not work so well in KOH. One needs either a small KOH concentration (below 10 wt%), or one has to add isopropyl alcohol to the solution. Both leads to very rough etched ⟨001⟩ surfaces with many pyramids, small ones (~5 μm) in the first case, fewer but large ones (50 μm) in the latter case.

The etching can also be stopped in an electrochemical set up. The etch rate of silicon n- or p- doped as a function of a bias is shown in Fig. 3.15a. The potentials indicated in the curves by "PP" denote the "passivation potential". At this potential there is a competition between etching silicon and the growth of an oxide layer. The oxide layer finally passivates the silicon against the etchant. This effect is exploited in the electrochemical etch stop.

The method involves growing an n-type epitaxial layer on a p-type substrate or to dope silicon accordingly by indiffusion or implantation to form a p/n junction. The aim is to etch away the substrate selectively by terminating the etching at the junction.

[3] Water 133 ml, ethylenediamine: 1l, pyrocatecol: 160 g, pyrazine: 6g

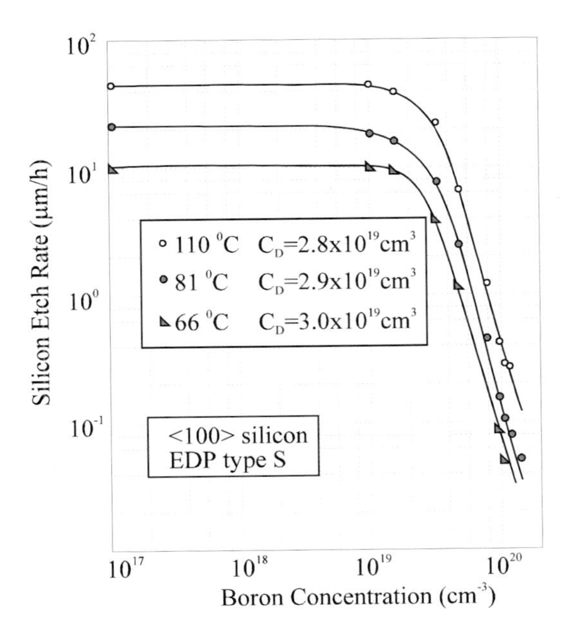

Fig. 3.14. Etch rate of ⟨100⟩ silicon as a function of boron concentration for various temperatures. EDP type S rate Reproduced from 'Silicon Micromachining', M. Elwenspoek and H. Jansen, 1999, by permission of Cambridge University Press (Seidel 1990)

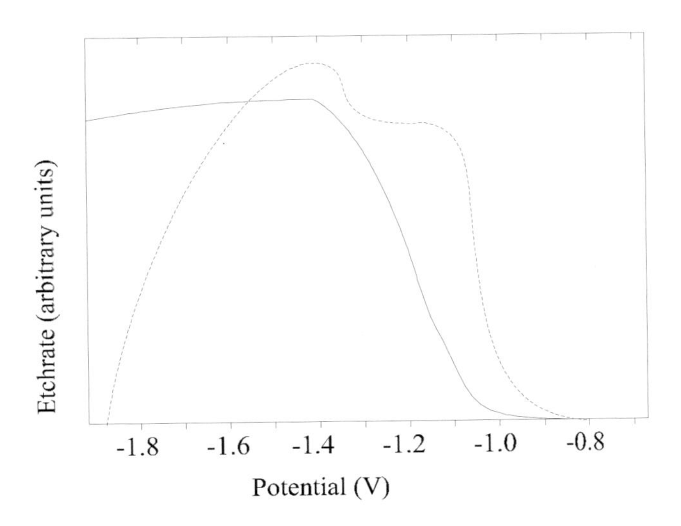

Fig. 3.15 a. Potential dependence of the etch rate of ⟨100⟩ silicon. Full line: p-type, broken line: n-type silicon in 40 % KOH at 60 degree C. (Kloeck 1989)

This is done in the following way. A positive voltage is applied to the epitaxial n-type layer (positive with respect to a reference electrode in the etch solution). This voltage maintains the epitaxial layer at a passivating potential to prevent its etching Simultaneously the p/n junction is reverse biased, and the substrate floats to the OCP (open circuit potential) and is etched. When the etch front reaches the p/n junction, the etching stops. The thickness of the remaining silicon membrane is thus defined exactly by the thickness of the epitaxial layer. Initial irregularities at the back surface have no influence on the final result. Thus the method is very precise, reproducible and reliable.

Best results are obtained when a four electrode setup is used, to force the potential of the substrate to the OCP (Kloeck 1987, 1989), Fig. 3.15b. Three of the four electrodes contact the n- and p-regions of the wafer and the etching solution, and the fourth is a reference electrode.

The disadvantage of this etchstop is the tricky technology. One needs special designed holders to make the necessary electrical contacts. The leakage of wafer holder for protection of parts that react with the etching solution (electric contacts) constitutes a notorious problem.

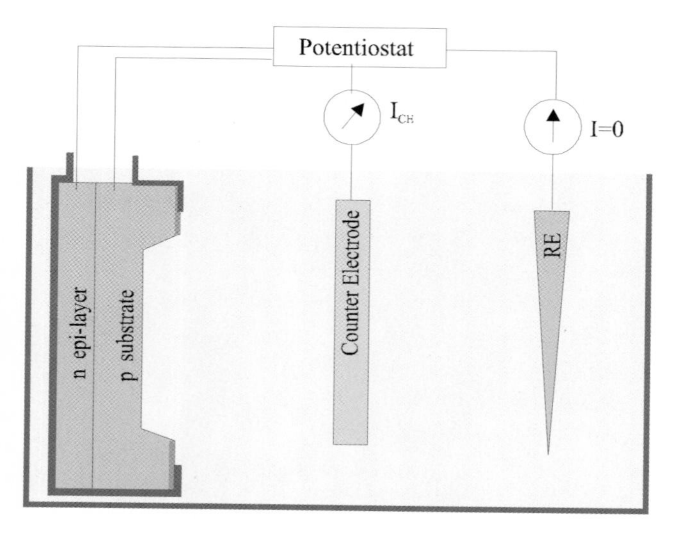

Fig. 3.15 b. Four electrode set up for the electrochemical etchstop

3.4 Wafer Bonding

Wafer bonding refers to the fixation of wafers on top of each other. It is an indispensable technique for silicon micromachining. Wafer bonding not only allows us to fabricate constructions with rather complex structures in three dimensions, but it also plays a role in packaging. Because whole wafers can be bonded containing many single components (here: sensors), tedious and expensive assembly can often be circumvented by wafer bonding.

Silicon is very smooth. Wafers, when contacted, do not touch each other at three small spots, as normally is the case, but on extended regions. Wafers, when clean and flat (not bent too much), spontaneously adhere to each other. This tendency can be enhanced using external forces, such as a weight on top of the wafer package or – much more efficient – electric fields. Bonding without the help of external forces between silicon wafers with or without thin films on top is called "direct silicon bonding", DSB, or "silicon fusion bonding", SFB (although the process has nothing to do with fusion). Using electric forces requires additional materials (glass), and is referred to as "anodic bonding".

3.4.1 Anodic Bonding

Anodic bonding was discovered in the sixties (Wallis 1969). The technology involves the bonding of a metal (electrical conducting) plate and a glass wafer, which becomes conductive at a high temperature. The electric field is then confined to a thin depletion layer in the glass and in the gap between the wafers. If the wafers are not too rough (maximum roughness must be below 1 μm), the resulting pressure is enough to deform the material at the interface so that the wafers join smoothly. Furthermore, the process temperature must be high enough for oxidation of the metal, so that covalent bonds form at the metal-glass interface.

The basic requirements for anodic bonding therefore are:

- the glass must be slightly conductive
- the glass and the metal must not inject charge carriers into each other (the field would break down)
- the roughness must be below 1 μm
- the thermal expansion coefficients must match sufficiently (otherwise the wafer package would break apart due to the thermal stress one gets when cooling the wafer package down to room temperature)

A suitable glass for silicon anodic bonding is Pyrex, but also Corning 7070 is OK (it is practically the same material). It is also possible to sputter a thin glass film from a Pyrex target on a silicon wafer and proceed with anodic bonding with a second silicon wafer (Hanneborg 1991).

A schematic set up is shown in Fig. 3.16. The wafer package is placed on a hot plate (process temperature 400–450 °C), and the glass wafer is connected to a high voltage source (required voltage in the range of 1000 V). The process can be monitored by the naked eye – the colour of the bonded region looks grey, quite different from unbonded regions. In Fig. 3.17 we show how voltage and temperature should be set, and how the current results. We see that typical process times are in the range of 20 minutes. It is also apparent that the process is finished if the current drops to zero.

When using a thin sputterdeposited Pyrex film, the voltage must be much smaller (20–50 V). Bonding has been demonstrated with Pyrex thin films ranging in thickness between 50 nm and 3 μm (Berenschot 1994).

Both processes are quite reliable, provided the wafers are dust free.

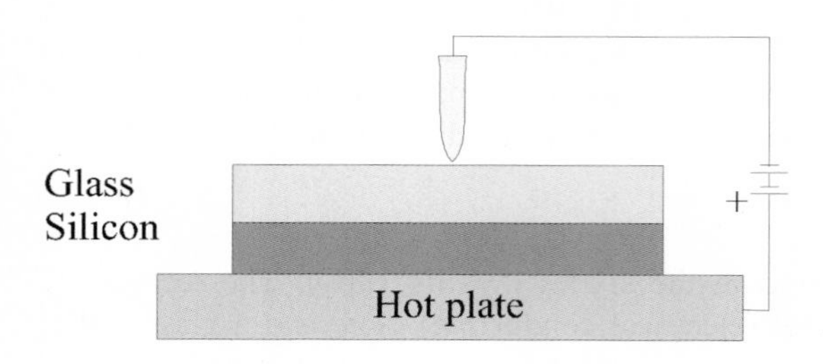

Fig. 3.16. Schematic of an anodic bonding set-up (Hanneborg 1991)

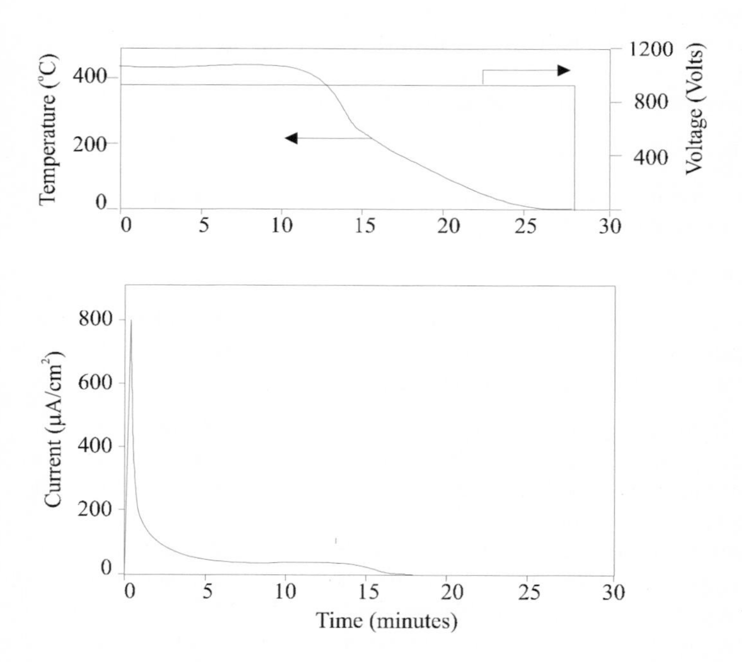

Fig. 3.17. Typical temperature, voltage and current density profiles as a function of the time during an anodic bonding process

3.4.2 Silicon Fusion Bonding

The interfacial energy between two bare silicon wafers covered with native oxide (hydrophilic wafers) is due to hydrogen bridging Stengl 1989). The resulting forces are orders smaller than forces that can be achieved with electric forces as in anodic bonding. The result is that direct silicon bonding is a much more difficult process. The microroughness of silicon wafers must be below 1 nm.

The oxide on the wafers can be removed (in HF) making the wafers hydrophobic. The interfacial energy of the contact is then due to van der Waals forces (Bäcklund 1992), which are still on order of magnitude smaller than forces due to hydrogen bridging. Nevertheless bonding is possible, though difficult, and this process even has advantages.

Normally one uses hydrophilic wafers, as they come out of all wet cleaning procedures. The wafers are contacted under clean room conditions. The adhesion is large enough to allow normal handling of the wafer package. The package is annealed in a wet oxidation oven at 700 °C (to our own experience, this is the optimal process temperature (Sanchez 1997)) or higher. Between 300 and 700 °C voids form, which result from water.

Prior to bonding the wafers can be processed (Harendt 1991), but each process step increases the roughness of the wafers. The process yield declines after too many steps. Some steps make DSB absolutely impossible, prominently wet etching of the whole surface (it is possible to etch cavities), deposition of too thick Si_3N_4 (more than 200 nm) or boron doping by indiffusion.

There is however a remedy. The wafer manufacturers polish silicon to the fantastic smoothness using a process called chemical-mechanical polishing (CMP, Fig. 3.18 (Gui 1998)). In CMP the wafer is placed on a rotating polishing

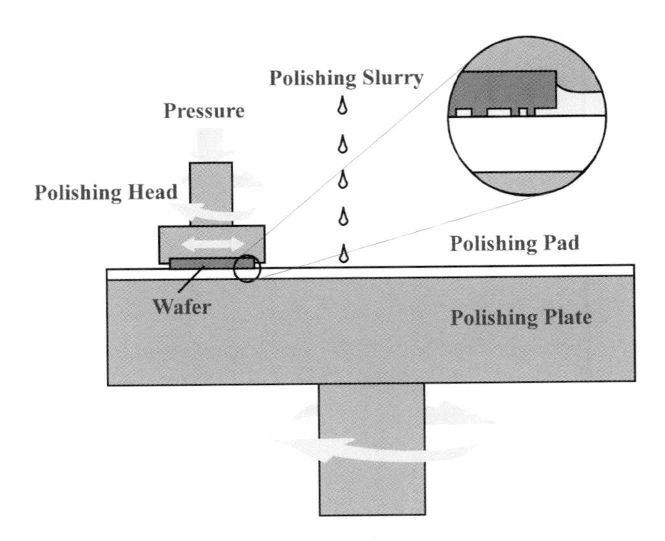

Fig. 3.18. Schematic view of a CMP machine (Gui 1998)

head and moved under a certain pressure over a polishing pad. A slightly alkalic slurry containing nm sized silica particles is added. The combined action of wear and etching yields the smooth surfaces. This process can also be applied to other materials among others silicon nitride, silicon oxide and poly silicon (Gui 1998). It appears that all materials that can be polished down to a roughness better than 1 nm are bondable (Haisma 1994).

In Fig. 3.19 we show a silicon nitride surface (1 mm thick stress free S LPCVD nitride) before and after polishing and in Fig. 3.20 we show the bonding interface (Gui 1998).

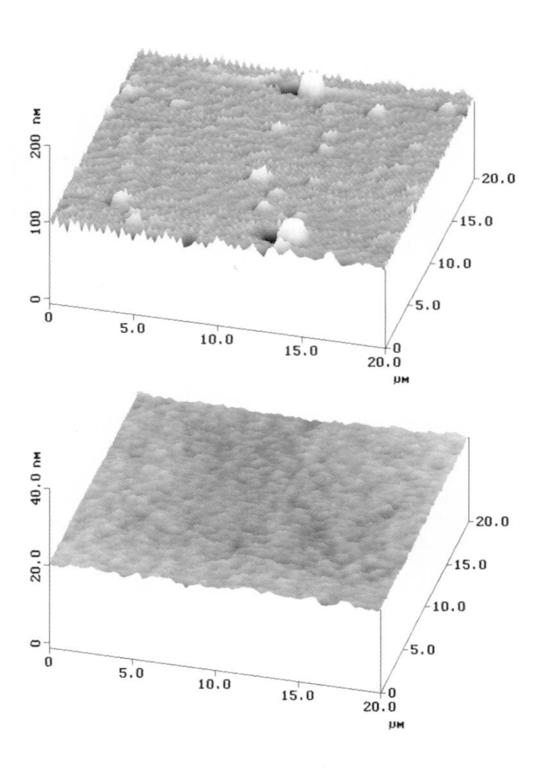

Fig. 3.19. Top: LPCVD $Si_{3+x}N_4$ surface before CMP. Root mean square roughness 3.6. nm. Bottom: LPCVD $Si_{3+x}N_4$ surface after CMP. RMS = 0.4 nm (Gui 1998)

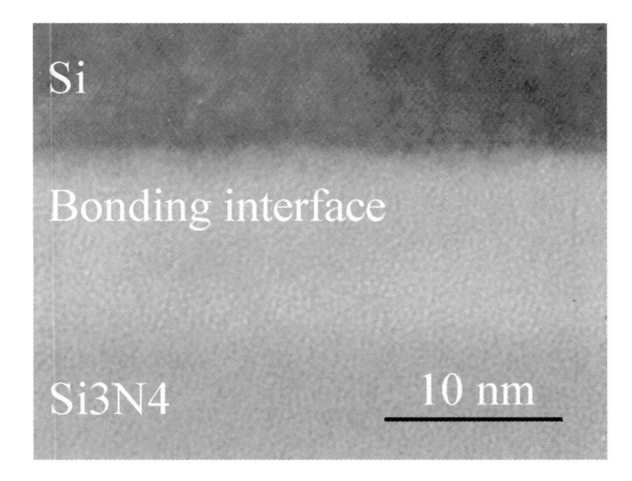

Fig. 3.20. TEM image showing the bonding interface of silicon and silicon nitride (Gui 1998)

3.5 Plasma Etching

Wet chemical etching is confined either by the ⟨111⟩ planes of single crystalline silicon or by poorly controllable diffusion fields. Both techniques severely limit the possible design space for mechanical construction. Plasma etching offers much more. In plasma etching ions impinge on the substrate, and neutral particles reach the substrate by diffusion. The combined action - which is quite complex, as will be seen - leads to anisotropic etching. The anisotropy here is not the result of crystallographic properties but a result of the direction of the ion flux towards the substrate. In particular reactive ion etching (RIE) and related techniques have been developed during the past years for micromachining. Originally developed for IC processing (etching of thin films, pattern transfer) for comparatively shallow etching (etch depth not more than 1 - 3 μm), MEMS has a need of great structure height, so etching must be deep, with neat sidewalls, vertically straight through the wafer. Several processes have been developed to make this possible.

3.5.1 Plasma

An overview of the main dry etch techniques is given in Fig. 3.21. This overview includes also the chemical etching from a vapour and laser assisted etch methods. Ions are created either in a glow discharge or an ion source. In the latter case we speak of ion beam techniques.

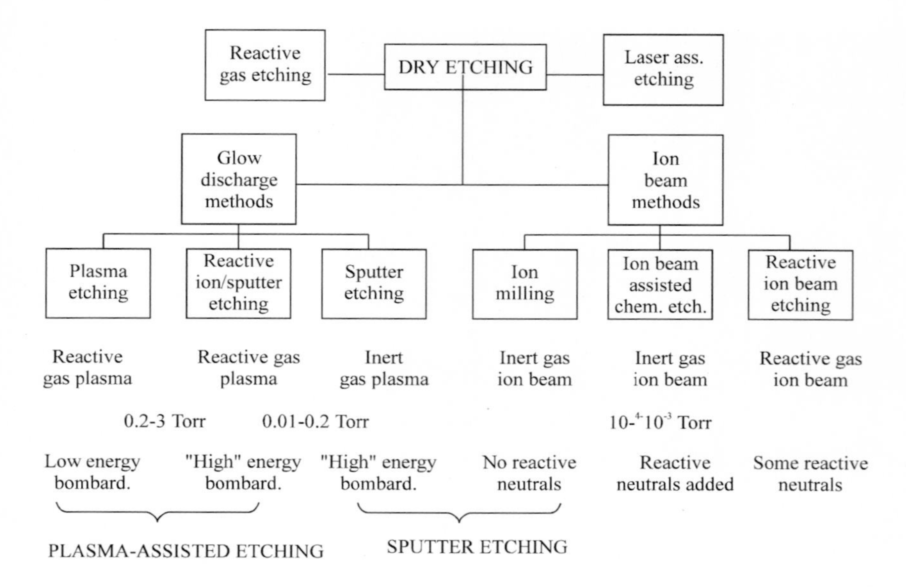

Fig. 3.21. The relationship between the various dry etching techniques (Lehmann 1991)

Chemical etching is isotropic in general. The etching is only possible if the reaction products are volatile. Further, because it is a chemical reaction, the etch rate depends strongly on the material to be etched: the etching is selective. In a plasma the etching species are radicals: atoms or parts of molecules with an unpaired electron. Radicals are very reactive. An important example for chemical plasma etching is the stripping of photoresist in an oxygen plasma, a process called “ashing”. In Fig. 3.22. we show an example for the result of isotropic etching of silicon in a fluor based plasma.

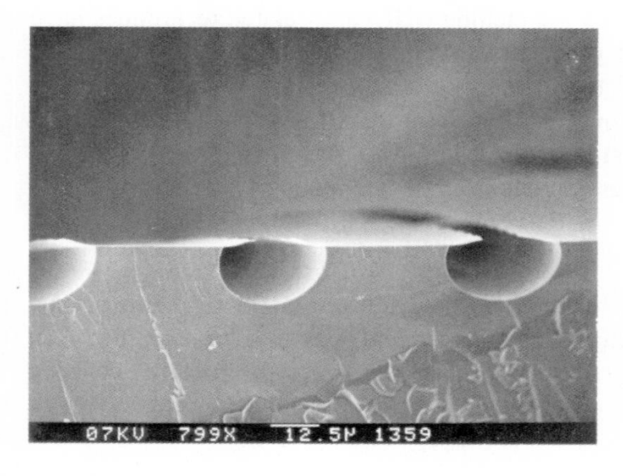

Fig. 3.22. Isotropic profile by reactive ion etching in SF_6/N_2 (Spiering 1993)

If there are only ions hitting the substrate while neutrals have no chemical effect the etching is said to be "physical". Using an ion beam we speak of "ion beam milling" (IBE), using a glow discharge we speak of "sputter etching". Both processes are physically identical. The main features are:

- the etching is anisotropic, as a result of the ion flux
- any material can be etched
- etch rates do not depend very much on the type of material
- the etched material is redeposited anywhere in the etch equipment
- maximum etch rates are in the order of 1 atom removed per impinging ion, about 1 keV is required
- there is a threshold energy for etching (tens of eV)
- the sidewalls are inclined: trenches become narrower the deeper they are etched

In Fig. 3.23 we show the result of IBE etched in a stainless steel foil.

The combination of chemical etching and physical etching makes plasma etching so interesting for micromachining. We call this mode "synergetic". In ion beam methods one speaks of Chemical Assisted Ion Beam Etching (CAIBE), in discharge reactors this is called reactive ion etching (RIE). The main difference is that in RIE the chemical etching is due to radical formed in glow discharge, while in CAIBE a reactive gas is fed into the reactor. Because a chemical reaction is involved in the process the etching is selective, and the etch products must be volatile.

3.5.2 Anisotropic Plasma etching modes

There are two distinct modes for the anisotropy in CAIBE and RIE. In one mode the anisotropy is due to a modification of the chemical-physical state after impact of an ion. This mode is called the ion energy induced mode. The temperature may be raised locally, chemical bonds may be ruptured, the physical site in which the atoms are may be altered, in one way or the other the chemical reaction rate is greatly enhanced by the ion - substrate interaction. Only where ions hit there is an appreciable etch rate. Ions do not hit the sidewalls in a trench (in a first approximation), so only the bottom of the trench is hit. The etching of silicon dioxide in CHF_3 plasmas follows this mode.

The other mode is more complex. If one of the reaction products is not volatile, inert against the radical or chemicals responsible for the chemical reaction, but are etched by ions, this material will be deposited everywhere in the reactor, in particular on the sidewalls of trenches and protect the sidewalls against chemical etching. The material is also deposited on the bottom of the trench but it is removed by ions. A schematic drawing of this process is shown in Fig. 3.24. In Fig. 3.25 we show a result in silicon using this method. This mode is called the inhibitor mode.

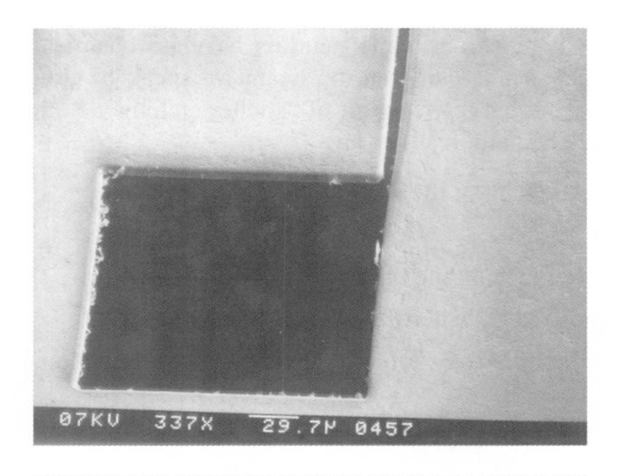

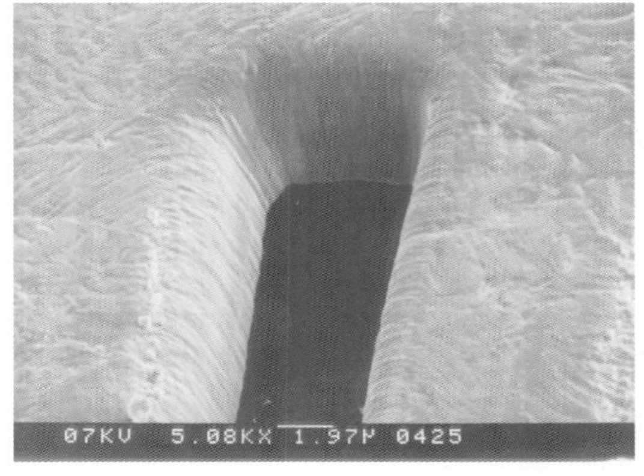

Fig. 3.23. Microstructure etched by IBE in stainless steel (Deheij 1996)

3.5.3 Configurations

A typical plasma etch configuration is shown in Fig 3.26. The plasma is created by an RF source coupled capacitively via a tuning network to the reactor chamber. The typical glow in the plasma is due to the relaxation of atoms in an excited state after collisions of electrons with atoms or molecules. Since the electrons have a much smaller mass than the ions, they move faster. Further in collisions with atoms the electrons do not loose much energy, so the electron gas is much hotter than the ions and neutrals. Table 3.1. gives the possible reactions in the plasma, together with an example, where we also give the energies ΔE involved in the reaction. The energies must be compared to the kinetic energy of the electrons, E_{kin} which is typically a few eV. The density of the various reaction products is proportional to exp-$\{\Delta E/kT\}$. The electrons have a certain mean free path. Electrons that hit the reactor walls are absorbed, so the plasma will be positively charged and there will be a voltage difference between the reactor walls and the plasma. The voltage drops over a region called dark space or sheath close to the reactor walls; this is a region of an extension of the order of the Debeye length, typically 0.1 mm, where only few electrons hit atoms. Since the plasma is conductive, there is no electric field inside the plasma, only in the sheath. Ions are accelerated in the sheath. In Fig. 3.27 we show a schematic drawing of the distribution of ions, radicals and electrons close to a substrate.

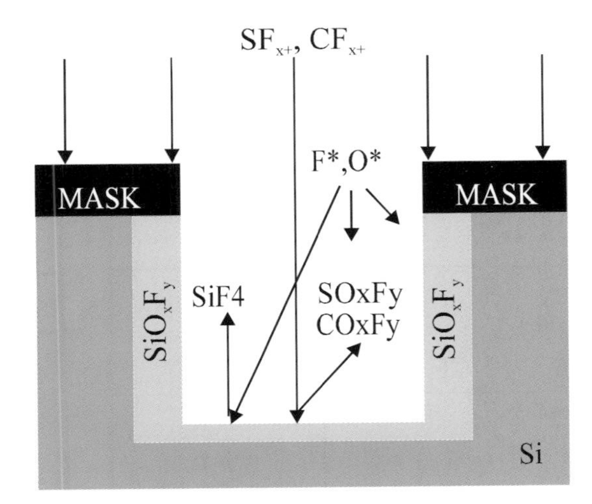

Fig. 3.24. Schematic overview of processes in a SF_6-O_2-CHF_3 plasma. Reproduced from 'Silicon Micromachining', M. Elwenspoek and H. Jansen, 1999, by permission of Cambridge University Press

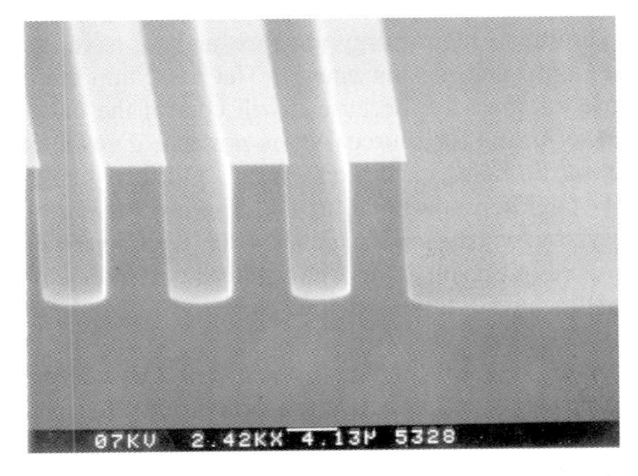

Fig. 3.25. Anisotropic profile by reactive ion etching in SF_6/O_2.Reproduced from 'Silicon Micromachining', M. Elwenspoek and H. Jansen, 1999, by permission of Cambridge University Press

Table 3.1. Examples of Electron Impact Reactions (Mucha 1983)

Reaction	Energy
Excitation (electronic, vibrational, rotational): $e + A_2 \rightarrow A_2^* + e$ $e + CF_4 \rightarrow CF_4^* + e$	4 eV
Ionisation: $e + A_2 \rightarrow A_2^{+*} + 2e$ $e + CF_4 \rightarrow CF_4^+ + 2e$	15.5 eV
Dissociation: $e + A_2 \rightarrow 2A + e$ $e + CF_4 \rightarrow CF_3 + F + e$	12.5 eV
Dissociative attachment: $e + A_2 \rightarrow A^- + A^+ + e$ or $A^- + A$ $e + CF_4 \rightarrow F^- + CF_3^+ + e$ or $F^- + CF_3$	7.3 eV
Dissociative ionisation: $e + A_2 \rightarrow A^+ + A + 2e$ $e + CF_4 \rightarrow CF_3^+ + F + 2e$	14.8 eV

Figure 3.28 shows how the electrical potential varies in the reactor. It is seen that the capacitively coupled electrode is on a large negative potential. Placing a substrate on this electrode will result in a high energy ion flux, and we have the possibility to control the voltage drop and the ion energy. The most important parameters for the ion energy are the pressure, because it will control the mean free path of the ions, and the power of the RF source. A low pressure gives ions more time to pick up energy from the field, so the ion energy, and the voltage drop across the sheath will very roughly proportional to 1/p. The mechanism of etching (physical, chemical or synergetic) depends mainly on the RF power but also on pressure. Low pressure tends to result in a large sputtering component: poor selectivity, redeposition. High pressure means high selectivity, isotropic etching. In between, typical at 1–100 mTorr, we have the synergetic mode.

To some extend ions also reach the sidewalls. In fact, if the trench becomes too deep, all ions have reached the sidewalls and the etching stops. Wider trenches etch faster than narrow ones. This phenomenon is known as "reactive ion lag", and demonstrated in Fig. 3.29. The RIE lag has consequences for the mask design. If a homogeneous depth of the etching is desirable it is important to define mask openings of a constant width.

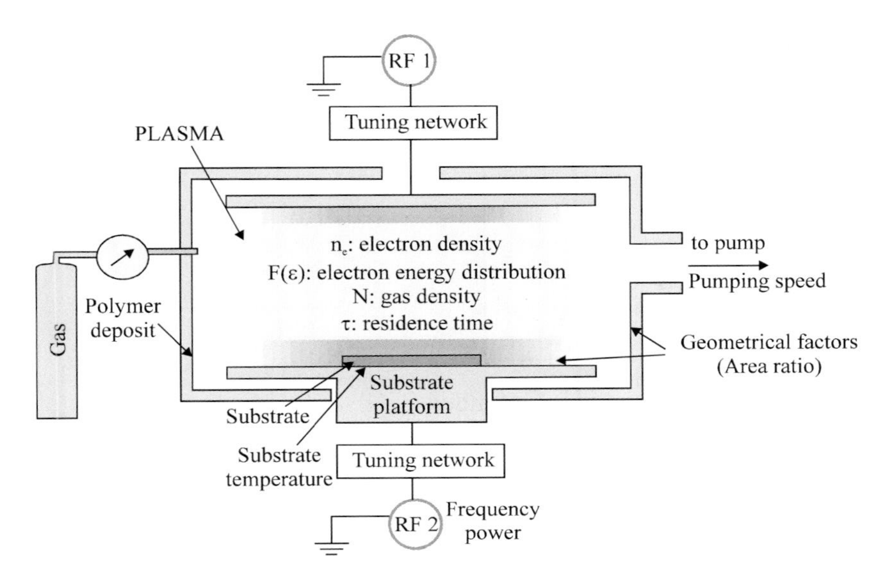

Fig. 3.26 Schematic cross-section of a parallel plate reactor showing experimental and fundamental plasma chemical parameters. Usually RF-voltage is applied either to the bottom electrode (in Reactive Ion Etching, RIE) or to the top electrode (in Plasma Etching, PE) (Lehmann 1991)

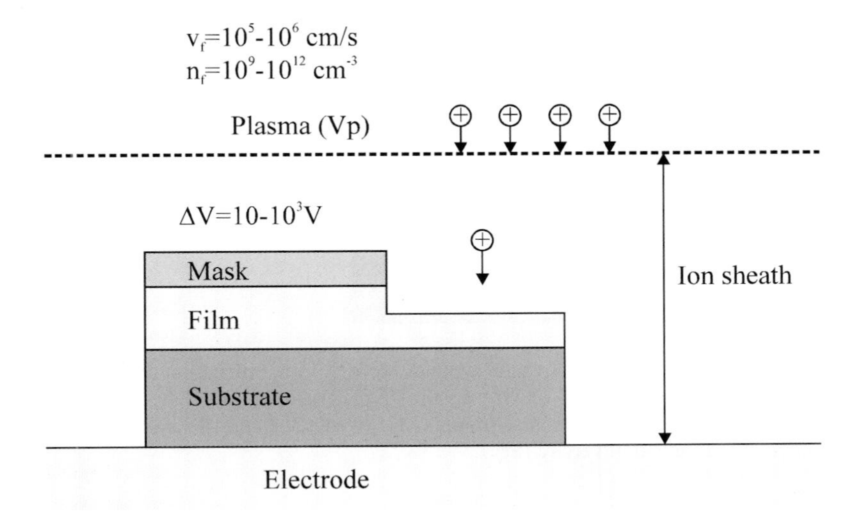

Fig. 3.27. Ion motion is random in the central glow but when a positive ion drifts to the sheath boundary (dotted line), the perpendicular electric field accelerates it towards the wall and wafer surfaces. Typical parameters are shown (Lehmann 1991)

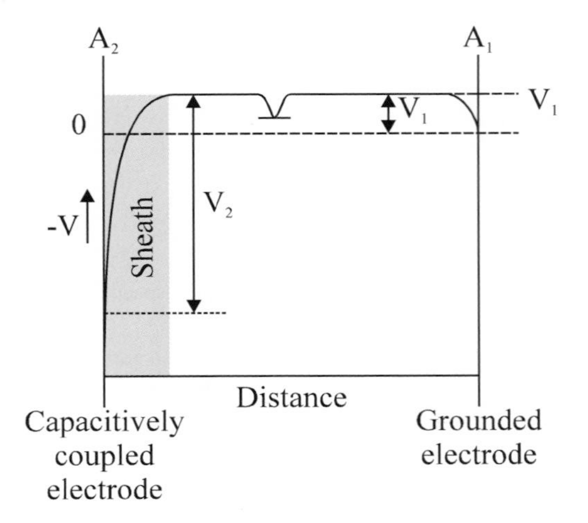

Fig. 3.28. Potential distribution in a parallel plate etcher with the grounded surface area larger than the area of the powered electrode. -V: RF-induced DC voltage (self-bias), V_p: plasma potential, $V_2 = | V + V_p |$ determines energy of ions striking the surface, V_f: floating potential (Mucha 1983)

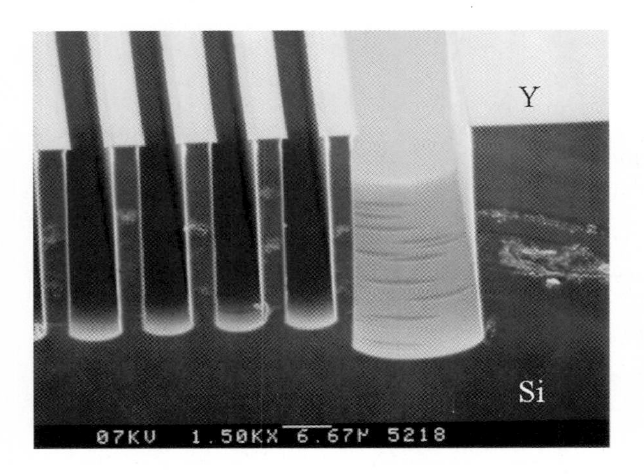

Fig. 3.29. Demonstration of the RIE-lag. Also the influence of ion bowing on the etched profile is visible. RIE settings: power 75 W, pressure 75 mTorr, SF_6:O_2:CHF_3 = 28:10:5 sccm. Reproduced from 'Silicon Micromachining', M. Elwenspoek and H. Jansen, 1999, by permission of Cambridge University Press

3.5.4. Black Silicon Method

In the inhibitor mode there is a competition of deposition and ion-induced etching of the passivating layer. If deposition prevails the side walls will tend to grow and the result is an inclination of the wall toward the trench. If etching prevails, there will be some underetching and the walls will be inclined towards the material. Jansen et al. found that it is possible to change the inclination by changing the etch conditions in a gas flow mixture containing SF_6, CHF_3 and O_2. In Fig. 3.30 we show structures with walls inclined in either direction. The conditions for vertical etching obviously must be found right in between the conditions resulting in the two structures shown. The interesting point now is that there is practically always some sputtering in the systems, so in particular, mask material will be redeposited. If the conditions lead to structures shown on the left hand side in Fig. 3.30, the redeposited material will act as a small mask, and features etched in silicon will emerge. These features are known as "micrograss" and are shown in Fig. 3.31. Macroscopically, micrograss reveals itself by absorbing light: the wafer covered by micrograss will look black. Once black wafers are seen in the reactor, the etch conditions are close to the optimal point for vertical etching. This is the basis of the "Black Silicon Method" (Jansen 1996).

The conditions leading to inclination of sidewalls to either side are shown in Fig. 3.32. Adding O_2 to the flow will reduce the underetching until the sidewalls incline towards the trench, and the wafer becomes black.

It must be noted that the locus of the lines shown in Fig. 3.32. are strongly temperature dependent. Since the chemical reactions are exothermic the substrate will be heated by the process and the temperature changes. The process will move away from the initial process line if the temperature is not controlled.

Fig. 3.30. The effect of oxygen addition to a SF_6 plasma. From left to right the oxygen content of the plasma is decreased, which leads to a change from a positively to a negatively tapered profile. Reproduced from 'Silicon Micromachining', M. Elwenspoek and H. Jansen,1999, by permission of Cambridge University Press

Fig. 3.31. Formation of "black silicon" (also frequently called "micro grass"). Reproduced from 'Silicon Micromachining', M. Elwenspoek and H. Jansen, 1999, by permission of Cambridge University Press

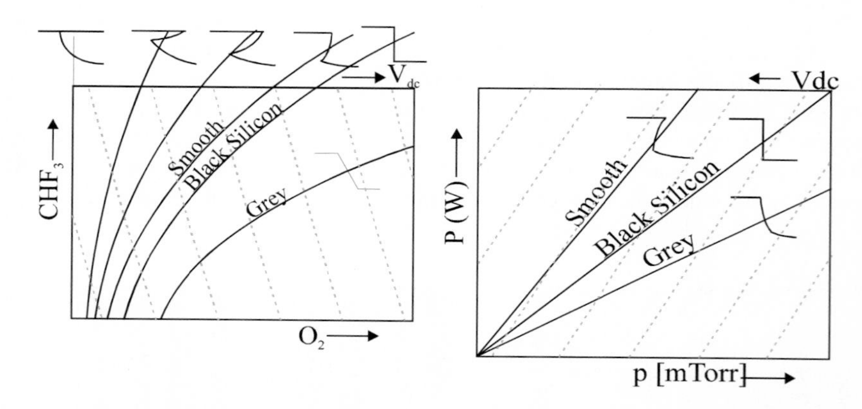

Fig. 3.32. (a) Influence of gas flow variations on etch profile, and (b) Influence of power and pressure variations on etch profile, for SF_6-O_2-CHF_3 plasmas. Reproduced from 'Silicon Micromachining', M. Elwenspoek and H. Jansen, 1999, by permission of Cambridge University Press

3.6 Surface Micromachining

Surface micromachining is a process to fabricate free standing and freely moving microstructures in a large two dimensional design space. The process idea is shown in Fig. 3.33. The trick is to deposit and pattern two thin films of materials that can be etched away selectively with respect to each other. The substrate now only plays a role as a mechanical carrier; the structure is made from a thin film material.

Although many combination of materials is possible for this process, the technology has developed to a high standard for the combination of silicon dioxide or PSG as the sacrificial material and polysilicon as the structural material. Here we restrict ourselves to this combination.

Users of the process must master two basic problems, thin film stress and sticking.

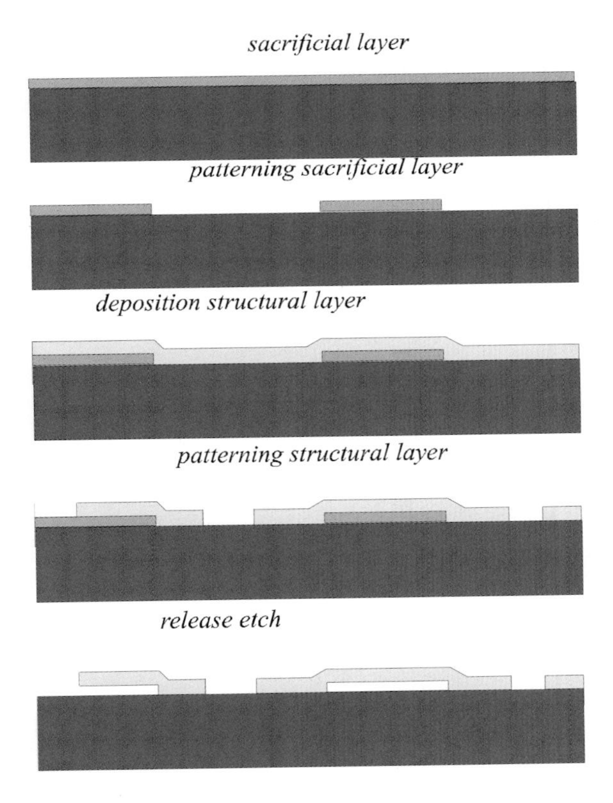

Fig. 3.33. Process scheme of surface micromachining.

3.6.1 Thin Film Stress

In Chap. 3.2 we commented on thin film stress of polysilicon films. We saw that annealing is the essential step. During processing one must incorporate test structures to monitor the stress of the structural layers. There are basically two types of stress. We have tensile or compressive lateral stress and there are stressgradients in the films. Test structures for lateral stress work on monitoring the buckling of structures. A buckled microbridge is seen in Fig. 3.34. For beams clamped at both sides we have

$$L_{\text{cr}} = 2\pi\sqrt{EI/\sigma A}$$

I is the moment of inertia of the beam, *A* it's cross section and σ is the strain.

In an array of microbridges such as those in Fig. 3.34 the shortest bridge that buckles gives one the value of the stress in the film. Tensile stress must be determined via a mechanical transformation (Guckel 1992). Stress gradients can be seen in cantilever beams. They curl up or down depending on the sign of the gradient, and the radius of curvature is proportional to the stressgradient.

The moment is related to the stress distribution $\sigma(z)$ (z is the co-ordinate normal to the surface of the beam) by

$$M_0 = \int_{-\frac{h}{2}}^{\frac{h}{2}} \sigma_x(z) z dz$$

and the radius of curvature R is given by

$$R = \frac{Eh^3}{12M_0}$$

Fig. 3.34. Buckled polysilicon structures demonstrating compressive stress of unannealed thin films (Howe 1988)

3.6.2 Sticking

The second problem that has to be mastered is sticking. When etching the sacrificial layer the free-standing structures have the tendency to touch the substrate and stay there. The cause is thought to be the surface tension of the liquid during evaporation, see Fig. 3.35. The liquid forms a droplet during drying between the microstructure and the substrate, and if the drying liquid wet both the droplet has an underpressure, and the result is a collapse of the microstructure, if it is not stiff enough.

There are solutions to the problem. If possible, the structures should be made stiff enough so that no sticking occurs. The critical length of a microbridge is (Legtenberg 1993)

$$L_{cs} > 1.059\left\{\frac{8Ed^2h^3}{\gamma(\cos\theta_1 + \cos\theta_2)}\right\}^{1/4}$$

Here γ is the surface tension of the rinsing liquid and θ_1, θ_2 are the contact angles the liquid makes with substrate and microstructure. Longer bridges will collapse.

Often the structures must be designed to that they are larger than the maximum length. Then other rinsing procedures can be used, such as critical point drying (Mulhern 1993) or freeze drying (Guckel 1989). Both methods avoid the free liquid surface. Another alternative is to introduce a surface roughness of the microstructure, because in the end sticking is just a waferbonding process. Special designed structures can be designed so that the surface which comes into contact with the substrate is quite small (Tas 1996).

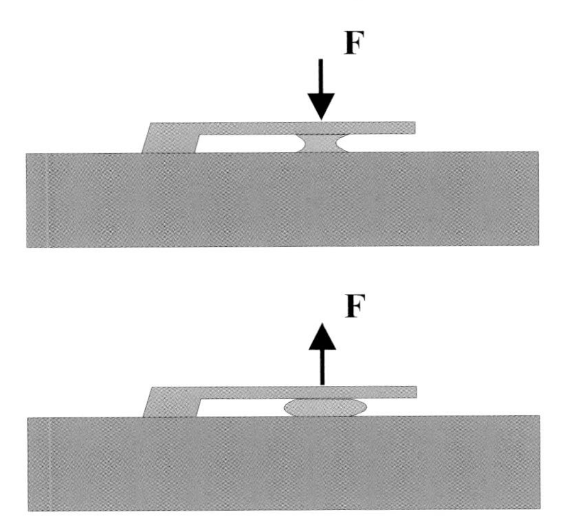

Fig. 3.35. Capillary forces during sacrificial layer etch

The surface micromachining process enables us also to seal structures in a vacuum. This special process is called reactive sealing, and due to Guckel (1986) and Tilmans (1993). In Fig. 3.36 we show the basic process. A narrow channel is left under the structural sealing layer to give access to the fluid for the sacrificial layer etch. After this process a silicon nitride is grown. The material grows also in the cavity until the channel is closed by the growing film. A fairly low pressure can be achieved in the cavity. The process has applications in absolute pressure sensors and in resonant strain gauges.

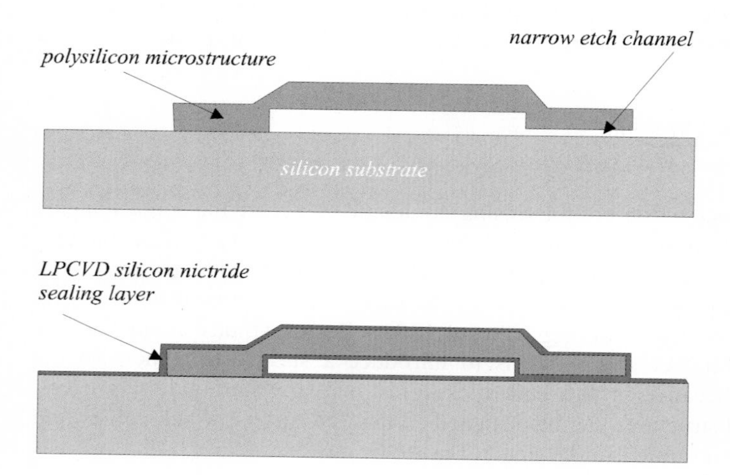

Fig. 3.36. Reactive sealing process

4. Mechanics of Beams and Diaphragms

In this chapter we discuss and in part derive the equations that describe the mechanical deformation of construction elements found in mechanical sensors, as far as they are needed for the basic design of mechanical sensors. Derivations are given in cases that illuminate the basic ideas and methods of thinking about mechanics. In particular, we shall give derivations of the energy of a bent beam, including an axial load and stretching, and we shall show how partial differential equations can be found from energy expressions. These differential equations will give us the static and dynamic behaviour of membranes and beams. In many cases a closed analytical solution of the differential equation cannot be given, therefore we need ways to find reliable approximations. Therefore we also discuss how to obtain simple approximations for deflection amplitudes by energy minimisation. Finally we shall consider mechanical stability.

Before we discuss the distributed mechanical systems we briefly review the mechanics of a mass connected to a spring. Then we turn to the simplest distributed mechanical system, the string. The discussion of beams follows. Beams are of considerable interest for mechanical microsensors because they form the basic spring elements, and they also form the basic element in the resonating strain gauge. Finally we discuss the mechanics of diaphragms.

4.1 Dynamics of the Mass Spring System

We shall use this simple and well-known system to illustrate a mathematical formalism we shall use later to derive partial differential equations. This formalism is based on the Hamilton principle which says that the time integral over the Lagrange function L – the difference between the kinetic and potential energy T and U, respectively, – is a minimum:

$$\int_{t_1}^{t_2} (T - U) \cdot \mathrm{d}t = \int_{t_1}^{t_2} L \cdot \mathrm{d}t = \text{minimum} \tag{4.1}$$

This integral is also called the action and Hamilton's principle is also called the principle of least action. The action is minimal for the actual trajectory of a mass, all other trajectories lead to a larger action.

The variation of this integral must be zero, and the result is the equation of Euler and Lagrange, which is identical to Newton's equation.

The formalism for variation is as follows. Variation is a type of differentiation where the argument is a function instead of a variable. Since L depends on the position $x(t)$ and velocity $\dot{x}(t)$ [1]of a mass and the time, the actual trajectory of the mass in the potential U makes the integral in (4.1) a minimum. Therefore, if the functions $x(t)$ and $\dot{x}(t)$ are varied around the actual trajectory the action is unchanged. We write

$$\delta \int_{t_1}^{t_2} L(x, \dot{x}, t) \cdot \mathrm{d}t = 0 \tag{4.2}$$

where the Greek δ now is reserved for the variation operation.

The endpoints of the trajectory in the variation are fixed, see Fig. 4.1. The variation operation can be carried out the following way. We observe that the Taylor expansion of L is given by

$$L(x, \dot{x}, t) = L(x_0, \dot{x}_0, t) + \frac{\partial L(x, \dot{x}, t)}{\partial x} \delta x + \frac{\partial L(x, \dot{x}, t)}{\partial \dot{x}} \delta \dot{x} + \ldots \tag{4.3}$$

$x_0(t)$ and velocity $\dot{x}_0(t)$ describe the actual trajectory. The variation therefore equals

$$\begin{aligned} \delta \int_{t_1}^{t_2} L \cdot dt &= \int_{t_1}^{t_2} [L(x, \dot{x}, t) - L(x_0, \dot{x}_0, t)] \cdot \mathrm{d}t \\ &= \int_{t_1}^{t_2} \left[\frac{\partial L(x, \dot{x}, t)}{\partial x} \delta x + \frac{\partial L(x, \dot{x}, t)}{\partial \dot{x}} \delta \dot{x} \right] \cdot \mathrm{d}t \end{aligned} \tag{4.4}$$

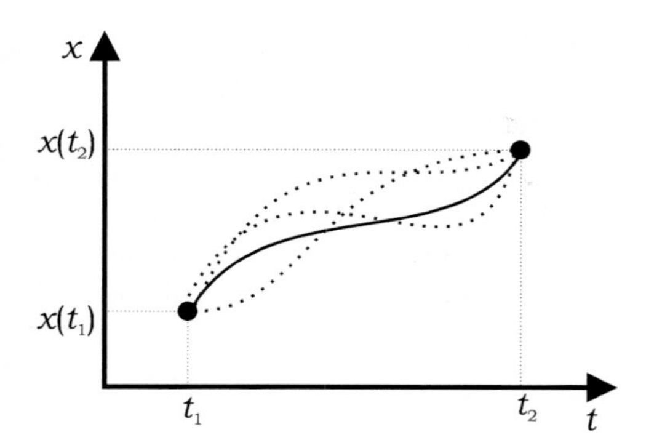

Fig. 4.1. Illustration of a variation of a trajectory

[1] The dot on top of a symbol denotes differentiation with respect to time, a prime differentiation with respect to a spatial coordinate

The first term in the Taylor expansion is sufficient because we can make the variation as small as we want to. The second term in (4.4) is partially integrated

$$\int_{t_1}^{t_2} \frac{\partial L(x,\dot{x},t)}{\partial \dot{x}} \delta\dot{x} \cdot \mathrm{d}t$$

$$= \frac{\partial L(x,\dot{x},t)}{\partial \dot{x}} \delta x \Bigg|_{t_1}^{t_2} - \int_{t_t}^{t_2} \frac{\mathrm{d}}{\mathrm{d}t}\left(\frac{\partial L(x,\dot{x},t)}{\partial \dot{x}} \right) \cdot \delta x \cdot \mathrm{d}t \tag{4.5}$$

$$= - \int_{t_t}^{t_2} \frac{\mathrm{d}}{\mathrm{d}t}\left(\frac{\partial L(x,\dot{x},t)}{\partial \dot{x}} \right) \cdot \delta x \cdot \mathrm{d}t$$

Here the first term equals zero because the endpoints of the motion are fixed:

$$\delta x(t_1) = \delta x(t_2) = 0 \tag{4.6}$$

Actually these equations represent the initial conditions. Normally, they are formulated by giving position and velocity of a mass at $t = t_0$.

We have now

$$\delta \int_{t_1}^{t_2} L \cdot \mathrm{d}t = \int_{t_1}^{t_2} \left[\frac{\partial L}{\partial x} - \frac{\mathrm{d}}{\mathrm{d}t}\frac{\partial L}{\partial \dot{x}} \right] \cdot \delta x \cdot \mathrm{d}t = 0 \tag{4.7}$$

This integral can be zero only if the term between the brackets vanishes. The equation of Euler follows from this:

$$\frac{\partial L}{\partial x} - \frac{\mathrm{d}}{\mathrm{d}t}\frac{\partial L}{\partial \dot{x}} = 0\,. \tag{4.8}$$

The kinetic energy of a mass with velocity $\dot{x}$ is

$$T = \frac{1}{2} m\dot{x}^2$$

and the potential energy of the mass connected to a spring

$$U = \frac{1}{2} kx^2$$

m and k are mass and spring constant, respectively.

We see that $\frac{\mathrm{d}}{\mathrm{d}t}\frac{\partial L}{\partial \dot{x}} = m\ddot{x}$ and $\frac{\partial L}{\partial x} = kx$. In other words, the first term in (4.8) gives the force (with a negative sign) and the second term the mass times the acceleration. (4.8) is identical to Newton's equations.

For the mass-spring system we have now

$$m\ddot{x} + kx = 0$$

Friction can be taken into account in this formalism in a somewhat artificial way. One uses the so-called dissipation function: $D(\dot{x}) = \frac{1}{2}\gamma\dot{x}^2$ (e.g. Goldstein, 1959) with the condition that derivative of D with respect to $\dot{x}$ must be added to the Euler equation. The differential equation reads then

$$\frac{\mathrm{d}}{\mathrm{d}t}\left(\frac{\partial L}{\partial \dot{x}}\right) - \frac{\partial L}{\partial x} + \frac{\partial D}{\partial \dot{x}} = 0 \tag{4.9}$$

and we have

$$m\ddot{x} + \gamma\dot{x} + kx = P(t)$$

with $P(t)$ an external driving force.

Below we use the following definitions:

$$\tau = \frac{m}{\gamma} \tag{4.10}$$

$$\omega_0^2 = \frac{k}{m} \tag{4.11}$$

and

$$Q = \omega_0 \tau \tag{4.12}$$

2τ is the lifetime of the free undamped oscillator, the energy stored in the oscillator decays with τ. ω_0 is the resonance frequency of the undamped oscillator. Q is known as the quality factor of the resonator. Loosely, it describes the number of oscillations of the system before it is damped out. In the literature damping is also described by the damping factor $c = \gamma / \sqrt{km}$ which is simply equal to Q^{-1}.

The solution of the homogeneous differential equation for the case $Q > \sqrt{1/2}$ is

$$x(t) = A\cos(\omega_r t + \phi)\exp-(t/2\tau) \tag{4.13}$$

with

$$\omega_r^2 = \omega_0^2 - \left(\frac{1}{2\tau}\right)^2 = \omega_0^2\left[1 - \frac{1}{Q^2}\right] \tag{4.15}$$

The stationary solution of (4.9) with an external force $P(t) = P_0\cos\omega t$ is

$$x(t) = x_0\cos(\omega t + \phi) \tag{4.16}$$

with

$$|x_0| = \frac{P_0}{m}\frac{1}{\sqrt{\left(\omega_0^2 - \omega^2\right)^2 + \omega^2/\tau^2}} \tag{4.17}$$

and

$$\tan\phi = -\frac{\omega / \tau}{\omega_0^2 - \omega^2}. \tag{4.18}$$

4.2 Strings

Most moving micromechanical structures are flexible. A spring and a mass, which individually can be identified in the real system, cannot describe them. These systems are distributed: the mass and the spring element are physically the same. The simplest mechanical example for this is the string, shown in fig. 4.2. The string is a rope under axial tension without bending rigidity. The equilibrium state is defined by the axial tension in the string. We start the analysis of distributed mechanical systems with this example.

In Fig. 4.3 we show a portion of the string. Below we indicate its equilibrium situation, and above in a deflected state. We describe the deflection as a function of x and t by $w(x,t)$. At the ends of the string portion the tensile forces act. Note that a resultant force pulling the string back to or away from its equilibrium position exists only if the portion of the rope is bent: would it be straight, the forces at the ends would cancel each other.

We assume that the tension in the string is constant, so $T_1 = -T_2$. The ends of the piece make angles θ_1 and θ_2 with respect to the equilibrium situation. We shall assume that the deflection and, therefore, the angles are small, so that $\sin\theta$ and $\tan\theta$ both equal θ. The normal force components at the end of the piece are

$$-T_1 \sin\theta_1$$

and

$$-T_2 \sin\theta_2$$

The total force is then

$$T_0\{\sin\theta_1 - \sin\theta_2\} \approx T_0\left\{\left(\frac{\partial w}{\partial x}\right)_1 - \left(\frac{\partial w}{\partial x}\right)_2\right\}$$

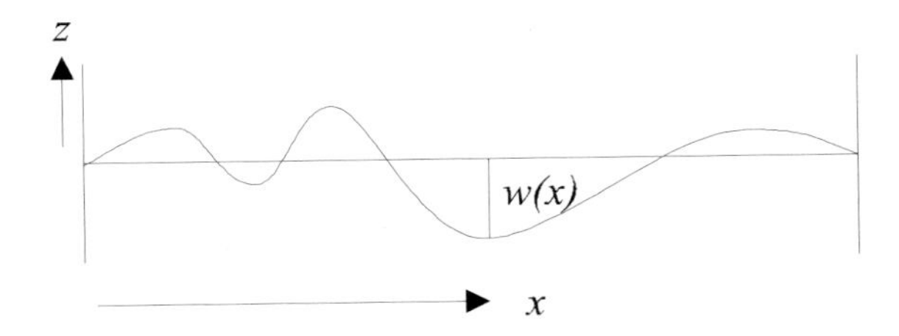

Fig. 4.2. Deflection of a string as a function of the position x

For a very small portion $\Delta x \to 0$, and $\{(\partial w/\partial x)_2 - (\partial w/\partial x)_1\}$ can be replaced by $\Delta x(\partial^2 w/\partial x^2)$. Using Newton's equations we have:

$$\Delta x \rho A \frac{\partial^2 w}{\partial t^2} = \Delta x \, T_0 \left(\frac{\partial^2 w}{\partial x^2} \right) \tag{4.19}$$

where the mass has been expressed in the length Δx times the density per unit length ρ. A is the cross-section of the string. The right hand side is the force that acts on the piece of string. It says that if the curvature is negative the force is negative and the string will be pulled down. This situation is sketched in Fig. 4.3. The opposite curvature would give an upward force. From (4.19) the wave equation results:

$$\frac{\partial^2 w}{\partial x^2} - \frac{\rho_0}{T_0} \frac{\partial^2 w}{\partial t^2} = 0 \tag{4.20}$$

This equation can be derived also using the variation approach. For this we must find the potential energy of the string as a functional of the deflection (the deflection is a function of x and t). This can be found by integrating the force as given in the right hand side (4.19) over the deflection. We then have the potential energy per unit length, and in order to get the total potential energy we have to integrate over the length of the string.

$$U(w) = \int_0^l U^*(w'(x,t)) \mathrm{d}x \tag{4.21}$$

with U^* the potential energy per length, and l the length of the string.

$$U^* = -\int_0^w T_0 \frac{\partial^2 w}{\partial x^2} \mathrm{d}w = \frac{1}{2} T_0 \left(\frac{\partial w}{\partial x} \right)^2 = \frac{1}{2} T_0 w'^2 (x,t) \tag{4.22}$$

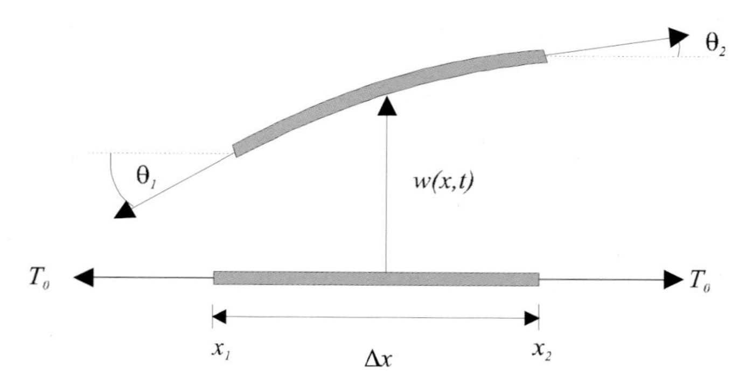

Fig. 4.3. Transversal deflection of a string. We show a small portion of the string of length Δx, with tensile forces T and deflection w.

The kinetic energy is also an integral over the length of the string:

$$T = \int_0^l T^*\left(\dot{w}(x,t)\right)\mathrm{d}x = \frac{1}{2}\rho A \int_0^l \dot{w}^2(x,t)\mathrm{d}x . \tag{4.23}$$

From (4.22) and (4.23) we see that the Legendre function is now an integral over Legendre functions per length:

$$L = \int_0^l L^*\left(\dot{w}, w', t\right)\mathrm{d}x \tag{4.24}$$

The variation of these two last expressions leads to a partial differential equation (4.20) as we shall see below.
First we derive the energy of a bent beam.

4.3 Beams

In the description of beams we have to include bending. We start with calculating the energy required to bend a beam with a certain radius of curvature. The aim is to get an expression for the potential energy similar to (4.22).

4.3.1 Stress and Strain

First we review some elementary notions for the deformation of solid bodies. We study a column under an axial load, as shown in Fig. 4.4. We pull with a stress σ,

$$\sigma = \frac{F}{A} \tag{4.25}$$

The body deforms, so that it becomes longer and thinner. We have

$$\begin{aligned} l &\to l + \Delta l \\ b &\to b - \Delta b \\ h &\to h - \Delta h \end{aligned} \tag{4.26}$$

The relative deformation $\Delta l/l$ is called the strain. For elastic materials the stress strain relations are

$$\sigma = E\frac{\Delta l}{l} = E \cdot \varepsilon \tag{4.27}$$

where E is the Young's modulus. E is a pure material property. The contraction in directions normal to the axial force is described by the Poisson ratio υ :

$$\frac{\Delta b}{b} = \frac{\Delta h}{h} = -\upsilon\frac{\Delta l}{l} = -\upsilon\varepsilon$$

In silicon υ depends on the crystallographic direction. We neglect the contraction effects in the following derivations.

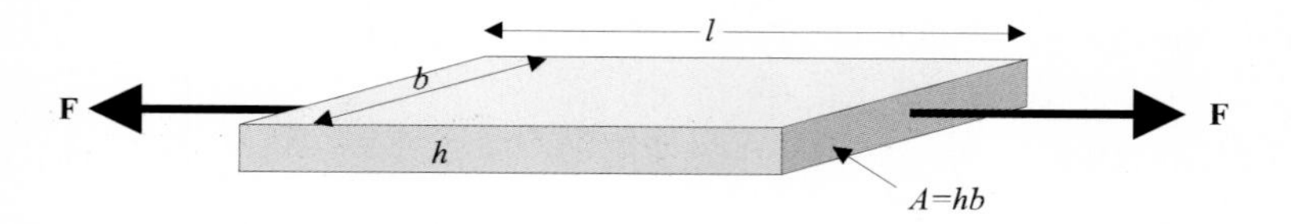

Fig. 4.4. A beam subject to a tensile force

The energy associated with the strain is given by

$$U_{stretch} = \int F\mathrm{d}x = \int AE\varepsilon\mathrm{d}(\varepsilon l) = \frac{l}{2}AE\varepsilon^2 \tag{4.28}$$

4.3.2 Bending Energy

Let us now take a beam with a centre line l, with l_0 the length in the unbent state. The beam is bent and it gets a radius of curvature R_0.

$$l(z) = \phi \cdot (R + z) \tag{4.29}$$

Here is z the distance from the centre line. If there is no tension in the beam, the length of the centre line stays unchanged after the deformation, so it is identical to the neutral line. Inspection of Fig. 4.5 shows that the upper layers of the beam become longer, the lower ones shorter. There is a strain in the beam that varies over its thickness. The relative change of the length, the strain, follows from elementary geometry

$$\frac{\Delta l}{l_0} = \frac{\phi \cdot z}{l_0} \tag{4.30}$$

If we use the relation of the angle ϕ and the radius of curvature,

$$\phi = \frac{l_0}{R} \tag{4.31}$$

we find the z-dependence of the strain in the beam:

$$\varepsilon = \frac{\Delta l}{l_0} = \frac{z}{R} \tag{4.32}$$

The potential energy of a thin slice of the beam with cross-section $\mathrm{d}A = b\mathrm{d}z$ is now given by the integral over the (external) force time the change of length. We express the integral in stress and strain:

$$\begin{aligned} \mathrm{d}U &= \int F \cdot \mathrm{d}\Delta l = \\ \mathrm{d}Al_0 \int \sigma \cdot \mathrm{d}\varepsilon &= E\mathrm{d}Al_0 \int \varepsilon \cdot \mathrm{d}\varepsilon = \frac{E\mathrm{d}A}{2} l_0 \varepsilon^2 \end{aligned} \tag{4.33}$$

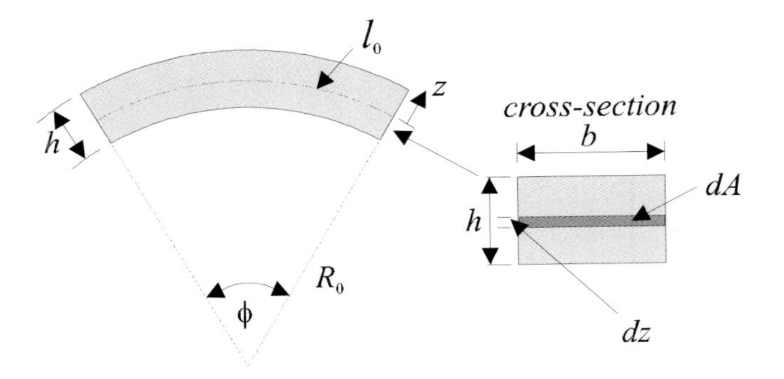

Fig. 4.5. A bent beam

The total energy is now obtained by integration over dA.

$$
\mathrm{d}U = \frac{Ebl_0}{2}\frac{z^2}{R^2}\mathrm{d}z
$$

$$
U = \frac{Ebl_0}{2}\int_{-\frac{h}{2}}^{\frac{h}{2}} \frac{z^2}{R^2}\cdot \mathrm{d}z \tag{4.34}
$$

$$
= \frac{Ebl_0}{6}\left(\frac{h^3}{2^3}+\frac{h^3}{2^3}\right)\frac{1}{R^2} = \frac{1}{2}\frac{Ebh^3 l_0}{12}\frac{1}{R^2}
$$

This is the energy of a piece of a beam with length l_0. This formula is correct for narrow beams. The energy expression for wide beams must be corrected by the factor $1/(1-\upsilon^2)$.

The curvature of a beam having an arbitrary deflection must be described by local radii of curvature, $R = R(x)$. If we divide U in eq. (4.34) by l_0, we have U^*, the bending energy per unit length. The total bending energy is then the integral of U^* over the length of the beam:

$$
U = \int \mathrm{d}U^* = \int_0^l U^*\,\mathrm{d}x \tag{4.35}
$$

4.3.3 Radius of curvature

Finally we have to find the relation between the radius of curvature and the deflection of the beam. For any angles we have

$$
\frac{\mathrm{d}\phi}{\mathrm{d}l} = \frac{1}{R} \tag{4.36}
$$

With the help of Fig. 4.6 we can see that

$$\tan\phi = \frac{\mathrm{d}w}{\mathrm{d}x} \rightarrow \phi = \arctan\frac{\mathrm{d}w}{\mathrm{d}x} \tag{4.37}$$

By differentiating we find

$$\frac{\mathrm{d}\phi}{\mathrm{d}x} = \frac{1}{1+\left(\frac{\mathrm{d}w}{\mathrm{d}x}\right)^2}\cdot\frac{\mathrm{d}^2w}{\mathrm{d}x^2} \tag{4.38}$$

We use Pythagoras' theorem

$$\mathrm{d}l = \sqrt{\mathrm{d}x^2 + \mathrm{d}w^2} \tag{4.39}$$

from which follows that

$$\frac{\mathrm{d}l}{\mathrm{d}x} = \frac{\sqrt{\mathrm{d}x^2 + \mathrm{d}w^2}}{\mathrm{d}x} = \sqrt{1+\left(\frac{\mathrm{d}w}{\mathrm{d}x}\right)^2}\,. \tag{4.40}$$

Rearranging (4.40) gives us

$$\frac{\mathrm{d}\phi}{\mathrm{d}l} = \frac{\mathrm{d}\phi}{\mathrm{d}x}\cdot\frac{\mathrm{d}x}{\mathrm{d}l} = \frac{\mathrm{d}\phi}{\mathrm{d}x}\bigg/\frac{\mathrm{d}l}{\mathrm{d}x}$$

Finally, using (4.36), (4.38) and (4.40) we have the result

$$\frac{1}{R} = \frac{\partial^2 w}{\partial x^2}\frac{1}{\left(1+\left(\frac{\partial w}{\partial x}\right)^2\right)^{\frac{3}{2}}} \cong \frac{\partial^2 w}{\partial x^2} \tag{4.41}$$

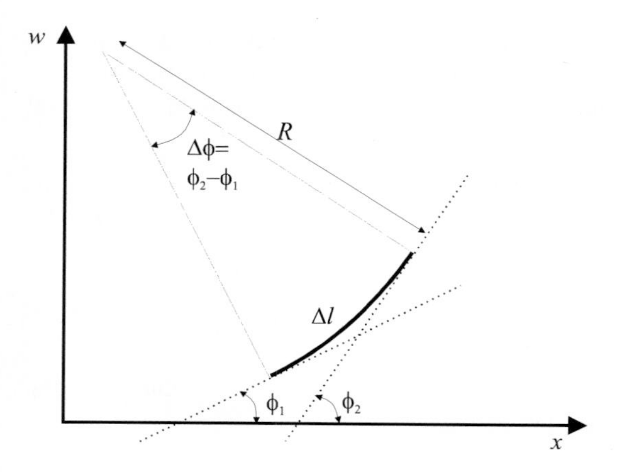

Fig. 4.6. Relation between curvature and deflection

Using this relation for U^*, we find

$$U^* = \tfrac{1}{2} EI \left(\frac{\partial^2 w}{\partial x^2} \right)^2 = \tfrac{1}{2} EI w''^2 \tag{4.42}$$

where we used the definition of the moment of inertia I

$$I = \int_{-h/2}^{+h/2} b z^2 \mathrm{d}z = \frac{bh^3}{12} \tag{4.43}$$

An external load P_0 also gives an energy contribution, which is

$$U_{ext} = -\int_0^l P_0(x) w(x) \cdot \mathrm{d}x \tag{4.44}$$

The deflection of the beam causes an elongation. However, this is a second order effect as we can see from (4.40). The change in the length of a small piece of a beam is obtained from (4.40) by a Taylor expansion:

$$\mathrm{d}\Delta l = \frac{1}{2} \left(\frac{\partial w}{\partial x} \right)^2 \mathrm{d}x$$

and the strain of this piece is then

$$\varepsilon = \frac{1}{2} \left(\frac{\partial w}{\partial x} \right)^2$$

So the energy associated with the stretching of the beam is, with the help of (2.28)

$$U_{stretch} = \frac{1}{2} \int_0^l EA\varepsilon^2 \mathrm{d}x = \frac{1}{8} \int_0^l EA w'^4 \mathrm{d}x \tag{4.45}$$

This energy term is proportional to the fourth power of the deflection, so in deriving differential equations we shall neglect this term. However, when we consider the phenomenon of buckling, this energy term is important. Note that in the approximation in (4.41) a term proportional to $w''^2 w'^2$ occurs. This also gives rise to a fourth order term in the energy. However, this term is much smaller than the stretch term, because it is proportional to $(h/l)^2$.

4.3.4 Lagrange Function of a Flexible Beam

The Lagrange function of a bent beam with an axial load and a normal load distribution is now

$$L=\frac{1}{2}\int_0^l \mathrm{d}x\left(\rho A\dot{w}^2-EIw''^2-T_0w'^2-P_0w\right)=$$
$$\int_0^l L^*(w,\dot{w},w',w'')\mathrm{d}x \tag{4.46}$$

For the overview we give a summary of the various energy terms:

$$U^*_{ext}=-P_0(x)w(x) \tag{4.47}$$

$$U^*_{ax}=\frac{1}{2}T_0w'^2(x) \tag{4.48}$$

$$U^*_{bend}=\frac{1}{2}EIw''^2(x) \tag{4.49}$$

$$T^*=\frac{1}{2}\rho A\dot{w}^2(x) \tag{4.50}$$

T^* denotes the kinetic energy per unit length. The Lagrange function is just the difference between the last term and the other ones. So the dependency of the Lagrange function on the deflection and its derivatives is represented in distinct terms.

4.3.5 Differential equation for beams

Next we show how a partial differential equation can be derived from this function by variation.

$$\delta\int_{t_1}^{t_2}\int_0^l L^*\left(w,w',w'',\dot{w},t\right)\cdot\mathrm{d}x\cdot\mathrm{d}t$$
$$=\int_{t_1}^{t_2}\int_0^l\left[L^*\left(w+\delta w,w'+\delta w',w''+\delta w'',\dot{w}+\delta\dot{w},t\right)-L^*\left(w,w',w'',\dot{w},t\right)\right]\cdot\mathrm{d}x\cdot\mathrm{d}t$$
$$=\int_{t_1}^{t_2}\int_0^l\left[\frac{\partial L^*}{\partial w}\delta w+\frac{\partial L^*}{\partial w'}\delta w'+\frac{\partial L^*}{\partial w''}\delta w''+\frac{\partial L^*}{\partial \dot{w}}\delta\dot{w}\right]\cdot\mathrm{d}x\cdot\mathrm{d}t \tag{4.51}$$

This integral must be expressed in terms of w, we must get rid of the derivatives of w with respect to the time and the position, just as we did earlier. This is done by partial integration:

$$\delta\dot{w}: \quad \int_{t_1}^{t_2}\int_0^l \frac{\partial L^*}{\partial \dot{w}}\delta\dot{w}\cdot dx\cdot dt = \int_0^l \frac{\partial L^*}{\partial \dot{w}}\delta w\Big|_{t_1}^{t_2}\cdot dx - \int_{t_1}^{t_2}\int_0^l \frac{\partial}{\partial t}\frac{\partial L^*}{\partial \dot{w}}\delta w\cdot dx\cdot dt \tag{4.52}$$

$$\delta w': \quad \int_{t_1}^{t_2}\int_0^l \frac{\partial L^*}{\partial w'}\delta w'\cdot dx\cdot dt = \int_{t_1}^{t_2} \frac{\partial L^*}{\partial w'}\delta w\Big|_0^l\cdot dt - \int_{t_1}^{t_2}\int_0^l \frac{\partial}{\partial x}\frac{\partial L^*}{\partial w'}\delta w\cdot dx\cdot dt \tag{4.53}$$

$$\delta w'': \quad \int_{t_1}^{t_2}\int_0^l \frac{\partial L^*}{\partial w''}\delta w''\cdot dx\cdot dt = \int_{t_1}^{t_2} \frac{\partial L^*}{\partial w''}\delta w'\Big|_0^l dt - \int_{t_1}^{t_2}\int_0^l \frac{\partial}{\partial x}\frac{\partial L^*}{\partial w''}\delta w'\cdot dx\cdot dt$$

$$= \int_{t_1}^{t_2} \frac{\partial L^*}{\partial w''}\delta w'\Big|_0^l dt - \int_{t_1}^{t_2} \frac{\partial}{\partial x}\frac{\partial L^*}{\partial w''}\delta w\Big|_0^l\cdot dt + \int_{t_1}^{t_2}\int_0^l \frac{\partial^2}{\partial x^2}\frac{\partial L^*}{\partial w''}\delta w\cdot dx\cdot dt \tag{4.54}$$

The total variation is:

$$\delta\int_{t_1}^{t_2} L\cdot dt$$

$$= \int_{t_1}^{t_2}\int_0^l\left[\frac{\partial L^*}{\partial w} - \frac{\partial}{\partial t}\frac{\partial L^*}{\partial \dot{w}} - \frac{\partial}{\partial x}\frac{\partial L^*}{\partial w'} + \frac{\partial^2}{\partial x^2}\frac{\partial L^*}{\partial w''}\right]\delta w\cdot dx\cdot dt \tag{4.55}$$

$$+ \int_{t_1}^{t_2}\left[\left(\frac{\partial L^*}{\partial w'} - \frac{\partial}{\partial x}\frac{\partial L^*}{\partial w''}\right)\delta w + \frac{\partial L^*}{\partial w''}\delta w'\right]_0^l\cdot dt = 0$$

This equation contains now all the secrets of bent beams. First of all, the variations δw are arbitrary. That means that the terms in both brackets must be equal to zero. The first bracket gives the Euler-Lagrange equation for the Lagrange function in (4.46). The other two terms in the second row are the boundary conditions of the bent beam. The Euler-Lagrange equation is

$$\boxed{\frac{\partial L^*}{\partial w} - \frac{\partial}{\partial t}\frac{\partial L^*}{\partial \dot{w}} - \frac{\partial}{\partial x}\frac{\partial L^*}{\partial w'} + \frac{\partial^2}{\partial x^2}\frac{\partial L^*}{\partial w''} = 0} \tag{4.56}$$

Using (4.47)–(4.50) we get the following partial differential equation:

$$EIw'''' - T_0 w'' + \rho A\ddot{w} = P_0(x) \tag{4.57}$$

We see that for the string, which has $EI = 0$, we find back equation (4.20) for $P_0 = 0$. Equation (4.57) can be derived in alternative ways, which can be found in standard textbooks on mechanics. In chapter 9 an alternative derivation is given. Solutions of (4.77) will be discussed in sections 4.3.7 through 4.3.9 and in chapter 9.

4.3.6. Boundary conditions for beams

Next we have to consider remaining set of expressions in equation (4.55). These tell us that at the ends of the beam (i.e. at $x = 0$, $x = l$) *at all times*

$$\left[\left(\frac{\partial L^*}{\partial w'} - \frac{\partial}{\partial x}\frac{\partial L^*}{\partial w''}\right)\delta w + \frac{\partial L^*}{\partial w''}\delta w'\right] = 0 \tag{4.58}$$

Using again (4.47)–(4.50) this can be rewritten

$$\begin{aligned} &\delta w(0)\left[T_0 w''(0) - EIw'''(0)\right] + \delta w'(0)\, EIw'''(0) = 0 \\ &\delta w(l)\left[T_0 w''(l) - EIw'''(l)\right] + \delta w'(l)\, EIw'''(l) = 0 \end{aligned} \tag{5.59}$$

These equations look quite complex, but they have a simple meaning. They tell us the conditions that must be valid for w and its spatial derivatives at the ends of the beam.

The δ's represent a variation, hence first we look for an interpretation of the equations $\delta w' = 0$ and $\delta w = 0$. They mean that the variation of the position and slope of the beam at both ends are fixed - but of course not necessarily zero. This is just what we might expect from a beam clamped at both sides: position and slope are fixed. If this is the case the remaining terms in equation (5.59) may have any value, which means that the higher derivatives (the second and the third) at the ends of the beam are free. The second derivative is proportional to the curvature, and proportional to the moment at the ends.

On the other hand, if w and w' both are free, we must have that w'' and w''' are equal to zero. Simply supported means that $\delta w = 0$, $\delta w' \neq 0$. From this we can conclude that $w''' = 0$ and w'' is free. These boundary conditions can be mixed at the respective ends of the beam. A cantilever beam as an example has at the clamped edge: $w = 0$ and $w' = 0$ (or constant in place of 0), and at the free end $w'' = 0$ and $w''' = 0$. In Fig. 4.7 we show examples for clamped–clamped beams where w and w' and the ends are non zero, but constant. Fig. 4.8. shows the other boundary conditions (clamped, free and simply supported).

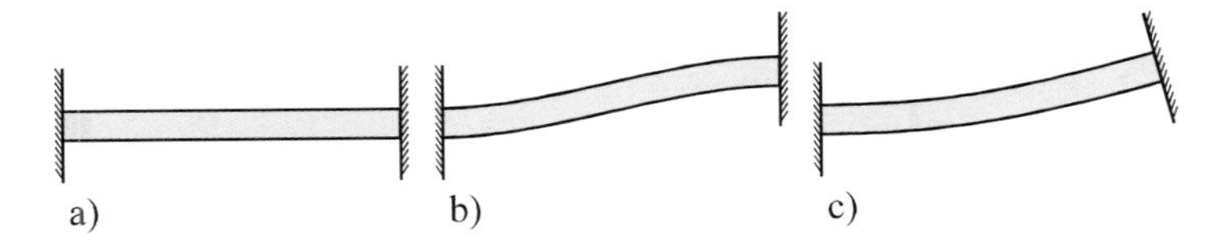

Fig. 4.7. Examples for clamped–clamped beams

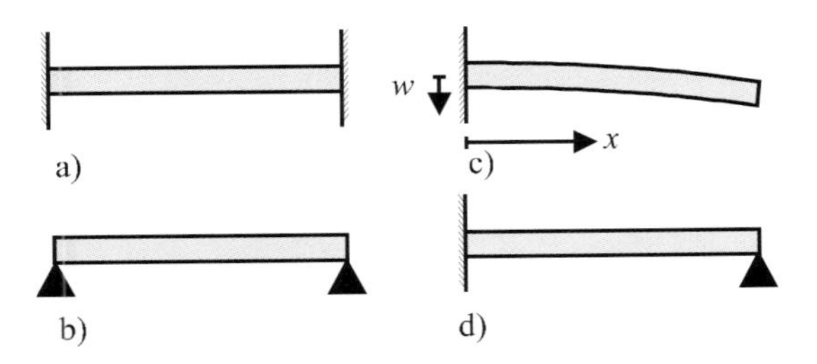

Fig. 4.8. Four combinations of boundary conditions: a) clamped–clamped (bridge) b) simply supported–simply supported c) clamped–free (cantilever beam) d) clamped–simply supported.

Table 4.1 Mechanical constructions and the corresponding boundary conditions

boundary	w	δw	w'	$\delta w'$	w''	w'''
free	free	free	free	free	0	0
clamped	defined	0	defined	0	free	free
simply supported	defined	0	free	free	0	0

4.3.7. Examples

For simple examples eq. (4.57) can be integrated directly. Let us show a procedure for the example of a bridge (a double clamped beam) loaded by its own weight in steady state (i.e. the static case). Then $P_0(x)$ is the weight per length of the beam, $-g\rho A$, with g the acceleration due to gravity. Let us denote $g\rho A = G$.

The differential equation now reads

$$w'''' = \frac{G}{EI} \tag{4.60}$$

The general form of the solution of eq. (4.60) is

$$w = a_0 + a_1 x + a_2 x^2 + a_3 x^3 + a_4 x^4 \tag{4.61}$$

Clamping at both sides means that

$$\begin{aligned}
&w'''' = \frac{\rho A}{EI} \rightarrow a_4 = \frac{\rho A}{24EI} \\
&w(0) = 0 \rightarrow a_0 = 0 \\
&w'(0) = 0 \rightarrow a_1 = 0 \\
&w(l) = 0 \rightarrow a_2 l^2 + a_3 l^3 + \frac{G}{24EI} l^4 = 0 \\
&w'(l) = 0 \rightarrow 2a_2 l + 3a_3 l^2 + 4\frac{G}{24EI} l^3 = 0
\end{aligned} \tag{4.62}$$

The combination of these equations gives us all constants a_i, and we can write

$$w = \frac{\rho A}{24EI} x^2 (x-l)^2 \tag{4.63}$$

The maximum deflection is in the centre of the bridge, and its value is

$$w\left(\frac{l}{2}\right) = \frac{1}{384} \frac{G\ell^4}{EI} \tag{4.64}$$

In a similar way we get the following forms of beams under a homogeneous load: both sides simply supported:

$$w = \frac{G}{24EI} x\left(x^3 - 2lx^2 - l^3\right) \tag{4.65}$$

one side clamped, other side free (cantilever beam):

$$w = \frac{G}{24EI} x^2\left(x^2 - 4lx - 6l^2\right) \tag{4.66}$$

one side clamped, other side simply supported:

$$w = \frac{G}{48EI} x\left(2x^3 - 3lx^2 - l^3\right) \tag{4.67}$$

Next we consider a beam subject to a point force, which we shall describe by the Dirac δ-function. This function $\delta(x-a)$ (more exactly: distribution) is everywhere zero except at $x = a$ and has the property that

$$\int_{-\infty}^{\infty} \delta(x-a)\mathrm{d}x = 1 \tag{6.68}$$

The differential equation now becomes

$$EIw'''' = F\delta(x-a) \tag{6.69}$$

We have everywhere at $x \neq a$ $w'''' = 0$ and integration over the length of the beam means that

$$EIw''' = F \tag{6.70}$$

The general solution (4.61) leads to the following results:
For a= $l/2$, $0 \leq x \leq l/2$, both ends clamped:

$$w = \frac{F}{48EI} x^2 (3l - 4x) \tag{6.71}$$

For a= $l/2$, $0 \leq x \leq l/2$, both ends simply supported:

$$w = \frac{F}{48EI} x \left(3l^2 - 4x^2\right) \tag{6.72}$$

For a cantilever beam with the point force at the free end:

$$w = \frac{F}{6EI} x^2 (3l - x)$$

These equations are important for micromechanics in general: they give the spring constant of the beams and bridges.

4.3.8. Mechanical Stability

A beam under a compressive load becomes unstable once a critical load is exceeded. This load depends on the boundary conditions. The differential equation for a beam under compressive stress is

$$EIw'''' + |T_0| w'' = 0 \tag{4.73}$$

A trivial solution is obviously $w = 0$; the beam will not deflect. However, there can be also a solution with $w \neq 0$. The small deflection theory will not give correct results for the deflection itself, but the conditions for a finite deflection can be obtained from (4.73).

The general solution of (4.73) is

$$w = a_0 + a_1 x + a_2 \sin kx + a_4 \cos kx, \qquad k = \sqrt{\frac{|T_0|}{EI}} \tag{4.74}$$

This is easily verified by inserting (4.74) into (4.73). For the case a) in Fig. 4.9. (w and w'' equal to 0 at both ends of the beam) we have that

$$w = a_2 \sin kx \tag{4.75}$$

Furthermore, $\sin kl = 0$. So, eq. (4.74) can only be satisfied for a particular value of T_0. Below $T = T_{cr}$ the deflection is equal to zero, at $T > T_{cr}$ the deflection is finite. Since we started with a small deflection theory we will get no answer on the size of the deflection; we shall consider this problem below. The critical force is:

$$T_{cr} = \frac{\pi^2 EI}{l^2} \tag{4.76}$$

Once the axial load exceeds T_{cr} the beam will buckle as shown in Fig. 4.9.

For the case of a beam clamped at both ends we find

$$T_{cr} = \frac{4\pi^2 EI}{l^2} \tag{4.77}$$

as given earlier when we considered scaling (chapter 2) and thin film stress (chapter 3). The factor 4 appears because now the cosine function describes the deflection of the beam, and we must have that 1-cos kl = 0, which leads to the result (4.77).

In order to describe the buckling amplitude of a beam under compressive axial stress one must include terms that lead to a non-linearity in the differential equation. The dominating effect is stretching the energy of which is given by eq. (4.45). The calculation comprises the total potential energy of the beam as a function of centre deflection, assuming a shape of the beam.

The total potential energy including stretching, bending and axial stress is

$$U_{tot} = \frac{1}{2}\int_0^l \left[\mathrm{AE}\frac{w'^4}{4} + EIw''^2 + Tw'^2 \right] \mathrm{d}x$$

It is assumed that the shape of the beam clamped at both ends is given by

$$w(x) = w_0\left(1 - \cos\frac{2\pi x}{l}\right)$$

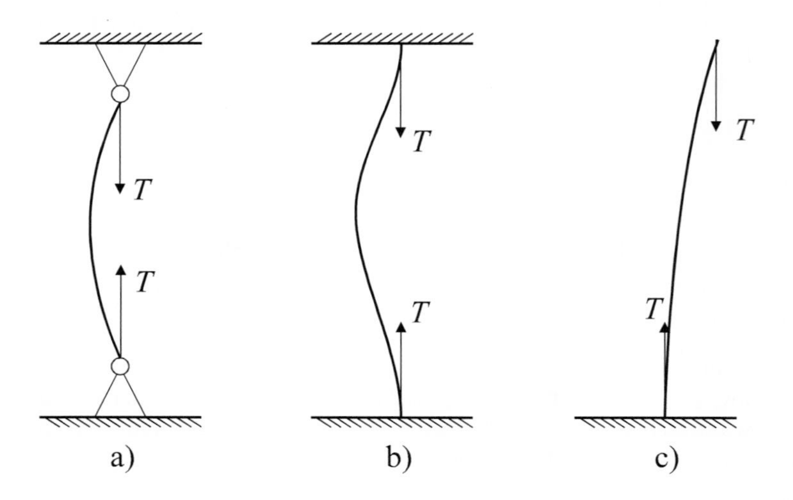

Fig. 4.9. Three possibilities to create an unstable situation

Minimisation of the potential energy with respect to the deflection amplitude w_0 gives (see appendix 4.1)

$$w_0 = \pm \frac{l}{\pi} \sqrt{\frac{2(|T| - |T_{cr}|)}{3EA}} \tag{4.78}$$

and $w_0 = 0$. w_0 is finite only if the compressive stress exceeds the critical stress.

4.3.9. Transversal Vibration of Beams

We restrict the discussion of the dynamic properties of a beam to the simple situation in which the beam is not axially loaded. The differential equation becomes

$$EIw'''' + \rho A \ddot{w} = 0 \tag{4.79}$$

To get the eigen modes we write a solution as

$$w_0 = w(x)\cos(\omega t + \alpha) \tag{4.80}$$

The solutions we find for ω are the eigenfrequencies. We insert (4.80) into (4.79) and get

$$w'''' = \kappa^4 w = \omega^2 \frac{\rho A}{EI} w \tag{4.81}$$

The general solution of this differential equation is

$$w = a_0 \cos\kappa x + a_1 \sin\kappa x + a_3 \cosh\kappa x + a_4 \sinh\kappa x \tag{4.82}$$

where the constants follow from the boundary conditions of the beam at the clamped edges, which are: $w(0) = 0, w(l) = 0, w'(0) = 0$ and $w'(l) = 0$. The result of a somewhat lengthy calculation is:

$$w(x) = (\sin\kappa l - \sinh\kappa l)(\cos\kappa x - \cosh\kappa x) - (\cos\kappa l - \cosh\kappa l)(\sin\kappa x - \sinh\kappa x) \tag{4.83}$$

for the modeshapes. Further, κ is found from the following transcendental equation:

$$\cos\kappa l \cosh\kappa l = 1 \tag{4.84}$$

With (4.83) and (4.80) we get the eigenfrequencies. The fundamental eigenfrequency is

$$\omega_1 = \frac{22{,}4}{l^2} \sqrt{\frac{EI}{\rho A}} \tag{4.85}$$

The numerical factor is a result of the solution if the transcendental equation (4.85). For simply supported beams the calculation is much simpler. The boundary conditions now lead to

$$w = a \sin\kappa x \tag{4.86}$$

and the eigenfrequencies follow from

$$\kappa = \frac{n\pi}{l} \tag{4.87}$$

with the smallest eigenfrequency

$$\omega_1 = \frac{\pi^2}{l^2}\sqrt{\frac{EI}{\rho A}} \tag{4.88}$$

Finally we give the result for the cantilever beam:

$$w = a_1\{(\cos\kappa l + \cosh\kappa l)(\cos\kappa x - \cosh\kappa x) - (\sin\kappa l - \sinh\kappa l)(\sin\kappa x - \sinh\kappa x)\} \tag{4.89}$$

with

$$\cos\kappa l \cosh\kappa l + 1 = 0 \tag{4.90}$$

$$\omega_1 = \frac{3.52}{\ell^2}\sqrt{\frac{EI}{\rho A}} \tag{4.91}$$

If there is an axial load on the beam, the matter becomes more complicated. We restrict ourselves to give the results. More details are given in Chap. 9. The characteristic equation giving the eigenfrequencies is:

$$\cos\lambda l \cos\mu l - \frac{1}{2}\left(\frac{\mu}{\lambda} - \frac{\lambda}{\mu}\right)\sinh\mu l \sinh\mu l = 1 \tag{4.92}$$

with

$$\lambda = \kappa\sqrt{\sqrt{a^2+1} - a}$$

$$\mu = \kappa\sqrt{\sqrt{a^2+1} + a}$$

$$a = T/2EI\kappa^2$$

κ is given in (4.81).

The modeshape is now given by the following complex function:

$$w(x) = \left(\frac{\lambda}{\mu}\sinh\mu l - \sinh\lambda l\right)(\cos\lambda x - \cos\mu x) + (\cos\lambda l - \cos\mu l)\left(\sin\lambda x - \frac{\lambda}{\mu}\sinh\mu x\right) \tag{4.93}$$

If we compare this expression to (4.82) the difference is modest if the parameter a is not too large. For the fundamental mode, both curves resemble the function $w(x) = w_0(1 - \cos 2\pi x/l)$ quite a lot, however at $T = 0$ the inflection points of the

beam are shifted a bit towards the edges ($x/l \approx 0.24$ and 0.76 in place of 0.25 and 0.75). Introducing a tensile axial load, the inflection points mover further towards the edges, but this is a small effect unless a >> 1.
The resonance frequencies are given approximately

$$\omega_n(T) = \omega_n(0)\sqrt{1 + \gamma_n \frac{Tl^2}{12EI}} \tag{4.94}$$

with

$$\omega_n = \alpha_n \frac{2\pi}{l^2}\sqrt{\frac{EI}{\rho A}} \tag{4.95}$$

Fig. 4.10 shows a plot of the first two eigenfrequencies as a function of the applied axial load.
The coefficients α_n and γ_n are given by integrals over the mode shapes. Bouwstra (1991) gives

n	α_n	γ_n
1	4.730	0.295
2	7.853	0.145
>2	$\approx\pi(n+1/2)$	$\approx 12(\alpha_n-2)/\alpha_n^3$

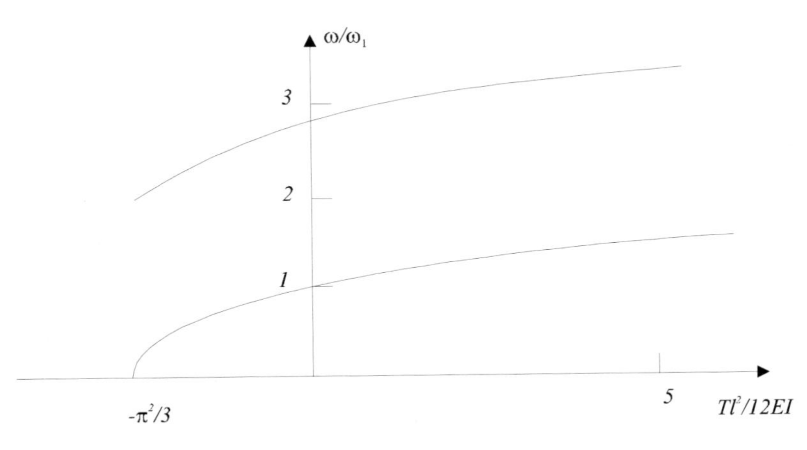

Fig.4.10. Dependency of the first two eigenfrequencies of a bridge loaded by axial forces

4.4. Diaphragms and Membranes

Diaphragms and membranes are the key elements in pressure sensors, but they have also important functions in other sensors, such as flow sensors and radiation sensors. Furthermore, they are widely used in pumps and valves.

We start with the energy due to bending of a diaphragm. In the discussion of bent beams we took into account only a curvature along the long axis of the beam. When considering diaphragms two principal radii of curvature must be taken into account.

Landau (1965) gives the energy of a diaphragm due to pure bending.

$$U_{bend} = \frac{Eh^3}{24(1-\upsilon^2)} \iint \mathrm{d}x\mathrm{d}y$$
$$\left[\left(\frac{\partial^2 w(x,y)}{\partial x^2} + \frac{\partial^2 w(x,y)}{\partial y^2} \right)^2 + 2(1-\upsilon) \left[\left(\frac{\partial^2 w(x,y)}{\partial x \partial y} \right)^2 - \frac{\partial^2 w(x,y)}{\partial x^2} \frac{\partial^2 w(x,y)}{\partial y^2} \right] \right] \tag{4.96}$$

From this energy term the differential equation can be found by variation. This calculation is rather cumbersome, and we refer to Landau (1965). The result is

$$\frac{Eh^3}{12\left(1-\upsilon^2\right)} \Delta^2 w(x,y) - P(x,y) = 0 \tag{4.97}$$

here, Δ is the two-dimensional Laplace operator, in Cartesian co-ordinates

$$\Delta = \frac{\partial^2}{\partial x^2} + \frac{\partial^2}{\partial y^2} \tag{4.98a}$$

and in circular co-ordinates

$$\Delta = \frac{1}{r} \frac{\mathrm{d}}{\mathrm{d}r} \left(r \frac{\mathrm{d}}{\mathrm{d}r} \right) \tag{4.98b}$$

There is no analytical solution of (4.97) for a rectangular membrane, and from Meleshko (1997) it can be seen that this forms a formidable mathematical problem. Therefore, for rectangular and square membranes we have to use approximate solutions.

4.4.1. Circular Diaphragms

Let us start the discussion of circular diaphragms of radius R. If there is a pressure difference p acting on the membrane, eq. (4.97) becomes

$$\Delta^2 w(r) = \beta \tag{4.99}$$

with $\beta = 12\left(1-\upsilon^2\right)p / Eh^3$. The origin of the co-ordinate system is in the centre of the diaphragm. The general solution of (4.99) is

$$w(r) = ar^2 + br^4 + cr^2 \ln r / R + d \ln r / R$$

At the edges $w(r = R)$ and $\mathrm{d}w / \mathrm{d}r|_{r=R}$ must vanish, and the deflection at $r = 0$ must be finite. It follows that c and d are equal to zero. The solution then becomes

$$w(r) = \frac{3}{16} \frac{1-\upsilon^2}{Eh^3} p[R^2 - r^2]^2 \tag{4.100}$$

From this we see that the center deflection is given by

$$w_0 = \frac{3}{16} \frac{\left(1-\upsilon^2\right)p}{Eh^3} R^4 \tag{4.101}$$

Some pressure sensors have a central boss, which is a part of the diaphragm which is much stiffer than the outer ring, see Fig. 4.11.

In this case a force acts on the central portion equal to the product of the pressure and the boss' surface. Then the boundary conditions at the clamped edges apply and at the boss the slope of $w(r)$ vanishes. The deflection of the boss then becomes:

$$w_0 = A_p \frac{p r_0^4}{Eh^3}$$
$$A_p = \frac{3\left(1-\upsilon^2\right)}{16} \left(1 - \left(\frac{r_b}{r_0}\right)^4 - 4\left(\frac{r_b}{r_0}\right)^2 log \frac{r_0}{r_b}\right) \tag{4.102}$$

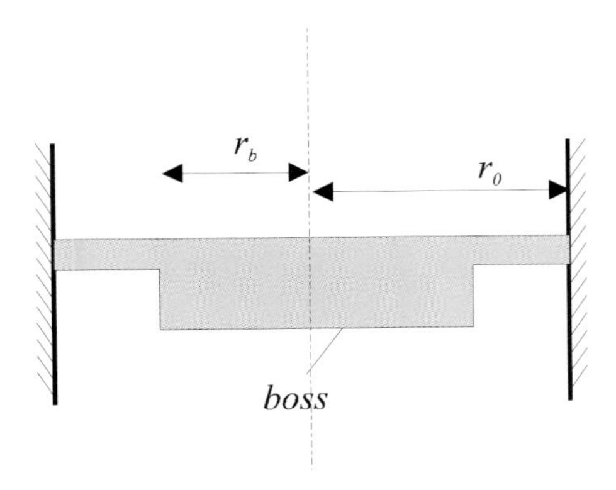

Fig. 4.11. A bossed diaphragm

A diaphragm which deflects with amplitudes comparable with and larger than its thickness suffers also form lateral stress due to stretching. Including stretching results, similar to what we found for beam, in non-linear differential equations. Technically, a membrane is a plate with vanishing bending stiffness under a large tension, which is caused by forces acting at the edges of the membrane. If the deflection of the membrane is small enough so that the additional tension due to the deflection can be neglected, the two-dimensional Poisson equation results:

$$T\Delta w + p = 0 \tag{4.103}$$

The solution of (4.103) for a circular membrane is a parabola,

$$w(r) = -R^2 \frac{p}{4T}\left(1 - \left(\frac{r}{R}\right)^2\right) \tag{4.104}$$

For the membrane without initial tension a deflection introduces a tension. This tension constitutes the restoring *non-linear* force. In this case, the deflection of a circular membrane resembles the surface of a sphere with a centre deflection given by di Giovanni (1982), p.143

$$w_0{}^3 \approx \frac{3(1-\upsilon)}{7-\upsilon}\frac{PR^4}{Eh} \tag{4.105}$$

It must be noted that there are other formulas for the Poisson ratio dependent coefficient in (4.105) found in the literature; see e.g. Timoshenko (1959)

The important thing here is that the deflection varies with the third root of the pressure. So the spring constant equals zero close to $p = 0$.

If stretching and bending terms both must be included in the deflection of a diaphragm, a description of the deflection amplitude by addition of (4.101) and (4.105) is rather satisfactory. Note that this addition is itself an approximation: the total differential equation is non-linear, and one cannot add the solutions of the various terms. The relation of the pressure and the deflection becomes

$$\frac{pR^4}{Eh^4} \approx f_1(\upsilon)\frac{w_0}{h} + f_2(\upsilon)\left(\frac{w_0}{h}\right)^3 \tag{4.106}$$

For $\upsilon = 0.3$, we have $f_1 = 5.86$ and $f_2 = 3.19$ (di Giovanni 1982), however, in the literature slightly different coefficients can be found.

4.4.2. Square Membranes

The following analytical form can describe approximately the deflection of a square diaphragm:

$$w(x,y) \approx w_0(1 - \cos 2\pi x / a)(1 - \cos 2\pi y / a) \tag{4.107}$$

(a is the side length of the plate) where w_0 follows a relation similar to (4.106) but with different coefficients. The relation is

$$\frac{pa^4}{Eh^4} = g_1(\upsilon)\frac{w_0}{h} + \frac{3.41a^2\sigma_i}{h^2E}\frac{w_0}{h} + g_2(\upsilon)\left(\frac{w_0}{h}\right)^3 \tag{4.108}$$

The second term in this equation is due to initial tension σ_i. Meleshenko (1997) gives

$$g_1 = \frac{4.13}{\left(1-\upsilon^2\right)}.$$

Based on finite elements simulation, Pan (1990) gives

$$g_2(\upsilon) = \frac{1.98(1-0.585\upsilon)}{1-\upsilon}.$$

(4.107) satisfies the boundary conditions (deflection and slope at the edges zero). It is important to note that the shape of the deflected membrane does not change with the amount of the deflection. This is a natural result of the approximation where a shape is assumed, which is independent of the deformation. A real diaphragm deflects such that the inflection line shifts towards the clamped edges if the amplitude increases. We have shown elsewhere that this effect is important for the estimate of the maximum stress in membranes (van Rijn et al. 1997). The maximum stress arises not from tension in the membrane but from the bending close to the clamped edges.

Appendix 4.1: Buckling of bridges

We have

$$U_{\text{tot}} = \frac{1}{2}\int_0^l \left[\text{AE}\frac{\text{w}'^4}{4} + EIw''^2 + Tw'^2 \right] \text{d}x$$

The deflection is assumed to have the following form:

$$w(x) = w_0(1 - \cos(2\pi x / l))$$

from which the derivatives follow:

$$w' = \frac{2\pi}{l} w_0 \sin(2\pi x / l);\ w'' = \left(\frac{2\pi}{l}\right) w_0 \cos(2\pi x / l)$$

The integration is straightforward and gives

$$U_{\text{tot}} = \frac{\pi^2}{l}\left\{ \left[EI(2\pi / l)^2 + T \right] w_0^{\ 2} + \frac{3}{16} EA(2\pi / l)^2 w_0^{\ 4} \right\}$$

In the first term in the curly brackets we recognise the critical axial load, T_{cr}=-$4\pi^2 EI/l^2$.

We require that

$$\frac{\partial U_{\text{tot}}}{\partial w_0} = 0$$

which gives us

$$2(T - T_{\text{cr}})w_0 + \frac{3}{4} EA\left(\frac{2\pi}{l}\right)^2 w_0^{\ 3} = 0$$

The three solutions are $w_0 = 0$ and

$$w_0 = \pm\frac{l}{\pi}\sqrt{\frac{2(|T| - |T_{\text{cr}}|)}{3EA}}.$$

5. Principles of Measuring Mechanical Quantities: Transduction of Deformation

This chapter deals with the problem of the transfer of a deformation in a mechanical construction by an external load to an electrical signal. This is called "transduction". The most important mechanisms for transduction of mechanical microsensors use the following effects: piezoresistivity, the dependency of the capacity on the geometric arrangement of conductors, piezoelectricity, optical resonance and optical interferometry. We concentrate here on the first two, most important, transducers. For flowsensors the thermal domain is of great importance, but we shall defer the discussion of transducers using heat to the chapter on flowsensors.

Historically the first transducers for the measurement of mechanical deformations are metal strain gauges, and we start the discussion with this subject.

5.1 Metal Strain Gauges

The electrical resistance of a piece of metal depends on its size and shape. For a rod with cross-section S and length l the resistance R is given by

$$R = \frac{l}{\lambda S} \tag{5.1}$$

with λ the specific conductivity. If as a consequence of an external load the resistor changes its dimensions the resistance will change. We have for the change of $R(l,S)$ therefore

$$\mathrm{d}R = \frac{\partial R}{\partial l}\mathrm{d}l + \frac{\partial R}{\partial S}\mathrm{d}S = \frac{\mathrm{d}l}{\lambda S} - \frac{l\mathrm{d}S}{\lambda S^2} \tag{5.2}$$

If the rod is subject to an axial stress $\sigma = F/S$ the change of the length and the cross-section change both lead to an increase (tensile stress) or a decrease (compressive stress) of the resistance. An illustration is given in Fig. 5.1.

The reason is that the change of the length and of the cross-section has different signs. We have $\mathrm{d}l = l\varepsilon = lE\sigma$ and

$$\mathrm{d}S = (y + \mathrm{d}y)(z + \mathrm{d}z) - yz \approx y\mathrm{d}z + z\mathrm{d}y = yz\left(\frac{\mathrm{d}z}{z} + \frac{\mathrm{d}y}{y}\right) = -2\upsilon\varepsilon S = -2\upsilon E\sigma S$$

z denotes the thickness of the rod (not drawn in Fig. 5.1). Using these equations in (5.2) we find for the relative change of the resistance

$$\frac{dR}{R} = (1+2\upsilon)\varepsilon \tag{5.3}$$

The ratio of the relative change of the resistance per unit strain is called the gauge factor G. Typical values of υ are 0.1 – 0.4, so for most materials the gage factor is between 1 and 2. The yield strain of metals mostly is well below 1% (this is about the yield strain of high quality steel), so changes in resistance of metal straingauges is maximal 1 to 2%.

Metal strain gauges are developed to a very high level. On the market there are strain gauges of all possible kinds, including temperature compensated strain gauges and strain gauges that match a great number of materials. To give an indication what traditional technology after 100 - 200 years development and experience is able to: There are loadcells based on metal straingauges with a precision of 1: 100,000, guaranteed in a temperature range between -40 and +80 °C. The load cell element for large loads (50 tons) can be made for a price of a few hundred dollar. Current micro technology cannot tip at this performance, and forget about price performance ratio of loadcells in this category for the relatively small market.

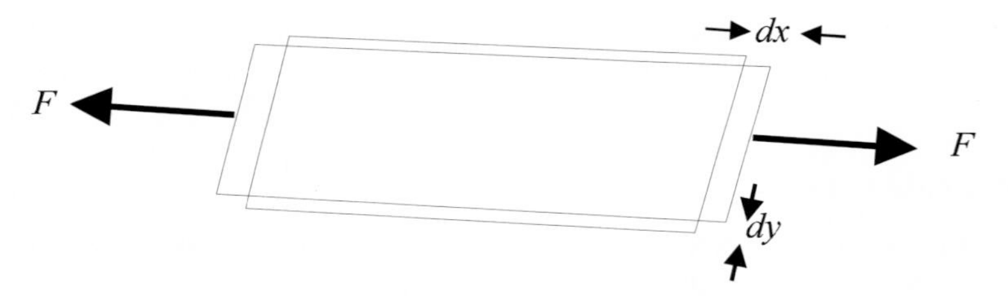

Fig. 5.1. The change of a prismatic body due to an axial force

5.2 Semiconductor Strain Gauges

The dominant effect in metal strain gauges is the change of the geometry. The materials property λ is independent of the strain. This is different in semiconductors. The effect is called piezoresistivity.

Conductivity, and therefore all effects connected to conductivity such as piezoresistivity in crystals are anisotropic. We do not dwell here into these complications, but refer to the literature; Kloek (1994) gives a good overview.

The piezoresistive coefficient π is defined as the relative change of the resistivity per stress. There is a longitudinal and a transversal effect: the resistivity changes in a direction parallel to the strain and normal to the strain.

$$\pi = -\frac{1}{p}\frac{\Delta\lambda}{\lambda} = \frac{1}{p}\frac{\Delta\rho}{\rho}$$

The effect is observed in single crystalline silicon and in polysilicon thin films. In single crystalline silicon the effect is exploited by doping p-regions in n-type silicon microstructures. Polysilicon thin films can be deposited on top of microstructures, doped and pattered.

5.2.1 Piezoresistive Effect in Single Crystalline Silicon

The gauge factor in silicon depends on the crystallographic orientation and on the doping. The maxima of the coefficients are:

$$p-Si:\pi_{l\langle 111\rangle} = 93.5\times 10^{-11}\,Pa^{-1}$$

$$n-Si:\pi_{l\langle 100\rangle} = -102.2\times 10^{-11}\,Pa^{-1}$$

If the electrochemical etch stop is used, the deformed member (a diaphragm in a pressure sensor, or a cantilever beam in an accelerometer) will be n-silicon, and p-type resistors will be indiffused or implanted into the upper layers of the member. Therefore, the larger π for n-silicon is of no use for micromachined sensors using this etch stop. Further, membranes or cantilever beams are machined in $\langle 100\rangle$ wafers, so there is no $\langle 111\rangle$ direction available. On a $\langle 001\rangle$ oriented p-silicon wafer the maximum piezoresistive effect is a little bit more than $70\times 10^{-11}\text{Pa}^{-1}$, both, for the transversal and longitudinal effects. The orientation of the maximum is in $\langle 110\rangle$, there are minima in $\langle 100\rangle$ directions.

The piezoresistive gage factor can be found easily from π. If we include the change of the specific resistivity ρ ($\rho = 1/\lambda$) in Eq. (5.3), we have

$$\begin{aligned}\frac{dR}{R} &= (1+2\upsilon)\varepsilon + \left(\frac{d\rho}{\rho}\right)_l \varepsilon_l + \left(\frac{d\rho}{\rho}\right)_t \varepsilon_t \\ &= (1+2\upsilon)\varepsilon + \pi_l E\varepsilon_l + \pi_t E\varepsilon_t\end{aligned} \tag{5.4}$$

The suffixes l and t refer to longitudinal and transversal piezoelectric coefficients.

The gaugefactor of p-Si$_{\langle 110\rangle}$ is 133, much larger than the geometric effects. Thus, change in geometry can be neglected.

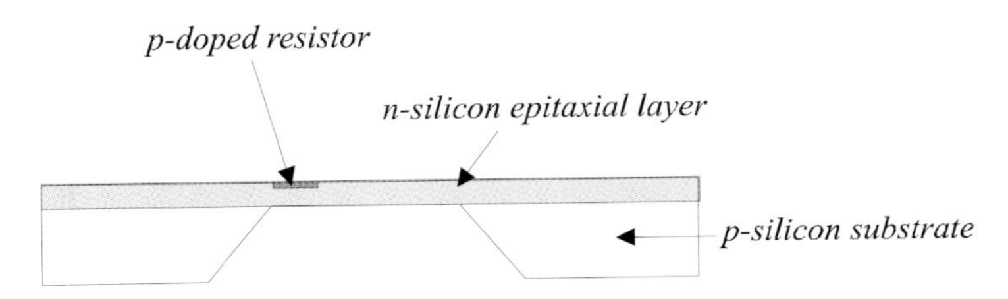

Fig. 5.2. Examples of a membrane with indiffused or ion implanted doped resistors

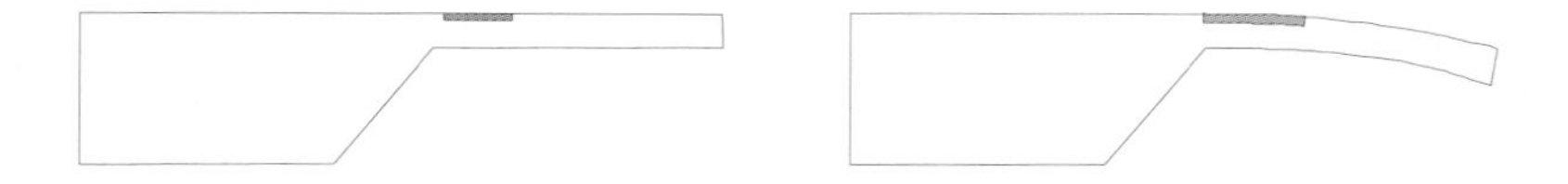

Fig. 5.3. Diffused resistors in silicon flexible structures

It is important to note that the piezoresistor must be placed in the upper part or on top of the flexible structure. Bending results in a strain distribution across the structure, and the strain will change its sign in the centre. A piezoresistor right across the flexible structure would not change its resistance upon bending.

The electrical isolation of the doped region relies on junction isolation. So the voltages are limited and there is some current leakage.

5.2.2. Piezoresistive Effect in Polysilicon Thin Films

Polysilicon thin films deposited by LPCVD processes on top of flexible structures can also be used as piezoresistive strain gauges. They have a few advantages compared to indiffused or implanted resistors: The technological possibilities allow a greater freedom of design and the choice of materials for membranes is greater, and the strain gauges do not rely on junction isolation, so there are less problems with leakage currents and greater voltages can be used. Electrical isolation from the substrate is done by means of a dielectric thin film (preferably stress free silicon nitride). Electrical contact can be made either by high-doped polysilicon leads, or better, because of higher conductivity, by metal leads.

Similar to single crystalline silicon, the piezoelectric coefficient in p-doped silicon is larger than in n-doped silicon. Using polysilicon it is possible to micromachine films with both types of doping, in contrast to implanted strain gauges. However, due to grain boundary effects and texture, the size of the piezoresistive effect is reduced. The maximum gauge factor in p-doped polysilicon is close to -40, in n-doped silicon to +20. Doping level must be in the range of a few 10^{19}cm^{-3}, deposition temperature 560°C, annealing at 1000-1100°C (French 1989).

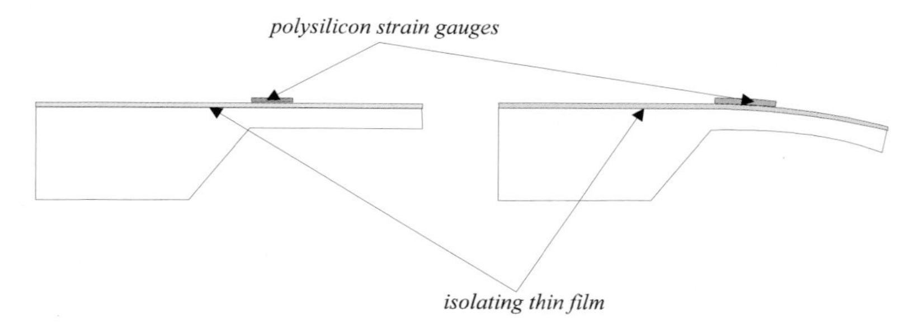

Fig. 5.4. Configuration of polysilicon this films on to of flexible silicon structures

Besides the possibility of doping by indiffusion and implantation, polysilicon can be doped during the deposition by feeding an additional gas for the doping through the reactor. Bouwstra et al. used in situ phosphorous-doped polysilicon by adding phosphine to the silane flow through the reactor (Bouwstra 1990). This way has some advantages, one of them is that the temperature coefficient of the gauge factor can me made zero. However the standard procedure is to dope the thin film after growth by ion implantation (Burger 1996).

5.2.3. Transduction from Deformation to Resistance

The piezoresistors are placed on top of flexible structures. Upon bending parts of the structures will be under compressive stress, other under tensile stress. Obviously it is important to know how the structure deforms under an external load to find the optimum positions for the strain gauges.

As we have seen in chapter 4, the strain in a bent member is

$$\varepsilon = \frac{z}{R} = zw'' \tag{5.4}$$

z is the distance normal from the neutral plane, R is the radius of curvature and w is the deflection of the neutral plane. The last equality holds for small deflections. This equation also holds locally. Generally the radius of curvature varies along the deformed construction, as can be seen by inspection of e.g. Eq. (4.63) for a bent bridge. Equation (5.4) means that per length $\mathrm{d}x$ at a distance z from the neutral plane there is a change of the length due to the bending $\mathrm{d}\Delta x$. $\mathrm{d}\Delta x/\mathrm{d}x$ is then the local strain. A piezoresistive element at a distance z from the neutral plane sees an average strain, or a total change in length of the element. This is given by the integral over $\mathrm{d}\Delta x$. So the average strain between two points x_1 and x_2 is given by

$$\begin{aligned}\bar{\varepsilon} &= \frac{1}{x_2 - x_1}\int_{x_1}^{x_2} \mathrm{d}\Delta x = \frac{1}{x_2 - x_1}\int_{x_1}^{x_2} \frac{z}{R}\,\mathrm{d}x \\ &= \frac{1}{x_2 - x_1}\int_{x_1}^{x_2} zw''\,\mathrm{d}x = \frac{z\left(w'(x_2) - w'(x_1)\right)}{x_2 - x_1}\end{aligned} \tag{5.5}$$

Thus, it is the derivative of the deflection, which is important for the average strain in a strain gauge.

This is simple for the cantilever beam: the sign of the curvature is constant, so the piezoresistor can be placed on top of the beam along the whole length. The total strain is easily calculated from (4.66) for a cantilever beam with length l:

$$\bar{\varepsilon} = \frac{20 P_0 \mathrm{h} l^2}{48 EI} \tag{5.6}$$

A bridge covered in its whole length by a piezoresistor would measure no effect. Since at both clamped edges the slope of the bridge is zero, there is no average strain. For an optimal effect, the strain gages must extend either from the edges to the two inflection points, or it must be placed in between them. The inflection points can be found from. (4.63), they are located at

$$x_{\text{inf}} = l\left(1/2 \pm \sqrt{1/12}\right) \tag{5.7}$$

The form of a circular diaphragm loaded by a constant pressure p_0 is (see eq. 4.100)

$$w(r) = \frac{3}{16}\frac{1-\upsilon^2}{Eh^3}p_0\left[r_0^2 - r^2\right]^2 \tag{5.8}$$

where r_0 is the radius of the diaphragm. It follows that the inflection line is at $r = r_0/\sqrt{3}$. A strain gage must be placed either at the edges, without exceeding the inflection line, or in the centre of the diaphragm between $\pm r_0 / \sqrt{3}$. The relative change of the resistance is given by

$$\frac{\Delta R}{R} = G\bar{\varepsilon} = \frac{1}{6}G\frac{p_0}{E}\left(\frac{r_0}{h}\right)^2 \tag{5.9}$$

For square diaphragms only the numerical factor differs from this equation. For large sensitivity, the diaphragm must be thin and wide.

5.3 Capacitive Transducers

5.3.1 Electromechanics

A capacitive transducer is a transducer of a very different type. The capacitor can store energy, and its properties are completely described once the energy function is known. There are two distinct ways to change the energy of a capacitor: by changing its charge, this would represent an electric work on the capacitor, or by changing the geometric arrangement, and this would represent mechanical work. The total energy change is then

$$\mathrm{d}W = u\mathrm{d}Q + \mathbf{F}_{\text{ext}}\mathbf{dr}\ . \tag{5.10}$$

with u the electrostatic potential and $\mathbf{F}_{\text{ext}}$ the force needed to realise the displacement d**r**. Let us restrict ourselves for the moment to a situation in which the geometrical change of the arrangement of the conductors is only given by the distance of the plates of a parallel plate capacitance, as shown in Fig. 5.5.

In this case, the mechanical work is $-F_{\text{int}}\mathrm{d}x$ and the total energy as a function of the charge and the distance of the plates is[1]

$$W = \frac{1}{2}\frac{x}{\varepsilon A}Q^2 + K(x - x_0)^2 \tag{5.11}$$

---

[1] Note that in a quasistatic process $\mathbf{F}_{\text{int}} = -\mathbf{F}_{\text{ext}}$

with A the overlap surface of the plates. We have added a spring to the system, with spring constant K. The position of the plate hold by the unloaded spring is x_0 (measured from the lower plate).

The total differential of W is given by

$$\mathrm{d}W = \left(\frac{\partial W}{\partial x}\right)_Q \mathrm{d}x + \left(\frac{\partial W}{\partial Q}\right)_x \mathrm{d}Q \tag{5.12}$$

from which we find by comparing with (5.10)

$$\left(\frac{\partial W}{\partial x}\right)_Q = F_{ext} = \frac{Q^2}{2\varepsilon A} + Kx \tag{5.13}$$

and

$$\left(\frac{\partial W}{\partial Q}\right)_x = u = \frac{Qx}{\varepsilon A} = \frac{Q}{C}. \tag{5.14}$$

We also see that the capacity and the spring constant are given by

$$\left(\frac{\partial^2 W}{\partial Q^2}\right)_x = \frac{1}{C} \equiv \frac{1}{C_x} \tag{5.15}$$

and

$$\left(\frac{\partial^2 W}{\partial x^2}\right)_Q = K \equiv K_Q . \tag{5.16}$$

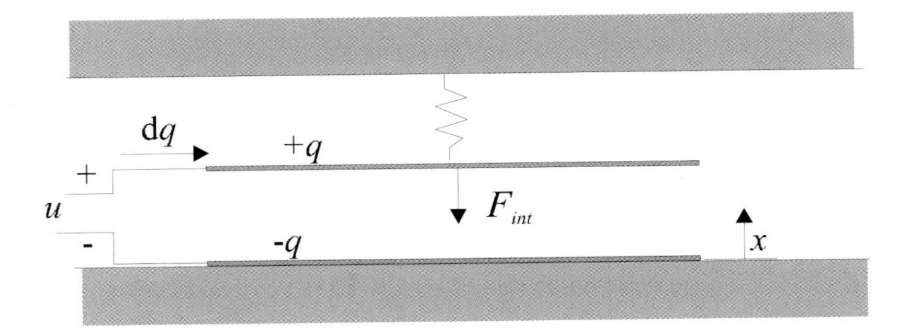

Fig. 5.5. A capacitor with a movable plate connected to a spring

In (5.13) the first term the right hand side represents the electrostatic force. Note that this force is independent of the plate separation. This might be surprising, one would expect that the force becomes smaller if the plates are farther apart. But this impression is based on the assumption that the voltage is constant when one changes the distance. In fact, the electrostatic field close to a plane surface is independent of the distance, so will be the forces on charges. Usually, in these systems the voltage is controlled. If the plate separation is changed while the voltage is constant, the charge must change. So a battery must also deliver work to the capacitance. We can see from Eq. (5.10) that the derivative of W with respect to x at variable charge does *not* give the force. In fact, the battery must do a work equal to $u\mathrm{d}Q = u^2\mathrm{d}C$.

There is a mathematical trick to circumvent this difficulty, which is called the Legendre transformation. One adds to the energy function the product $-uQ$, and one gets the so-called co-energy:

$$W' = W - uQ \tag{5.17}$$

We get immediately:

$$\mathrm{d}W' = \mathrm{d}W - u\mathrm{d}Q - Q\mathrm{d}u = -Q\mathrm{d}u + F \quad \mathrm{d}x\,. \tag{5.18}$$

With the help of $Q = Cu$ we can express W' by u and x:

$$W' = -\frac{1}{2}Cu^2 + K(x - x_0)^2 = -\frac{\varepsilon A u^2}{2x} + K(x - x_0)^2 \tag{5.18}$$

and we get now

$$F_{\mathrm{ext}} = \left(\frac{\partial W'}{\partial x}\right)_u = \frac{1}{2}\frac{\varepsilon bl}{x^2}u^2 + Kx \tag{5.19}$$

Now we do have an x-dependency of the force. But note that this is exactly the same force as in (5.13), as can be seen if one eliminates u in (5.19). The important point is that the stiffness of the capacity is different. If we calculate the spring constant from (5.19) we get a voltage dependency:

$$K_{\mathrm{u}} = \left(\frac{\partial^2 W'}{\partial x^2}\right)_u = K - \frac{\varepsilon A u^2}{x^3}\,. \tag{5.20}$$

There is a voltage u_{c} at which the spring constant K_{u} vanishes. This voltage is the so-called pull-in voltage: the capacitance collapses. When using the capacitance as a sensor for an external load, it is clear that one has to stay away from u_{c} because of uncontrollable instabilities. On the other hand, a small K_{u} means a sensitive sensor.

There is a second way to change the geometry, which is of importance for mechanical microsensors. This is the change of the overlap of the electrodes at constant separation of the plates. We have given an example of a micromachined structure for this in Chap. 2, Fig. 2.3. Besides in actuators, these types of electrodes are used in accelerometers, gyroscopes and resonators. Schematically, the device is shown in Fig. 5.6.

In this case the energy is given by

$$W = \frac{Q^2 d}{2\varepsilon a(b-x)} + K(x-x_0)^2 \tag{5.21}$$

Here, the surface of the capacitor plate is $A = ab$, and the overlap is $a(b\text{-}x)$. d is the separation of the plates. The force is now

$$F_{\text{ext}} = \left(\frac{\partial W}{\partial x}\right)_Q = \frac{Q^2 d}{2\varepsilon a(b-x)^2} + Kx \tag{5.22}$$

and we see that in this case the electrostatic force (first term in eq. (5.22)) is dependent on the position. It also follows that the electromechanical spring constant is dependent of the charge of the capacitor. The second derivative of (5.21) is

$$K_{\text{Q}} = K + \frac{Q^2 d}{\varepsilon a(b-x)^3} \tag{5.23}$$

and we do not get an instability. On the other hand, the co-energy is now

$$W' = -\frac{u^2 \varepsilon a(b-x)}{d} + K(x-x_0)^2 \tag{5.24}$$

and at constant voltage the electrostatic force is independent of the overlap:

$$F = \left(\frac{\partial W'}{\partial x}\right)_u = \frac{u^2 \varepsilon a}{d} + Kx \tag{5.25}$$

and the spring constant is simply K.

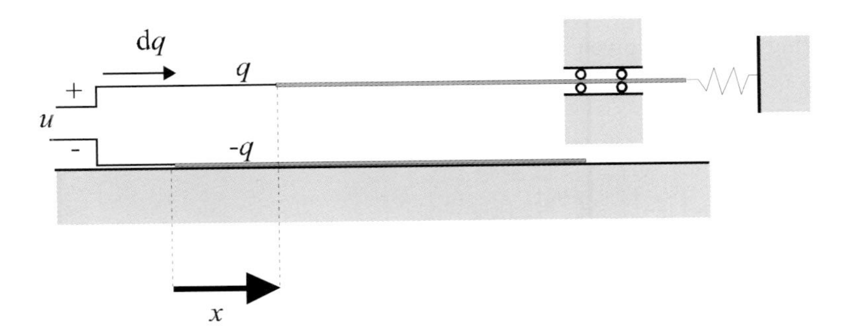

Fig. 5.6. Schematic of the capacitor with variable overlap b-x. b is the width of the plate

In practice, the distance between the plates is kept constant by means of a flexible mechanical construction, which is much stiffer than the spring in Fig. 5.6. However, when controlling the combdrive by voltage, there is the same instability as described by Eq. (5.20). Therefore, one is limited in the voltage of these electromechanical transducers. See Fig. 5.7.

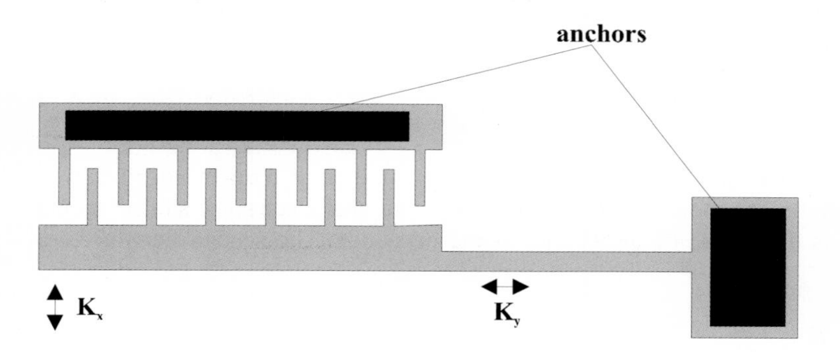

Fig. 5.7. Comb drive suspended by a spring with spring constants K_x (vertical movement) and K_y (horizontal movement). The latter spring constant is important to keep the fingers right in place. Generally, $K_x << K_y$. K_y is decisive for the largest deflection

5.3.2 Diaphragm Pressure Sensors

There are three types of mechanical constructions for mechanical diaphragm pressure sensors. A simple diaphragm bending under the pressure load (Fig. 5.8a), a partially collapsed capacitance (Fig. 5.8b) and a diaphragm with a stiff portion, a so-called boss (Fig. 5.8c).

The configuration in Fig. 5.7c is quite similar to the schematic in Fig. 5.5. Here the spring constant is given by di Giovanni (1982)

$$K \approx \frac{Eh^3}{\pi r_b{}^2 r_0{}^4} \frac{1}{A_p}$$
$$A_p = \frac{3\left(1-\upsilon^2\right)}{16}\left(1-\left(\frac{r_b}{r_0}\right)^4 - 4\left(\frac{r_b}{r_0}\right)^2 \log\frac{r_0}{r_b}\right) \tag{5.26}$$

Here, r_b is the radius of the boss and r_0 the radius of the membrane, and we have assumed that the pressure acts only on the boss, with is acceptable if $r_b \approx r_0$. h is the thickness of the flexible part of the structure. Keep in mind that this spring constant is the one that belongs to operation with charge control. For voltage control the spring constant must be corrected by the term in (5.20).

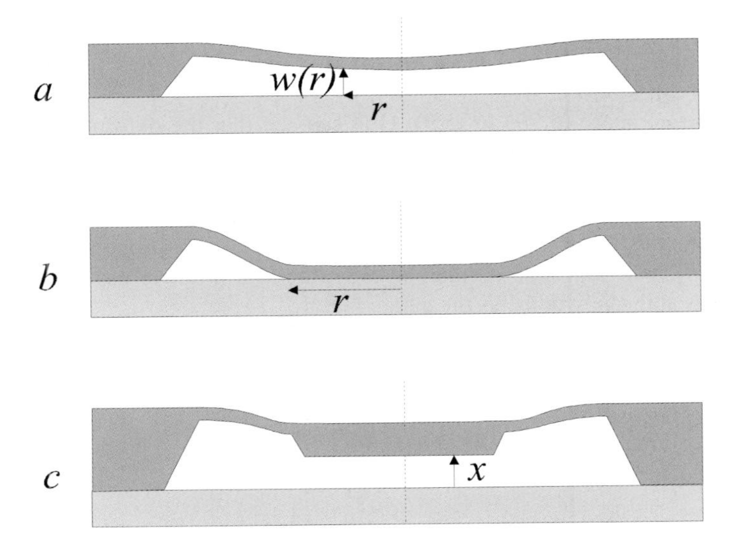

Fig. 5.8. Three configurations for pressure sensors

The relative change of the capacity is in this case

$$\frac{\delta C}{C}=\frac{\delta x}{x}=\frac{F}{xK}=\frac{p}{E}\frac{r_0^{\ 4}}{h^3 x}A_p . \tag{5.27}$$

For the case shown in Fig. 5.7a, the analysis of the relative change of the capacitance leads to a formula similar to this one. The capacity is now given by

$$C=2\pi\int_0^{r_0}\frac{\varepsilon}{d-w(r)}r\mathrm{d}r\approx 2\pi\int_0^{r_0}\frac{\varepsilon}{d}\left(1-\frac{w(r)}{d}\right)r\mathrm{d}r \tag{5.28}$$

We assumed a small deflection. d is the plate separation of the unloaded diaphragm. The deflection of a circular diaphragm under a pressure is given by (5.8) and the integration leads to

$$\frac{\delta C}{C}=\frac{1}{32}\frac{p}{E}\frac{r_0^{\ 4}}{dh^3} \tag{5.29}$$

The response of the capacity is non-linear. In the analysis here we linearised all equations to make clear the most essential features of the capacitive sensors. The example in Fig. 5.7b is an attempt due to Rosengren (1992). The result of their analysis is that once the membrane has touched the bottom electrode (one needs of course an electrical isolation), the capacitive change is dominated by the change of the contact area, and this turns out to be much closer to linearity than the capacitance change in the case 5.7a. Also the linearity of the bossed membrane is better (e.g. Mallon 1990).

It is interesting to compare (5.29) to (5.9). One can see that the deformation has quite different consequences for the relative changes of capacity and resistance. Typically we have in micromachined capacities $d \approx h$, so the sensitivity of capacitive pressure sensors is proportional to (r_0/h) to the fourth power, while in piezoresistive pressure sensors the sensitivity is only proportional to the square of (r_0/h). Sine this ratio is small anyway, the fourth power has quite drastic consequences and easily outrules the gauge factor.

6. Force and Pressure Sensors

There are several techniques to measure forces and pressures. Very often, the force to be measured is converted into a change in length or height of a piece of material, the *spring element*. The change in dimensions is subsequently measured by a *sensor element*, e.g. a (piezo-) resistive or resonant strain gauge or a changing capacitance. This is illustrated in Fig. 6.1. Sometimes the sensor element and the spring element can not be distinguished, i.e. the sensor element itself is also the spring element. For example, in piezoelectric force transducers, the deformed crystal both supports the load and supplies the output signal. More sophisticated systems incorporate an electronic feedback to balance the external force or pressure by an equal but oppositely directed counterforce or pressure. The obvious advantage of such a system is that the spring element can be omitted, thus eliminating problems like linearity, creep and hysteresis related to the spring element. However, application of such systems is limited to relatively small forces and pressures because of the limited size of the counterforce or pressure that can be exerted.

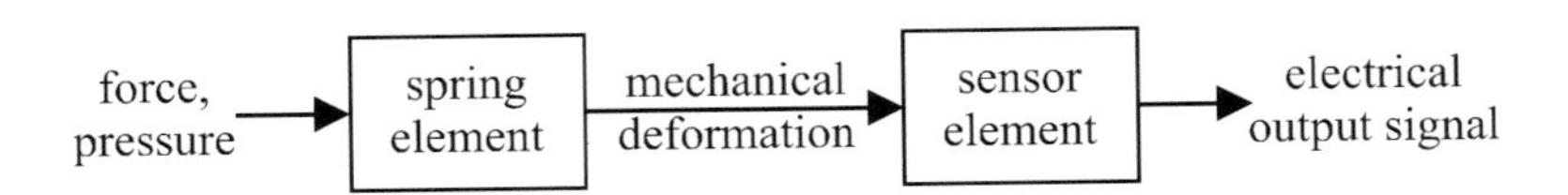

Fig. 6.1. Most force and pressure sensors consist of a spring element and a sensor element

Silicon micromachined force and pressure sensors are commonly based on the first mentioned technique and the spring and sensor elements can easily be distinguished. In this chapter, we will start with a discussion on force sensors. We will first show some simple bending structures that can be used to measure relatively small forces. Due to the small wafer thickness of silicon bending is not an option for measuring large forces. Therefore, in Sect. 6.1.1 we will discuss a special type of force sensor based on the compression of silicon.

Silicon pressure sensors are commercially available in various shapes and sizes and with significant differences in performance. In Sect. 6.2 an overview will be given. Three categories are distinguished based on the read out mechanism: piezoresistive, capacitive and resonant sensors. Also some examples will be given of capacitive sensors employing force feedback by means of an electrostatic counterforce. The chapter will be concluded with two types of sensors that are closely related to pressure sensors: silicon microphones and tactile imaging arrays.

6.1 Force Sensors

A simple sensor structure to measure forces is indicated in Fig. 6.2. The structure consists of a silicon beam with an integrated strain gauge, which can either be a (piezo-)resistive strain gauge or a resonant strain gauge. The strain at the position of the strain gauge as a result of the applied force is defined by the Young's modulus and cross sectional area of the beam:

$$\varepsilon = \frac{F}{Ebh} \tag{6.1}$$

Fig. 6.3 shows a possible implementation of the sensor using a piezoelectrically driven resonant strain gauge as the sensor element (Van Mullem, 1991). Fig. 6.4 shows a photograph of the device. As will be discussed in chapter 9, resonant sensors have several advantages compared to (piezo-) resistive strain gauges and very high resolutions of 100 ppm or better can be obtained.

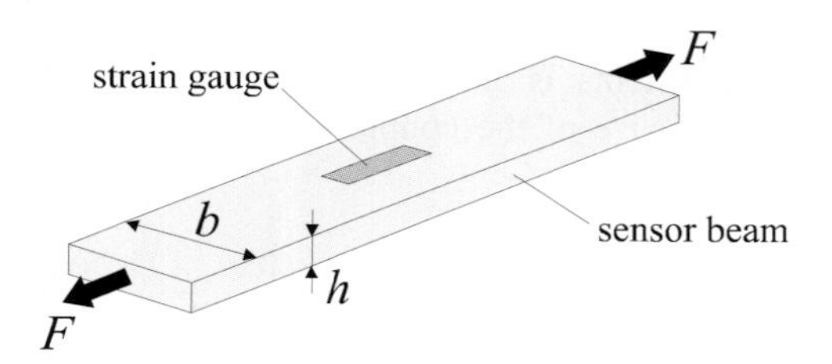

Fig. 6.2. Silicon beam force sensor

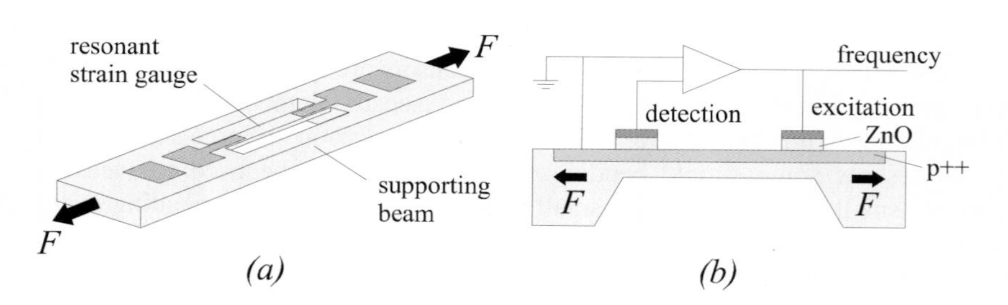

Fig. 6.3. Silicon beam force sensor with piezoelectrically driven resonant strain gauge (a) and schematic of oscillator circuit which uses the resonator as the frequency defining element (b)

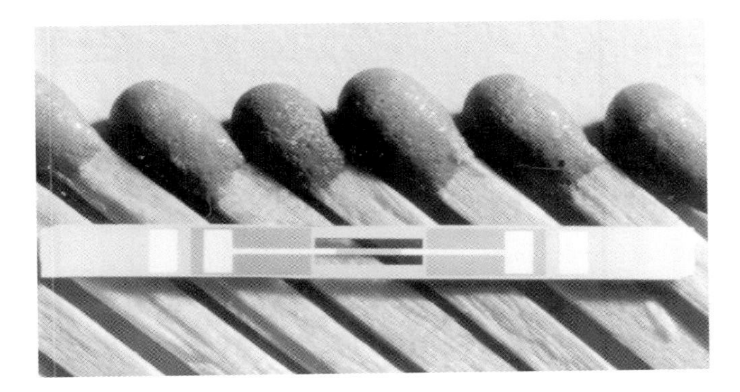

Fig. 6.4. Photograph of the piezoelectrically driven resonant force sensor from Fig. 6.3

Instead of loading the silicon beam force sensor by a tensile force, it can also be used in a bending configuration. For example, the four-point bending beam structure indicated in Fig. 6.5 can be used to measure compressive forces. An important drawback associated with this structure, however, is that it is not very robust. For application in a practical force sensor it is preferred that all parts are firmly fixed to each other. Furthermore, only tensile stresses occur at the place of the strain gauge(s). Structures with both tensile and compressive stress have the important advantage that the strain gauges can be operated in a differential fashion, which provides first order cancellation of temperature effects. Such a structure is indicated in Fig. 6.6. It consists of a beam (or membrane) which is supported at the ends and has a rigid boss in its center. When a force is applied to the boss the beam will deform in an S-shape resulting in both compressive and tensile strains at the positions of the strain gauges as indicated in Fig. 6.6(b).

In chapter 4 we have seen that the deflection w of a clamped-clamped beam due to a point force F at the center of the beam as a function of the position x along the beam is given by (see eqn. (4.71)):

$$w = \frac{F}{48EI} x^2 (3L - 4x) \qquad \left(0 \le x \le \frac{L}{2}\right) \tag{6.2}$$

Where $I = bh^3/12$ is the second moment of inertia of the beam.

The largest deflection occurs in the middle of the beam at the position of the boss ($x = L/2$):

$$w_{\max} = \frac{F}{16Eb}\left(\frac{L}{h}\right)^3 \tag{6.3}$$

Thus the spring constant K of the force sensor is given by:

$$K = 16Eb\left(\frac{h}{L}\right)^3 \tag{6.4}$$

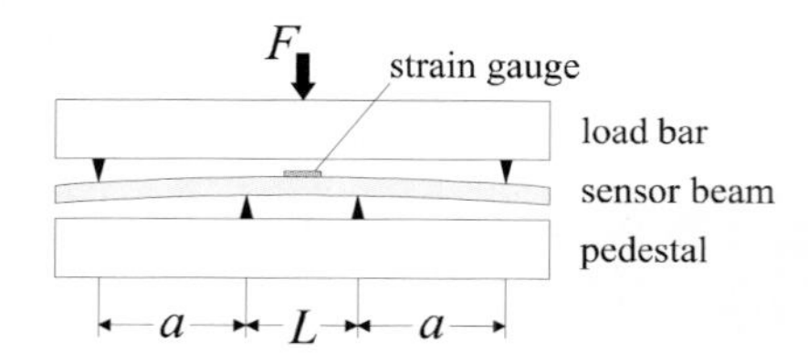

Fig. 6.5. Four point bending beam force sensor

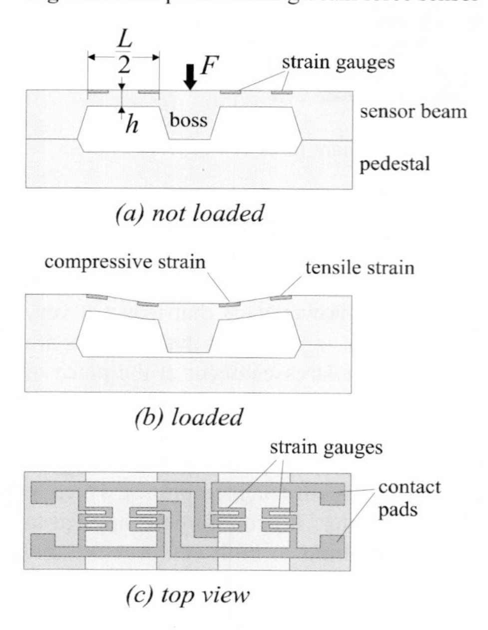

Fig. 6.6. Quasi-monolithic bending beam force sensor

As we have seen in chapter 5, the strain at the surface is proportional to the second derivative of the deflection (eq. (5.4)):

$$\varepsilon = \frac{h}{2} w'' = \frac{Fh}{16EI}\left(L - 4x\right) \tag{6.5}$$

Thus the strain changes linear with x and has maxima $\frac{FhL}{16EI}$ and $-\frac{FhL}{16EI}$ at $x = 0$ and $x = L/2$, respectively. For optimal sensitivity, the strain gauges should start at these points. For strain gauges with a length l, the measured average strain is (see also Sect. 5.2.3):

$$\bar{\varepsilon} = \frac{1}{l}\int_0^l \frac{h}{2} w'' dx = \frac{Fh}{16EI}\left(L - 2l\right) \tag{6.6}$$

So we see that the measured strain decreases with an increase in length of the strain gauges. This is why in most cases a meandering shape is used, as indicated in Fig. 6.6(c).

Some additional force reduction and an improved vertical alignment of the force is obtained by using a second beam on top of the structure as shown in Fig. 6.7. This structure was proposed by Tilmans et al. in combination with resonant strain gauges (Tilmans 1993). The structure can be dimensioned for a large range of maximum forces varying from 0.03 N up to 1500 N.

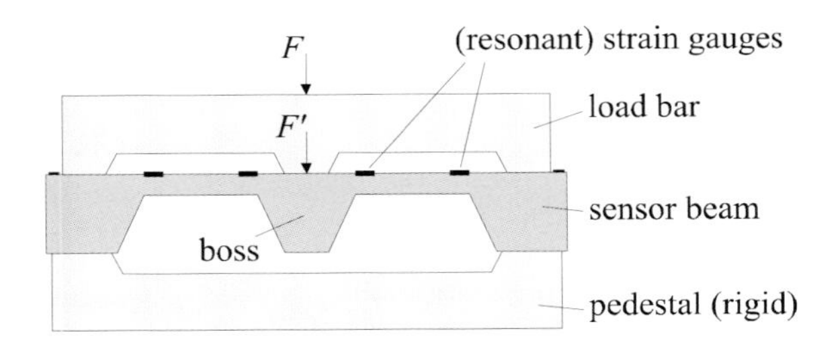

Fig. 6.7. Force sensor consisting of a sensor beam sandwiched between a load bar and a pedestal (Tilmans 1993). The load bar serves for force reduction and suppression of shear forces. The boss provides an easy means for overload protection

6.1.1 Load Cells

Load cells are force sensors which are used in weighing equipment. In most conventional load cells the spring element is made from steel or aluminum and resistive strain gauges are used as the sensor element. For small loads, i.e. less than 1000 kg, these load cells are fabricated in large quantities and at very low prices. For loads above 1000 kg this is different. Expensive high-grade steels and labor-intensive fabrication methods are required to minimize the influence of hysteresis and creep. Silicon does not suffer from hysteresis and creep and, therefore, a load cell made from silicon might be a good alternative to traditional load cells made from steel. However, the bending beam type force sensors discussed above are not suitable for very high loads. It can be shown that bending beam structures may be used for loads up to 150 kg (Tilmans 1993), but certainly not for loads higher than 1000 kg. For these high loads a load cell has to be based on the compression of silicon as indicated in Fig. 6.8. This sensor consists of two bonded silicon wafers. The edge of the sensor chip is compressed under the load and the amount of compression can be measured by measuring the change in capacitance between two capacitor plates located in the center. A problem with this design is that it is extremely difficult to design a package that applies the load homogeneously on the edge of the chip (Wensink 1998). Much better results are obtained when the single large capacitor in the center is replaced by an array of smaller capacitors distributed over the entire chip as illustrated in Fig. 6.9 (Wiegerink 1999).

Fig. 6.10 shows a cross-section of the distributed capacitive load cell and a bottom view of the top wafer. The top-wafer contains poles which bear the load. The bottom-wafer contains an electrode pattern forming an array of capacitors with the top wafer as a common electrode. On application of a load the poles will be compressed and the distance between the metal electrodes and the top-wafer at the position of the capacitors will decrease thereby increasing the capacitance.

The change in height of each pole as a result of the applied load can be expressed as (assuming that the force distribution is homogeneous over the surface of a single pole):

$$\Delta l_{i,j} = \frac{l_{\text{pole}}}{E \cdot A_{\text{pole}}} F_{i,j} \tag{6.7}$$

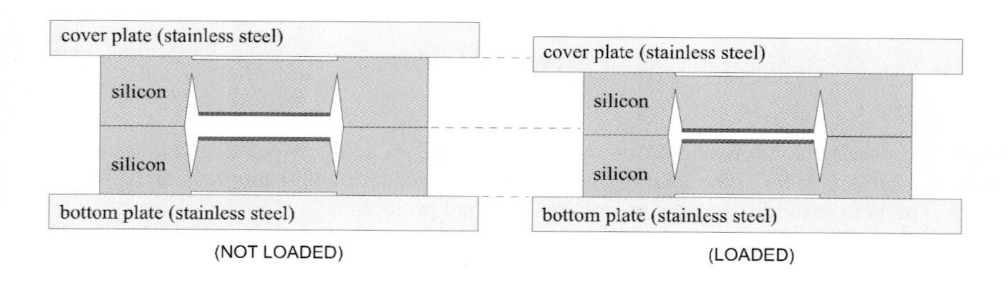

Fig. 6.8. Principle of a load cell based on compression of silicon

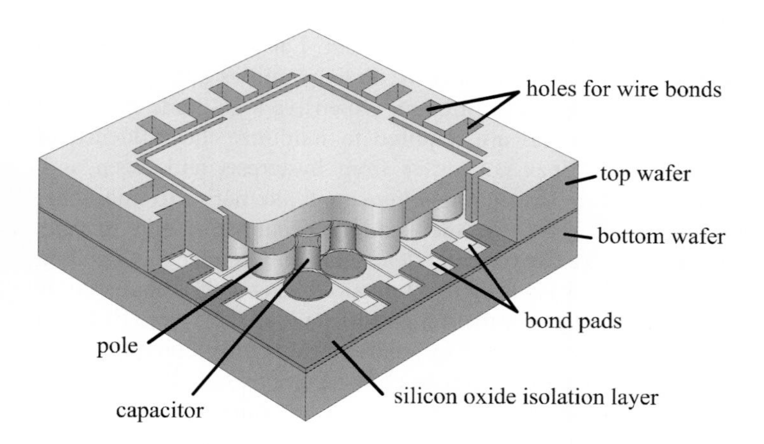

Fig. 6.9. Distributed capacitive load cell (Wiegerink 1999)

where l_{pole} and A_{pole} are the height and surface area of the poles, respectively, E is the Young's modulus of silicon and $F_{i,j}$ is the force acting on the pole. The indices i and j indicate the position of the pole in the array.

For each capacitor, the capacitance is given by:

$$C_{i,j} = \frac{\varepsilon A_{\mathrm{cp}}}{d_0 + \Delta d_{i,j}} \tag{6.8}$$

where ε is the dielectric constant, A_{cap} is the surface area of each capacitor, d_0 is the distance between the capacitor plates with no load applied and $\Delta d_{i,j}$ is the change in distance as a result of the load.

Approximating the change in distance $\Delta d_{i,j}$ by the average change in height of the surrounding poles, (6.8) can be written as:

$$C_{i,j} = \frac{\varepsilon A_{\mathrm{cap}}}{d_0 - \frac{1}{4}\left(\Delta l_{i,j} + \Delta l_{i+1,j} + \Delta l_{i,j+1} + \Delta l_{i+1,j+1}\right)} \tag{6.9}$$

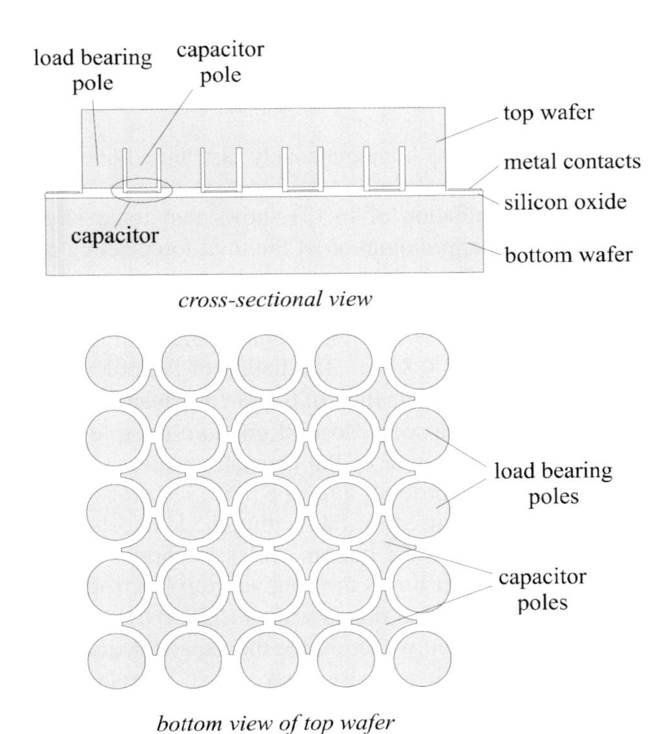

Fig. 6.10. Structure of the distributed capacitive load cell chip

or:

$$\frac{1}{C_{i,j}} = \frac{d_0 - \frac{1}{4}\left(\Delta l_{i,j} + \Delta l_{i+1,j} + \Delta l_{i,j+1} + \Delta l_{i+1,j+1}\right)}{\varepsilon A_{\mathrm{cap}}} \tag{6.10}$$

which is, according to (6.7), proportional to the forces acting on the poles.

Summation of (6.10) for all capacitors gives:

$$\sum_{i=1}^{n-1}\sum_{j=1}^{n-1}\frac{1}{C_{i,j}} = (n-1)^2\frac{d_0}{\varepsilon A_{\mathrm{cap}}} - \sum_{i=1}^{n-1}\sum_{j=1}^{n-1}\frac{\frac{1}{4}\left(\Delta l_{i,j} + \Delta l_{i+1,j} + \Delta l_{i,j+1} + \Delta l_{i+1,j+1}\right)}{\varepsilon A_{\mathrm{cap}}} \tag{6.11}$$

Substituting (6.7) into (6.11) gives:

$$\sum_{i=1}^{n-1}\sum_{j=1}^{n-1}\frac{1}{C_{i,j}} \approx (n-1)\frac{d}{\varepsilon A_{\mathrm{cap}}} - \frac{l_{\mathrm{p}}}{E\varepsilon A_{\mathrm{cap}}A_{\mathrm{p}}}F_{\mathrm{tt}} \tag{6.12}$$

where F_{tot} is given by:

$$F_{\mathrm{tot}} = \frac{1}{4}\sum_{i=1}^{n-1}\sum_{j=1}^{n-1}\left(\Delta F_{i,j} + \Delta F_{i+1,j} + \Delta F_{i,j+1} + \Delta F_{i+1,j+1}\right) \tag{6.13}$$

When the total force acting on the chip is homogeneously distributed over the poles (all $F_{i,j}$'s are equal) F_{tot} is exactly equal to the total force. In case the force distribution is not homogeneous, examination of (6.13) shows that for a large number of poles F_{tot} is still a very good approximation of the total force acting on the chip.

A first prototype has been designed and fabricated using a relatively small number of 5x5=25 poles and 16 capacitors. Each pole has a diameter of 2 mm, resulting in a total area bearing the load of 0.8 cm^2. The height of the poles was approximately 200 μm, resulting in a change in height of 0.2 μm at a load of 1000 kg. The distance between the capacitor plates was chosen 1 μm, in order to obtain a 20 % change at the maximum load. The area of the capacitor plates was 0.5 mm^2, resulting in capacitance values in the order of a few pF.

Figure 6.11 shows the process sequence used for fabricating the load cell chip. Both wafers have to be highly conductive, the bottom wafer to form a stable ground contact and the top wafer because it forms the common top electrode for the capacitors. Therefore, we used highly boron-doped wafers (0.01-0.018 Ωcm). In the bottom wafer, first the gaps are etched which define the capacitor distance d_0 using Reactive Ion Etching (RIE) with a resist mask. Next, wet thermal oxidation is used to obtain a 1.5 μm thick SiO_2 layer for electrical isolation from the substrate. The SiO_2 layer is stripped from the backside which will be used to make the ground contact. Finally, aluminum is deposited and structured on both sides of the wafer. The process for the top wafer starts with RIE etching (with a resist mask) the shallow gaps in the backside, which ensure that the load is applied at the position of the poles and not at the capacitors. Next, wet thermal oxidation is used to obtain an 800 nm thick SiO_2 layer, which will later be the mask when etching the 200 μm deep grooves which separate the poles from the capacitors. After structuring the SiO_2 layer a polymer layer is applied on the backside to allow etching through the entire wafer in the subsequent RIE steps [1]. A cryogenic RIE process is used to etch the first part of the contact holes using resist as the mask material and finish etching the contact holes and etching the grooves using the previously structured SiO_2 layer as the mask. Finally, the wafers are connected by low temperature silicon direct bonding using a bonding temperature of 450 °C. Prior to bonding, the bottom wafer is treated with an O_2 plasma and the top wafer with a piranha cleaning. Fig. 6.12 shows a photograph of the finished load cell chip. Besides the 16 sensor capacitors the chip contains 4 reference capacitors located at the corners.

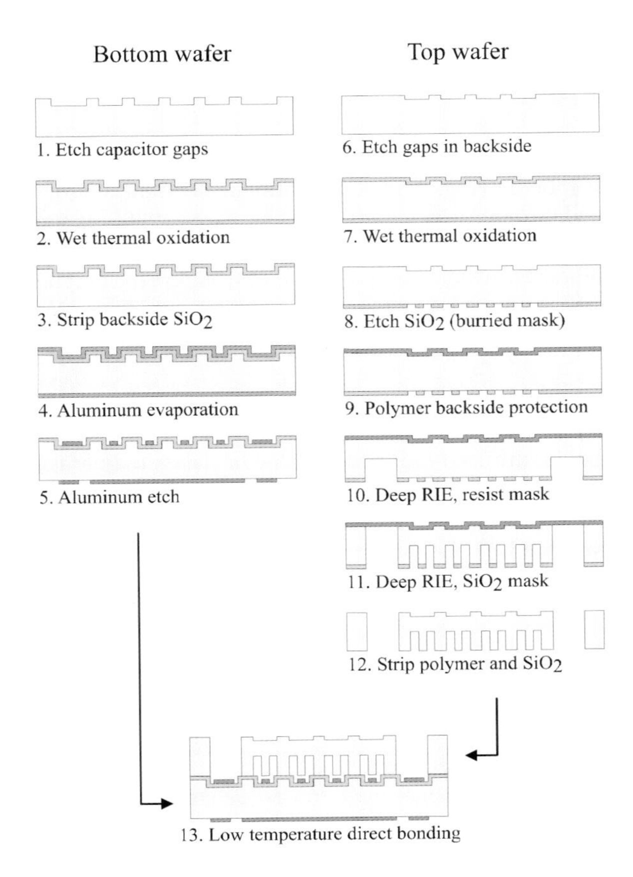

Fig. 6.11. Process sequence for the fabrication of the load cell chip

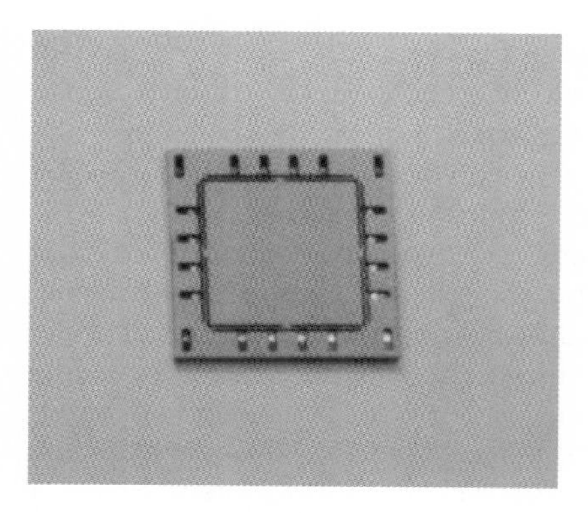

Fig. 6.12. Photograph of the fabricated load cell chip. The area in the center bears the load and has dimensions of 1 cm x 1 cm. The entire chip measures 1.5 cm x 1.5 cm

6.2 Pressure Sensors

Closely related to force sensors are pressure sensors. In pressure sensors the spring element is always a membrane (Fig. 6.1). Conventional pressure sensors used metal membranes. A breakthrough was achieved in the early 1980's when micromechanics was introduced and the metal membranes were replaced by (monocrystalline) silicon membranes, which suffer much less from creep, fatigue and hysteresis. Furthermore, the combination of small size and the high elastic modulus and low density of silicon results in sensors with a very high resonance frequency.

The first silicon pressure sensors were based on a piezoresistive read-out mechanism. At the moment, piezoresistive pressure sensors are still the most widely used. Piezoresistors may be diffused in the membrane or deposited on top of the membrane. Usually, the resistors are connected in a Wheatstone bridge configuration for temperature compensation (see Chap. 10). The main advantages of a piezoresistive read-out mechanism are the simple fabrication process, the high linearity and the fact that the output signal is conveniently available as a voltage. The main problems are the large temperature sensitivity and drift. Furthermore, because of the low sensitivity of piezoresistors, piezoresistive devices are not suitable for accurate measurement of very low pressure differences.

Capacitive read-out mechanisms are inherently less sensitive to temperature variations and an extremely low power consumption can be obtained. However, the capacitance to be measured is usually very small and an electronic interface circuit is required, which either has to be integrated on the sensor die or at least has to be positioned very close to the sensor chip. Compared to piezoresistive sensors the obtained sensitivity is significantly higher. Usually a capacitance change of 30 to 50 % can easily be obtained while the change in resistance of piezoresistive devices is limited to 2 to 5 %. Capacitive structures also offer the possibility of force-feedback as the electrostatic force between the capacitor plates can be used to compensate the external pressure. A few examples of such structures are discussed in this chapter. A discussion on the required electronic circuitry can be found in Chap. 10.

The highest accuracy is obtained using resonant sensors. These sensors have an output signal in the form of a change in resonance frequency of a vibrating element. In this chapter we will focus on resonant structures used for pressure measurement. An extensive discussion on the general theoretical background of resonant sensors can be found in Chap. 9. A problem with resonant sensors is the complexity of the fabrication process. Furthermore, in the common case that the vibrating element is integrated on a deflecting membrane, problems may arise from the mechanical coupling between the resonator and the membrane.

6.2.1 Piezoresistive Pressure Sensors

As mentioned above, most currently available silicon pressure sensors use a piezoresistive read-out mechanism. Fig. 6.13 shows a typical example of a bulk micromachined piezoresistive pressure sensor. The resistors may be diffused in the membrane or deposited on the membrane with an intermediate isolation layer (usually SiO_2). Especially at higher temperatures (above about 120 °C) deposited strain gauges perform better, as diffused gauges suffer from additional drift and noise originating from the parasitic reverse-biased pn-junction to the substrate. On the other hand, diffused strain gauges provide a larger sensitivity. An interesting technique to attach monocrystalline strain gauges to an SiO_2 isolation layer was presented by Petersen et al. (Petersen 1990). First, the piezoresistor pattern is realized in a sacrificial wafer, which is subsequently fusion bonded to the wafer containing the SiO_2 isolation layer. Next, the sacrificial wafer is removed by preferentially etching away the n-type bulk region, leaving behind the ion-implanted p-type resistors.

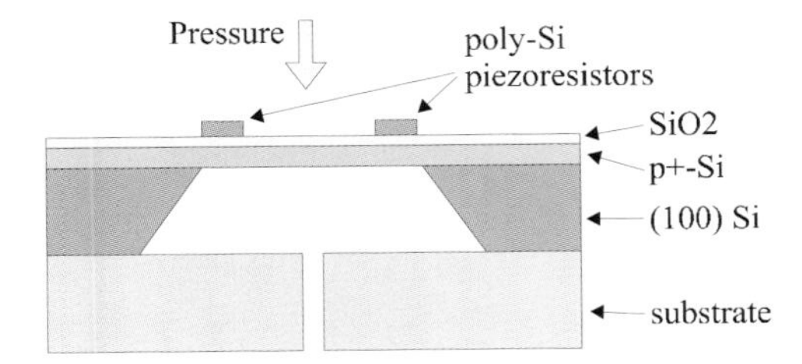

Fig. 6.13. Schematic cross section of a typical bulk micromachined piezoresistive pressure sensor

The membrane is etched from the backside of the wafer and is typically several tens of micrometers thick. All etch stop mechanisms described in Sect. 3.3.3 are used to realize the membrane. A time etch stop is simple and has the advantage that it does not require doping of the membrane with boron. However, the reproducibility of the membrane thickness is rather poor. In chapter 4 we have seen that the deflection of a membrane is dependent on the third order of the membrane thickness, thus a 1% change in thickness will result in a 3% change in deflection. A boron etch stop gives good control over the membrane thickness, however the high doping level prohibits the use of diffused strain gauges. Therefore, often an electrochemical etch stop is used with a more lightly doped membrane.

Because the membrane is etched from the backside of the wafer it is very well possible to combine this with a standard IC fabrication process. One of the first sensors with on-chip electronics was developed at Toyota (Sugiyama 1983). They combined a piezoresistive pressure sensor with a bipolar electronic circuit to provide temperature compensation and to convert the output voltage into a frequency (which can be easily interfaced with digital electronics). Other examples were developed at Hitachi (Yamada 1983) and NEC (Tanigawa 1985). Fig. 6.14(a) shows a schematic cross section of the latter, where an NMOS operational amplifier is integrated with the sensor for amplification and temperature compensation of the strain gauge signal. First the piezoresistors are realized using a boron implantation, followed by the standard NMOS process. Finally the membrane is realized using a time etchstop. More recently, a combination with a standard CMOS process was proposed (Kress, 1991). In this case the CMOS circuit was realized in an n-type epitaxial layer as shown in Fig. 6.14(b). In this way an electrochemical etch stop can be used using the epitaxial layer as the stop layer. The standard p-well is used for the strain gauge resistors, thus eliminating the need for a separate implantation step.

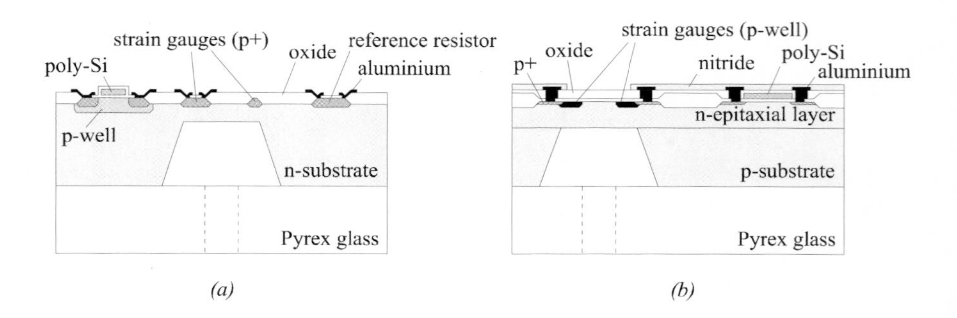

Fig. 6.14. Combination of a piezoresistive pressure sensor with:
(a) an NMOS process (Tanigawa 1985), and
(b) a CMOS process (Kress 1991)

A problem with pressure sensors having the membrane etched from the backside is that the chip surface required for the sensor is rather large. An interesting technique to reduce the required chip surface and still fabricate a single crystalline membrane was proposed by Petersen et al. (Petersen 1988). Fig. 6.15(a) shows an outline of the fabrication process. First, square cavities with the size of the membrane are etched in a wafer. Next, the wafer is fusion-bonded (see Section 3.4) to a p-type wafer with an n-type epitaxial layer. The bulk of the second wafer is removed using an electrochemical etch stop at the epitaxial layer, leaving membranes with the thickness of this layer. A complete pressure sensor is realized by ion-implanting piezoresistive strain gauges, etching contact vias and depositing and etching metal interconnect. Finally, the backside of the wafer is ground and polished back to a thickness of about 140 µm. The main advantage of this technology is that the support of the membranes can be much smaller, as indicated in Fig. 6.15(b) which may be critical in some applications, e.g. in ultra-miniature catheter-tip sensors for in vivo pressure measurement. Obviously, a smaller chip size also results in a larger number of chips per wafer, thus reducing the fabrication costs. A simple modification of the process, i.e. etching very shallow cavities as shown in Fig. 6.16, provides an effective overpressure protection (Christel 1990).

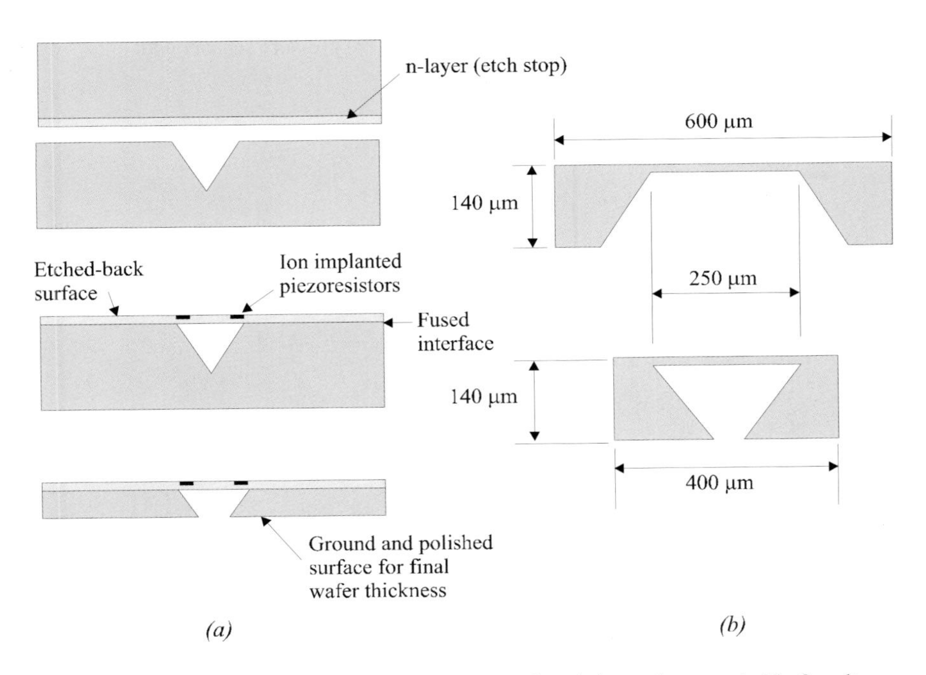

Fig. 6.15. Fabrication sequence of a silicon fusion bonded membrane suitable for ultra-miniature piezoresistive pressure sensors (a) (Petersen 1988). The main advantage of this technology is that the support of the membrane can be much smaller (b)

Ultra-miniature and much thinner membranes can be made by under-cut etching from the front side of the wafer or by using surface micromachining. However, membranes fabricated this way cannot consist of monocrystalline silicon. Sugiyama et al. (Sugiyama 1986) proposed to use a silicon nitride membrane. Fig. 6.17 shows an outline of their fabrication process. The membrane and a reference pressure chamber are formed by undercut etching through one or more small holes. The etch holes are sealed by depositing a 1 μm thick silicon nitride layer using plasma CVD. The main problem with this process was the low yield due to the production of hydrogen bubbles during the etching of the cavities. A solution was found in using surface micromachining with a sacrificial polysilicon layer (Sugiyama 1993). This results in a much shallower cavity but with a much higher production yield.

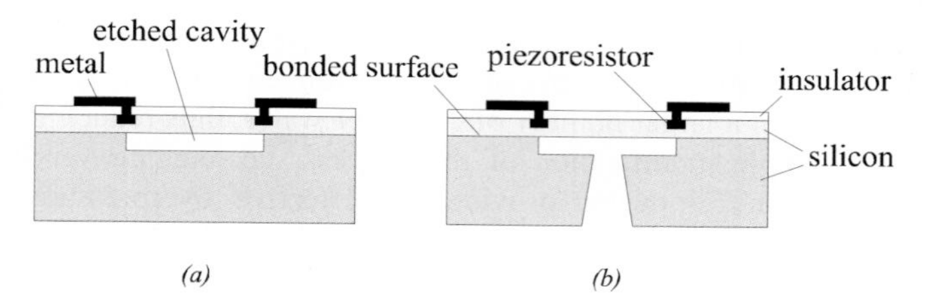

Fig. 6.16. Fusion bonded absolute (a) and differential (b) pressure sensors with overpressure protection (Christel 1990)

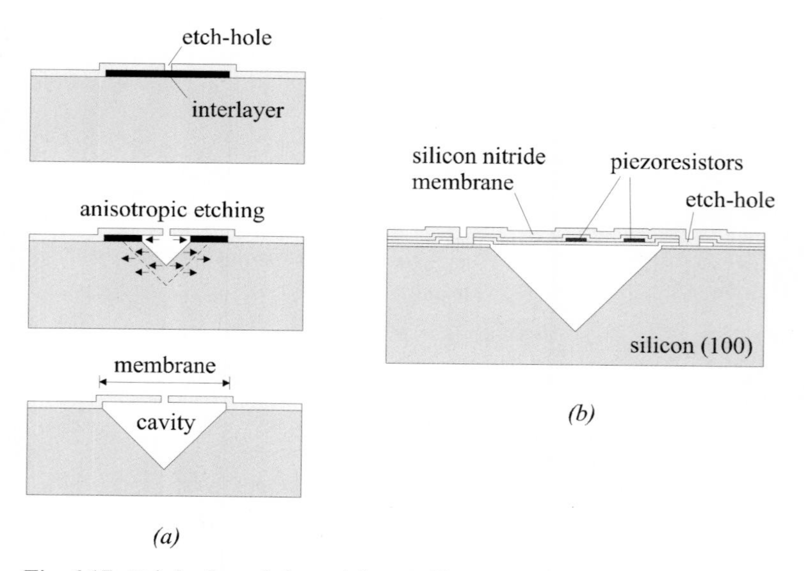

Fig. 6.17. Fabrication of ultra miniature silicon nitride membranes by under-cut etching from the front side of the wafer (Sugiyama 1986): fabrication principle (a) and cross section of the finished device (b)

Important work on piezoresistive pressure sensors using surface micromachined polysilicon membranes has been done by Guckel. Fig. 6.18 shows a schematic drawing of a typical device (Guckel 1993). An SiO_2 layer is used as the sacrificial layer. The polysilicon membrane is about 2 μm thick and is deposited using low-pressure chemical vapor deposition (LPCVD). Deposition of a polysilicon layer with reproducible mechanical properties is not easy, as the mechanical properties of polysilicon are related to the film morphology and LPCVD polysilicon can have a wide morphological range (Guckel 1988). Furthermore, the built-in strain has to be controlled carefully, as a compressive strain may cause buckling of the membrane and a large tensile strain reduces the sensitivity. Therefore, after deposition an anneal step is used to convert the initially compressive strain into a tensile strain (Guckel 1988).

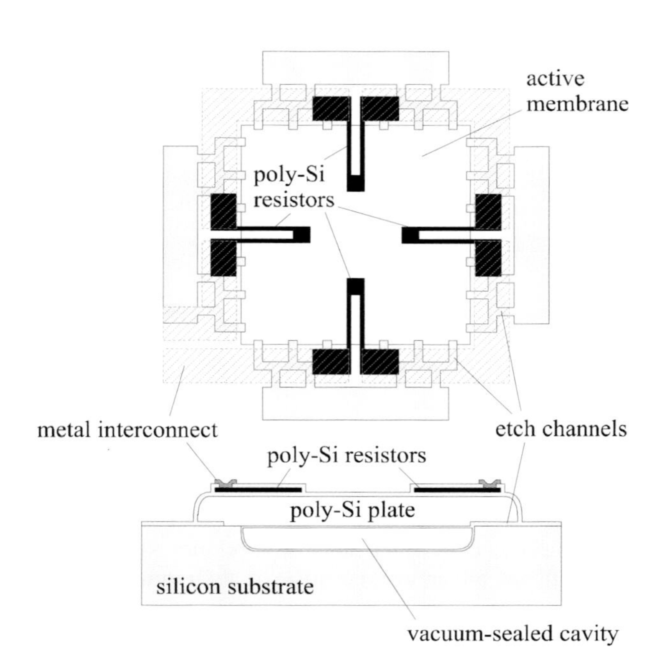

Fig. 6.18. Surface micromachined pressure sensor with polysilicon membrane (Guckel 1993)

In the low pressure range (e.g. a full-scale pressure range of 500 Pa), flat-diaphragm pressure sensors suffer from low accuracy, poor sensitivity and a high degree of non-linearity. In particular the non-linearity with values of a few percent is unacceptably high. The non-linearity is a result of stretching of the middle plane which becomes significant at large deflections. In particular high aspect ratio diaphragms that are used in low pressure sensors suffer from this effect. Specialized geometries, such as diaphragms with a rigid center or boss, aimed at increasing the stiffness and thus limiting the maximum deflection of the diaphragm have been developed for enhanced linearity. Fig. 6.19 shows some typical examples (Mallon 1990). A center bossed diaphragm also proves convenient in providing an overpressure protection, e.g. (Christel 1990).

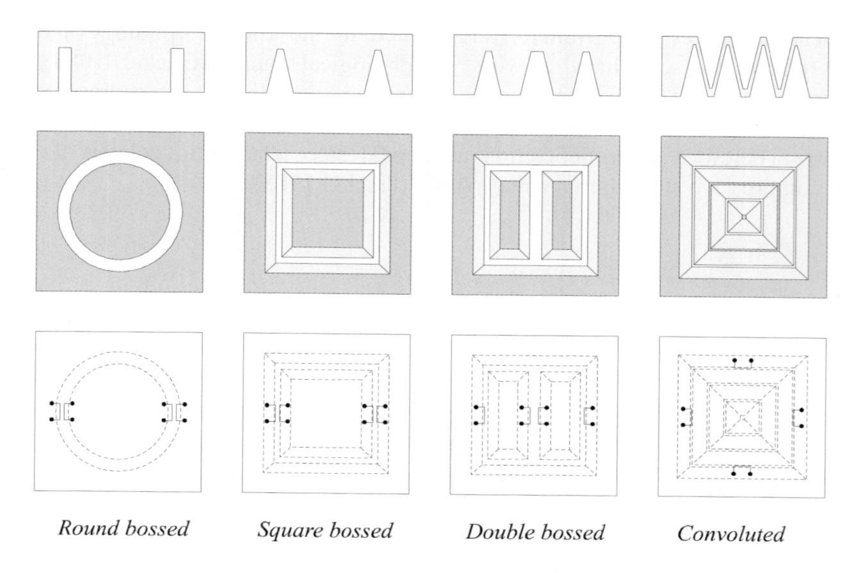

Fig. 6.19. Typical linearity-enhancing geometries for low-pressure sensors (Mallon 1990)

6.2.2 Capacitive Pressure Sensors

As mentioned before, capacitive pressure sensors have several advantages compared to piezoresistive sensors. Especially at low pressure differences capacitive sensors offer a significant advantage because of their much higher sensitivity. Furthermore, capacitive sensors are inherently less sensitive to temperature variations and offer a much better long-term stability. On the other hand, capacitive sensors are also inherently non-linear and the measurement of small capacitance changes requires much more complicated electronic interface circuits than piezoresistive sensors. Furthermore, the interface circuit has to be integrated on the sensor chip or at least be placed very close to the sensor chip to avoid influence of stray capacitance.

The basic structure of a capacitive pressure sensor consists of two parallel plates with surface A and separated by a distance d. The capacitance at rest is given by:

$$C_0 = \frac{\varepsilon A}{d} \tag{6.14}$$

where ε is the dielectric constant of the medium between the plates. In a simple sensor structure where one of the electrodes is a deflecting membrane the capacitance is given by the following integral:

$$C = \iint \frac{\varepsilon}{d - w(x,y)} \mathrm{d}x\mathrm{d}y\,, \tag{6.15}$$

where $w(x,y)$ is the deflection of the membrane as a function of x and y. From (6.15) it is clear that the relation between the applied pressure and the change in capacitance is nonlinear. In the case that the membrane contains a boss and the capacitance comprises only the area of the boss, (6.15) can be simplified to

$$C = \frac{\varepsilon A_{\mathrm{boss}}}{d - \Delta d}\,, \tag{6.16}$$

where Δd is the pressure dependent deflection of the boss and A_{boss} is the surface area of the boss. Now the reciprocal value of the capacitance $1/C$ is linearly dependent on the deflection and -for small deflections- also linearly dependent on the pressure. However, in practice this linear relationship will be disturbed by parasitic capacitance C_{p}. In that case the total capacitance is given by:

$$C = \frac{\varepsilon A_{\mathrm{boss}}}{d - \Delta d} + C_{\mathrm{p}}\,, \tag{6.17}$$

and the reciprocal value is no longer linearly dependent on the deflection.

Rosengren compared several sensor structures with respect to linearity (Rosengren 1992). One interesting structure is to let the membrane touch the surface beneath it as described in Sect. 5.3.2. In that case the increase in capacitance is not the result of a decrease in plate distance but of an increase in the "touching" surface area. The latter effect appears to be much more linear.

One of the first published capacitive pressure sensors was developed at Stanford University for use in cardiology (Sander 1980). Fig. 6.20 shows a schematic drawing of this device. The fixed counter electrode (Al) is deposited on a Pyrex wafer. The deflecting electrode is realized in a silicon wafer using an n-doped epitaxially grown. The membrane of a thickness of 25 μm is etched from the backside by anisotropic etching using the electrochemical etchstop. The two wafers are attached by anodic bonding. A bipolar integrated circuit is integrated on the chip to convert the capacitance change into a pulse period output. An improved design contained more advanced electronics for temperature compensation and a dummy reference capacitor (Smith 1986).

A similar design of a capacitive pressure sensor using anodic bonding of a silicon wafer to a Pyrex wafer has been used by Puers et al. (Puers 1990). They did not integrate the electronics in the sensor chip but used a separate CMOS chip to invert the hyperbolic capacitance versus pressure relation and suppress the influence of parasitic capacitance.

Ko et al. introduced the idea to combine a pressure dependent capacitor and a reference capacitor in a single cavity (Ko 1982). Because the reference capacitor is located at the edge of the membrane, it is almost independent of the pressure.

Kim et al. proposed to use an interdigitated electrode structure as indicated in Fig. 6.21 (Kim 1997). They optimized the design such that the nonlinear part in the relation between the applied pressure and the change in the sensor capacitance is compensated by a corresponding nonlinear pressure dependence of the reference capacitor.

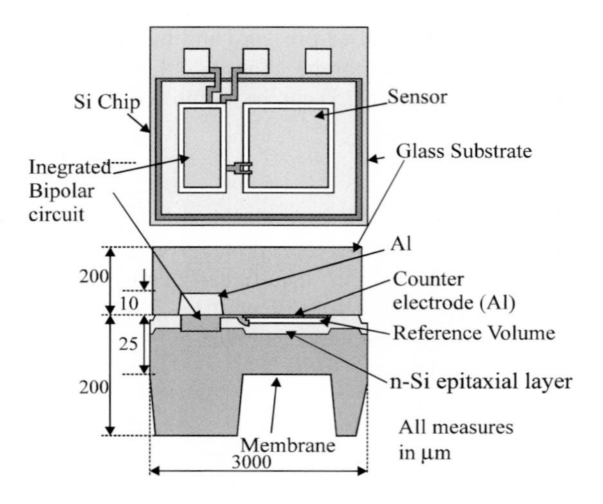

Fig. 6.20. Schematic of an integrated capacitive pressure sensor (Sander 1980)

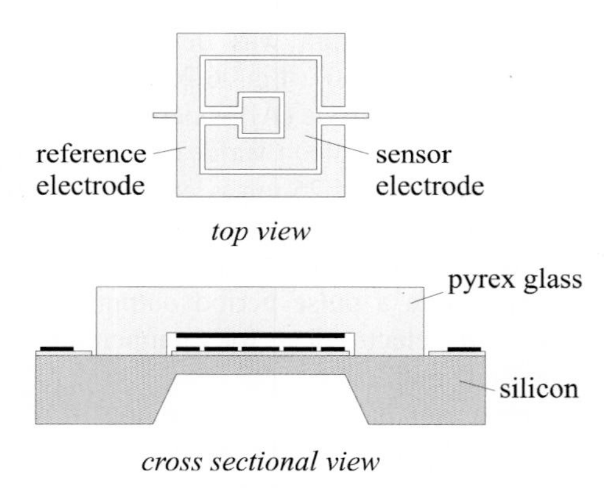

Fig. 6.21. Capacitive pressure sensor with interdigitated electrode structure for linearity compensation (Kim 1997)

A completely different design was proposed by Chau et al. (Chau 1988). They fabricated a pressure sensor small enough for use in a cardiovascular catheter, with an outer diameter of 0.5 mm. Fig. 6.22(a,b) shows an overview of the catheter and the sensor chip. The silicon chip with the deflecting membrane was bonded on a glass substrate. It was micromachined with a concentrated HF solution using a gold mask in order to realize 30 μm deep groves for lead attachment and to realize the metallization. The process sequence for the fabrication of the membrane is given in Fig. 6.22(c). First a 3 μm deep cavity is etched, which defines the distance between the capacitor electrodes. Next, two boron indiffusion steps are used to define an etchstop at 12 μm for the support of the membrane and an etchstop at 1.2 μm for the membrane itself. A dielectric is deposited on the membrane to protect the membrane against shorts. After bonding the silicon wafer to the glass wafer all undoped silicon is etched away in EDP, only leaving the membrane and its support.

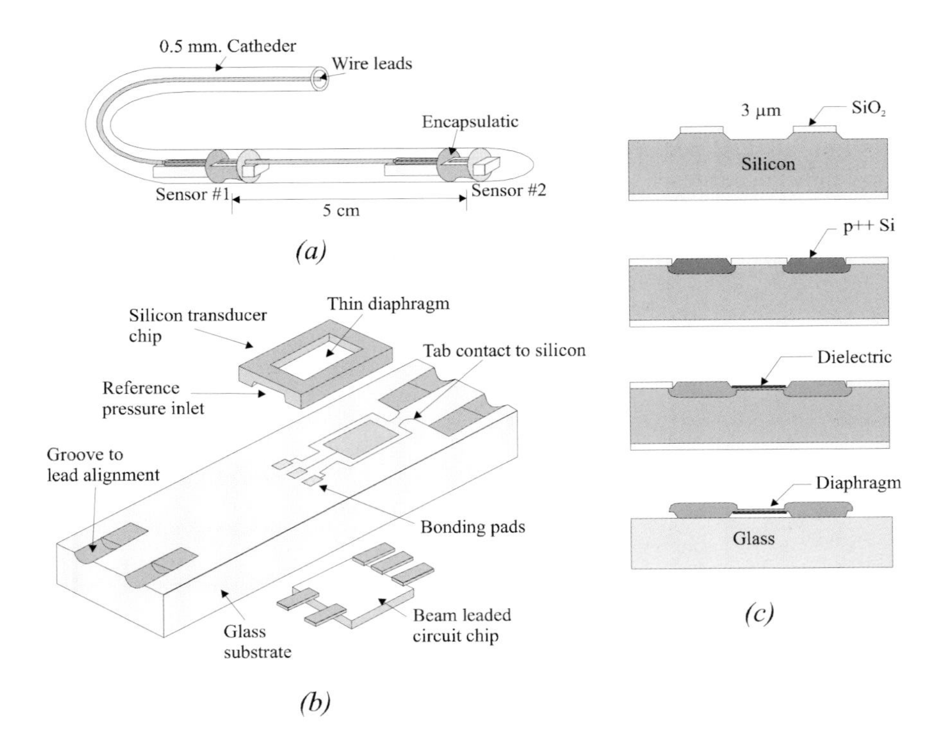

Fig. 6.22. Overall view of a the catheter (a), construction of the pressure sensor (b), and fabrication sequence of the silicon transducer chip (c) (Chau 1988)

A similar fabrication process but with a totally different sensor structure was used by Park et al. (Park 1999). Fig. 6.23 shows the basic structure and operating principle of their sensor. The pressure sensitive capacitance is between a flap or skirt-like extension of the membrane that reaches past the sealed cavity and an electrode patterned on the substrate directly below this extension. The sensor is realized as follows. First shallow cavities are etched in a silicon wafer defining the distance between the membrane and the glass substrate. Next, a boron indiffusion is used which defines the thickness of the membrane. After bonding the silicon wafer to the glass substrate the bulk of the silicon wafer is etched away, leaving only the sensor membrane. The main advantage of having the electrodes outside the sealed cavity is that it is not necessary to make an electrical contact to the inside of the cavity.

Hanneborg and Ohlckers proposed to replace the commonly used Pyrex 7740 support chip by a silicon chip with a thin sputtered layer of Pyrex. In this way two common sources of errors are avoided: thermal sensitivity due to thermal mismatch between the bonded wafers and poor long-term stability due to slow relief of large built-in mechanical stresses. Figure 6.24 shows a cross section of their device (Hanneborg 1990). The silicon chips as well as the pressure port tube are sealed together by anodic bonding. Experimental results showed a remaining thermal zero shift of 0.01 % of full scale per °C, and a long term drift of less than 0.01 % of full scale per week. The thermal zero shift can be fully explained by the temperature dependence of the dielectric constant of the Pyrex layer as approximately ¼ of the zero-pressure capacitance is due to the inactive bonded area with Pyrex as the dielectric.

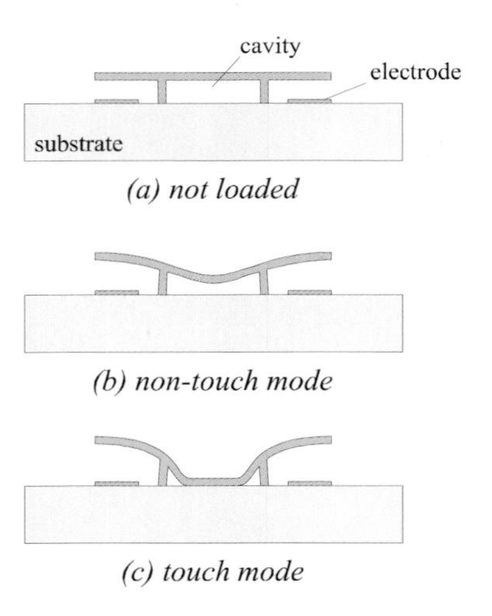

Fig. 6.23. Capacitive pressure sensor with the electrodes located outside the cavity (Park 1999)

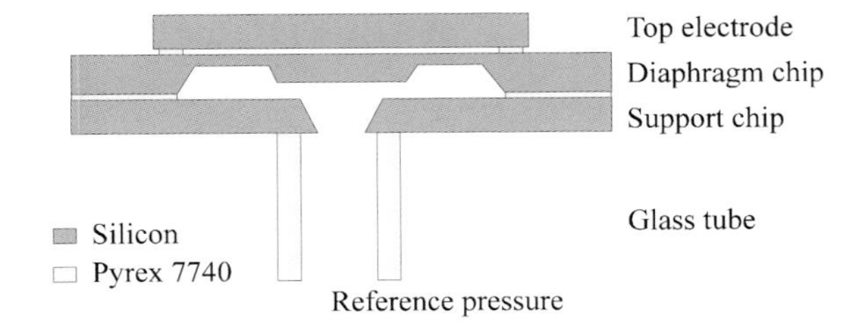

Fig, 6.24. Schematic of a capacitive pressure sensor using a bossed diaphragm and sputtered thin films of Pyrex for anodic bonding (Hanneborg 1990)

Another sensor which is entirely made of silicon was presented by Bäcklund *et al.* (Bäcklund 1990). Instead of a thin intermediate Pyrex layer, they used direct silicon wafer bonding (fusion bonding). Fig. 6.25 shows a schematic cross section and Fig. 6.26 shows a microphotograph of the device. The device is planned to be implanted in the human eye for continuous monitoring of the intraocular pressure. The pressure-dependent capacitance changes the resonance frequency of a passive LC resonator. The resonance frequency is measured with an external, inductively coupled oscillator. The authors report on difficulties combining a boron etch stop with direct bonding. The indiffusion process apparently roughens the surface too much. As a consequence the authors used a time etch stop in the following way. Before bonding of the wafers, V-grooves were etched in the top wafer with a known depth. After bonding, the top wafer was thinned using a KOH-etch which was manually stopped as it reached the tip of the V-grooves, thus defining the thickness of the membrane.

All capacitive pressure sensors discussed so far use a bonding step in which each wafer contains one electrode of the capacitance. This wafer bonding step can be eliminated when surface micromachining is used and the capacitor is realized by etching away a sacrificial layer. In this way it is also possible to combine the sensor with a standard CMOS or BiCMOS process. One of the first surface micromachined capacitive pressure sensors was proposed by Dudacevs *et al.* (Dudaicevs 1994). Fig. 6.27 shows a summary of the fabrication sequence they used. First one electrode of the capacitor is realized as an n+ region in the substrate by ion implantation. Next, a thin silicon nitride isolation layer is

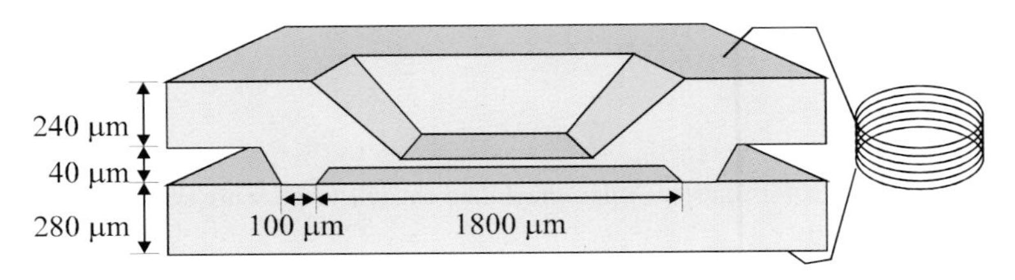

Fig. 6.25. Schematic cross section of the intra ocular pressure sensor (Backlund 1990)

Fig. 6.26. SEM-photo of the membrane showing the bonded region (The region marked with 100 μm in Fig 6.25, upside down)

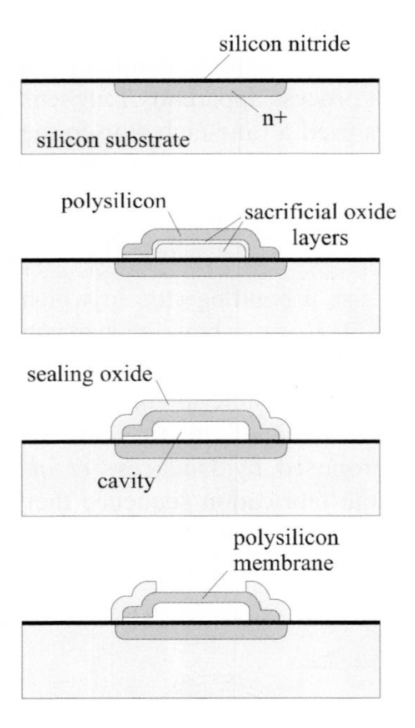

Fig. 6.27. Fabrication steps for a surface micromachined capacitive pressure sensor (Dudaicevs 1994)

deposited to avoid short circuits when the top plate of the capacitor touches the substrate. Sacrificial SiO_2 layers are used to define the membrane area and the etch channels. Polysilicon is deposited and structured to form the membrane. After etching away the sacrificial layer through the etch channels, the etch channels are sealed by depositing a new layer of SiO_2, which is removed again from the top of the membrane. A reference capacitor is conveniently realized by leaving the oxide on the membrane. All processing steps are slightly modified CMOS process steps so the sensor can be realized with standard CMOS equipment and CMOS circuitry can be integrated with the sensor.

Another example of a surface micromachined capacitive pressure sensor was developed recently at Siemens (Scheiter 1998). They used the standard layers in a BiCMOS process to realize the pressure sensor structure. Compared to the standard BiCMOS process only one photolithography step and a sacrificial layer etching step are added.

6.2.3. Force compensation pressure sensors

An important property of capacitive devices is that a capacitance can also be used to exert an electrostatic force. Thus, while using the same basic structures a force-feedback sensor can be realized. In this way, nonlinearity due to the membrane deflection can be completely eliminated as the membrane always remains in its rest position. Fig. 6.28 illustrates the operating principle of such a configuration. A small displacement of the membrane due to the external pressure P_o is detected and amplified. The output signal is used to exert an electrostatic counter balancing pressure P_e. If the loop gain is sufficiently high, the membrane deflection will be virtually zero and the output signal will be independent of the mechanical properties of the membrane and the displacement detector. The output is completely defined by the actuator characteristic, therefore the actuator structure should be reproducible and independent of material properties as is the case in a parallel plate electrostatic actuator.

Fig. 6.29 shows the structure of a force balancing pressure sensor proposed by Wang and Esashi (Wang 1998). The deflection of a bossed silicon membrane is measured by sense capacitor C_s and compensated by a voltage applied to the actuator capacitor C_a.

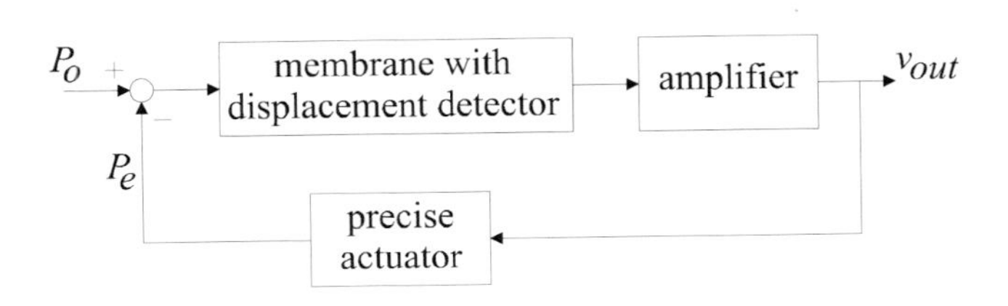

Fig. 6.28. Principle of force balancing pressure sensor

The required actuation voltage is quite high, typically in the 100–300 V range for atmospheric pressure ranges. A force multiplication scheme can be used to reduce the required actuation voltage (Gogoi 1999). In this scheme, the pressure sensing area and the force balancing actuator area are decoupled. The external pressure is sensed by a small membrane while the electrostatic restoring pressure is applied to a large plate as indicated in Fig. 6.30.

The external pressure P_o is exactly balanced by an electrostatic pressure P_e if:

$$P_e \cdot A_d = P_o \cdot A_s \tag{6.18}$$

The electrostatic pressure due to a drive voltage v_{drive} is given by:

$$P_e = \frac{\varepsilon_0}{2}\left(\frac{v_{drive}}{d}\right)^2 \tag{6.19}$$

where d is the distance between the plates and ε_0 is the permittivity of vacuum.

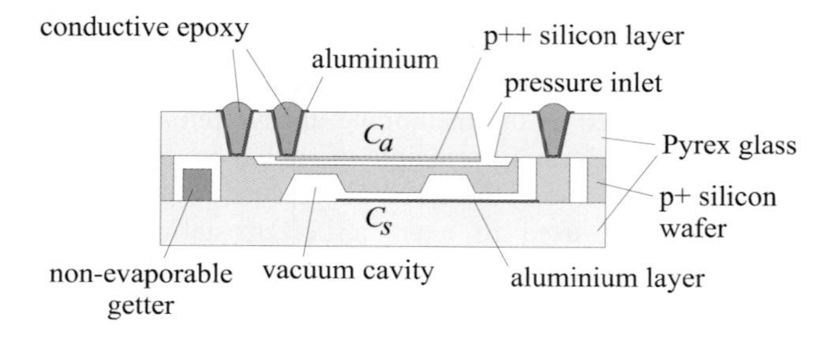

Fig. 6.29. Structure of a force balancing pressure sensor proposed by Wang and Esashi (Wang 1998)

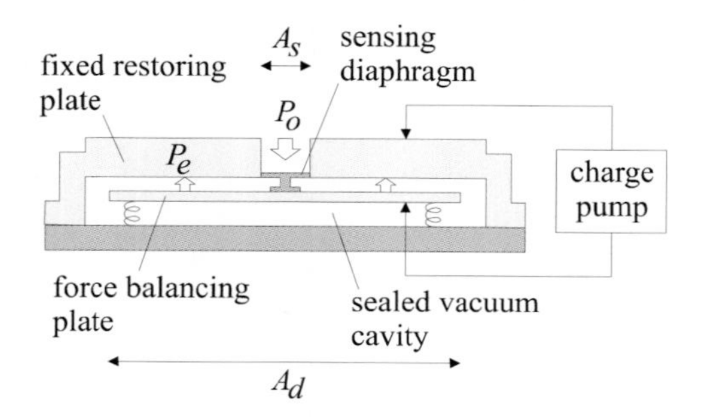

Fig. 6.30. Multiplying force balancing scheme for low voltage pressure sensing with hermetically sealed actuator (Gogoi 1999)

Substituting (6.19) into (6.18) and rearranging gives:

$$v_{\text{drive}} = \sqrt{\frac{2P_{\text{o}}A_{\text{s}}}{\varepsilon_0 A_{\text{d}}}}d \qquad (6.20)$$

Gogoi reports a required actuation voltage of 12 volts at a full-scale pressure of 1 atmosphere for a sensor with A_d/A_s=100 and d = 0.5 μm.

6.2.4 Resonant Pressure Sensors

Two types of resonant pressure sensors can be distinguished. The first type consists of a vibrating membrane. In this case the resonance frequency is dependent on the pressure difference across the membrane. The other type uses a vibrating structure on top of the membrane: the membrane deflects due to a pressure difference and the resonance frequency of the vibrating structure changes as a result of the changing strain at the membrane surface.

Vibrating membrane pressure sensors have been described in various versions. The common element is an anisotropically etched membrane, which can be fabricated by using a time controlled etch stop or a boron etch stop. Several mechanisms can be used to excite and detect the vibration. For example, a piezoelectric thin film can be sputtered on top of the membrane (Smits 1983). A voltage across the thin film leads to a change of the dimensions of the film. Both the thickness and the lateral dimensions of the film change. The latter gives rise to a bending moment in the membrane that can be used to excite bending vibrations of the membrane. Another excitation mechanism is to use the thermal expansion of a resistor which is integrated on the membrane (Lammerink 1985; Bouwstra 1989). A detailed description of various excitation and detection mechanisms can be found in Chapter 9.

A common problem of vibrating membrane pressure sensors is that the resonance frequency is not only dependent on the pressure but also on the mass of the gas in the vicinity of the membrane. Hence, the resonance frequency is also dependent on the sort of gas and its temperature. Furthermore, the gas is allowed to interact directly with the resonator. Absorption of chemicals and dust, as well as corrosive effects, change the mass of the resonator and therefore cause a drift in the readout of the sensor. Therefore, other sensors have been developed in which the membrane itself does not vibrate, but instead a resonator has been integrated on the membrane. One of the first devices operating this way was developed by Greenwood (Greenwood 1988; Greenwood 1993) and is now commercially available from Druck Ltd. Fig. 6.31 shows a SEM photograph of the device. It consists of a "butterfly"-shaped resonator on top of a 6 μm thick membrane. The structure is realized by a boron etch stop. The silicon wafer is bonded to a Pyrex wafer containing electrodes for electrostatic excitation and detection of the vibration. Fig. 6.32 shows a schematic cross section of a complete sensor.

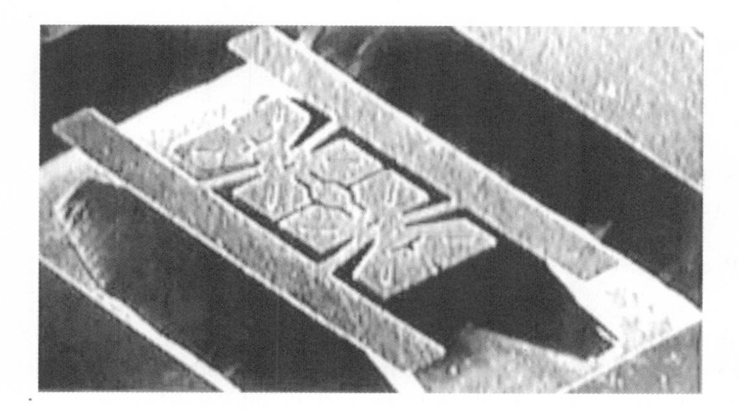

Fig. 6.31. SEM photograph of the Druck resonant pressure sensor chip (Greenwood 1988)

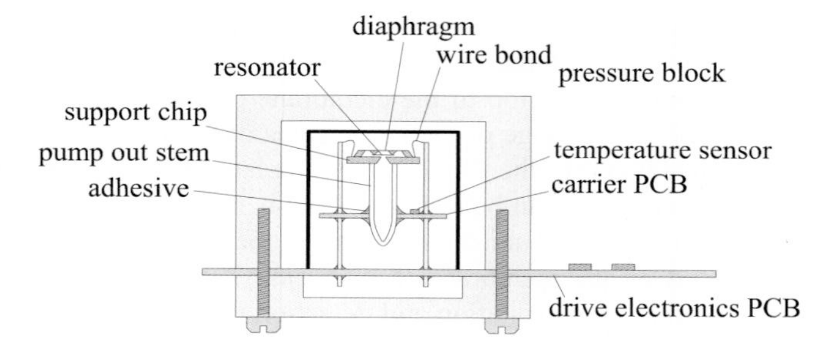

Fig. 6.32. Cross section of the complete Druck pressure sensor (Greenwood 1993)

Another resonant pressure sensor which is now commercially available was developed by Ikeda *et al.* at Yokogawa Electric Corporation (Ikeda 1990). They integrated single crystalline resonant strain gauges in the membrane surface using epitaxial growth of doped silicon and the dopant selectivity of anisotropic etching in electrochemical etching. The resonator is protected by a cap and operating in a local vacuum which results in low air damping and, thus, a high quality factor. It consists of two parallel beams which are mechanically connected in the middle. The operating principle is illustrated in Fig. 6.33. A constant magnetic field is applied by means of a permanent magnet. The resonator is excited by a current flowing through one of the beams, which exerts a Lorentz force. The induced voltage at the second beam is used to detect the vibration. An electronic feedback circuit is used to maintain a stable oscillation (see Chapter 10).

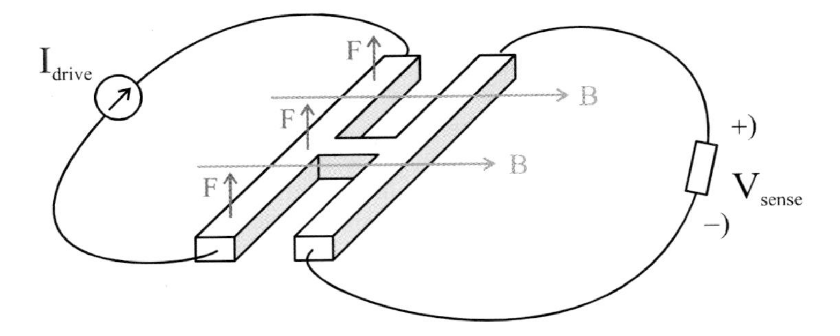

Fig. 6.33. Operating principle of an electrodynamically driven H-shaped two-port resonator

The fabrication process is shown in Fig. 6.34. The process starts with a HCl dry etch at 1050 °C in an epi reactor through a hole in an oxide mask. Selective epitaxial growth of silicon of the following doping: p+, p++[1], and p+, p++ as follows. These steps are carried out in the same epi-reactor, just by changing the concentration of B_2H_6. The next step is stripping of the oxide in HF, followed by selective electrochemical etching in hydrazine. In this etching step, the n-substrate is pacified against the etchant by the electrochemical potential, and the p++ structures by the boron etchstop. After this process one has a microbridge covered by a cap, both from single crystalline heavily B-doped silicon. The cap is sealed by growth of an epitaxial layer, and annealing in an nitrogen atmosphere leads to a pressure inside the cavity of 1 mTorr (j).

In order to obtain a structure that works as a pressure sensor, one has to etch a membrane from the backside of the wafer, a step rather trivial after the process just described.

The concept of integrating resonators in the membrane surface, operating in a local vacuum, appeared to be very succesful. However, for a resonator consisting of monocrystaline silicon a problem is the complexity of the fabrication process. Therefore, other designs have been proposed using surface micromachined *polysilicon* resonators, e.g. (Guckel 1990; Tilmans 1993). These resonators are described in more detail in Chap. 9. One example will be shown here, which is the sensor proposed by Tilmans (Tilmans 1993). Fig. 6.35 shows a schematic drawing of this sensor. A differential pressure Δp causes a cross section of the annular diaphragm to deform like an S-shape. This causes a change in the resonant frequencies of the resonators. The resonant frequency of the resonators close to the boss will decrease and the resonant frequency of the resonators close to the edge will increase. By taking the frequency difference as the output signal, common mode errors like the influence of temperature are kept small. More than one resonator pair is incorporated for reasons of symmetry and to provide built-in

---

[1] Here p^+ and p^{++} refer to medium and high doping levels, respectively

redundancy. As mentioned in Sect. 6.2.1 the use of a bossed diaphragm results in better linearity (Mallon 1990). In this particular design, the bossed structure offers an elegant way to reduce the degrading influence of the sealing caps on the sensitivity. Important design issues are the position of the resonators relative to the boundaries of the annular diaphragm and the in-plane dimensions of the annulus and the boss. To achieve the required dimensional control, front-side wafer processing must be used. Possible micromachining technologies are sacrificial layer etching of locally oxidized wells (as suggested by the rounded shapes in Fig. 6.35) (Guckel 1992), silicon fusion bonding (Christel 1990) and the dissolved wafer process (Cho 1992).

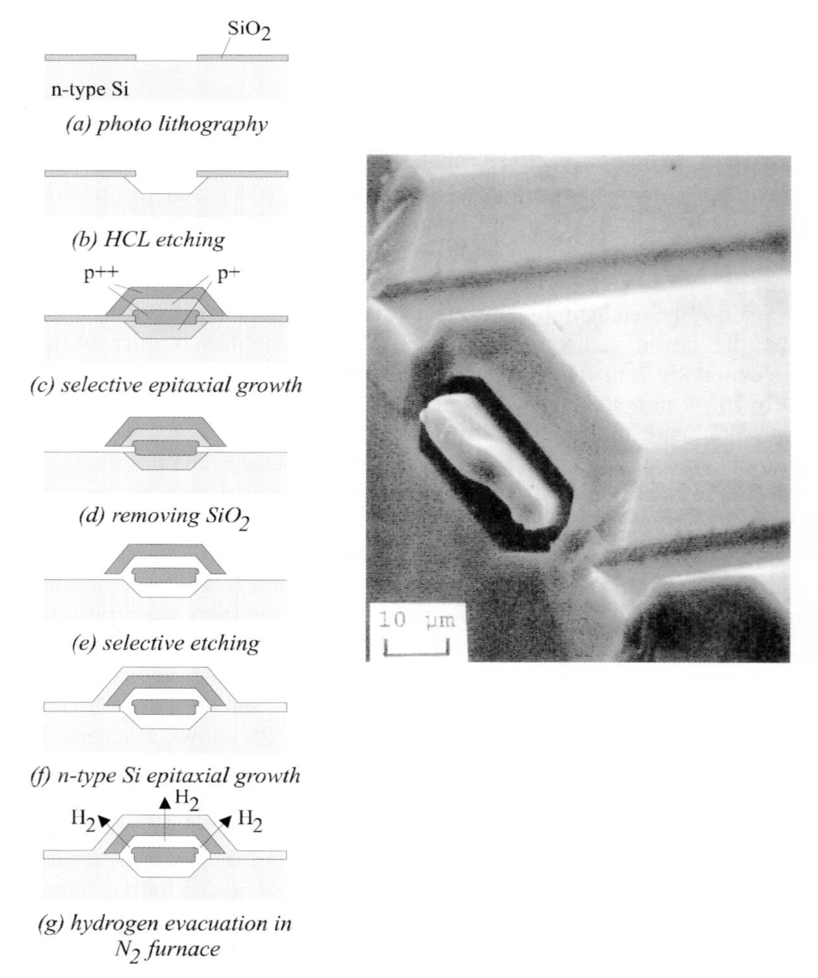

Fig. 6.34. Fabrication process and photograph of a vacuum sealed resonating microbridge (Ikeda 1990)

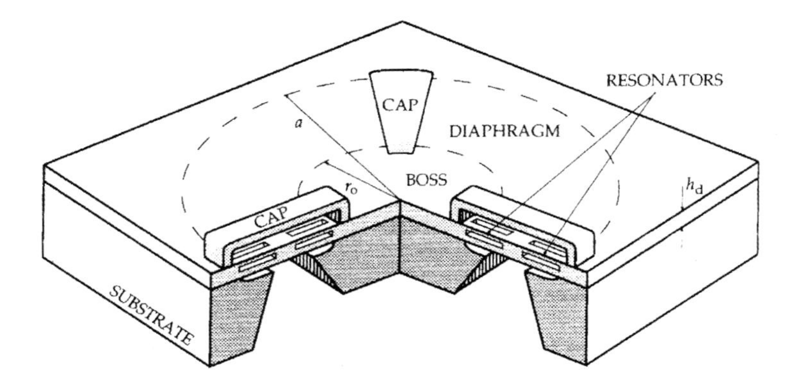

Fig. 6.35. Differential pressure sensor using encapsulated polysilicon resonant strain gauges. (Tilmans 1993). For a full scale range of 1000 Pa the sensor has an outerradius a of 600 μm and an inner radius r_0 of 300 μm. The diaphragm thickness is 3 μm

An important problem with resonant sensors integrated on a flexible membrane is that there may be an undesired coupling between the two resonators and between each resonator and the membrane. This manifests itself as unwanted resonance peaks in the frequency spectra of the resonators and leads to difficulties for the electronic oscillator to lock-in at the desired frequency (Tilmans 1993). The effects of coupling are more pronounced for matched resonators. Mechanical interference will be largely eliminated by introducing a strong mismatch between the two resonators, or, by utilizing decoupling or mechanical isolation techniques. Two approaches can be used for decoupling: (i) using *tuning-fork-type* resonators and (ii) using a *mechanical isolator* to decouple the resonator from the support. The second approach is less attractive since it makes the design and fabrication very complex. The characteristic feature of tuning-fork-type resonators is that different branches vibrate in antiphase fashion, so that reaction moments and forces are cancelled at the mounting edges. This leads to a reduction of the energy losses into the mount, thereby raising the mechanical quality factor and moreover providing a form of inherent isolation. Fig. 6.36 indicates how the design of Fig. 6.35 could be modified to eliminate resonator coupling. The single beam resonators have been replaced by quadruple beam resonators where the two outer beams resonate in antiphase with the inner beams. Furthermore, a stiffener is used to concentrate the end moments and shear forces of the individual beams into a single point. Moreover, the stiffener forces the edges of the individual beams to be displaced by the same amount, thus ensuring equal loading of the individual beams. The stiffener is located at the inflection points of the annulus and thus hardly needs to bend, but must be able to rotate easily about an axis perpendicular to radial axis, otherwise the important S-shape will be seriously distorted. This is why the stiffener cannot extend along the entire annulus: a circular stiffener does not rotate easily, resulting in an awkward deformation.

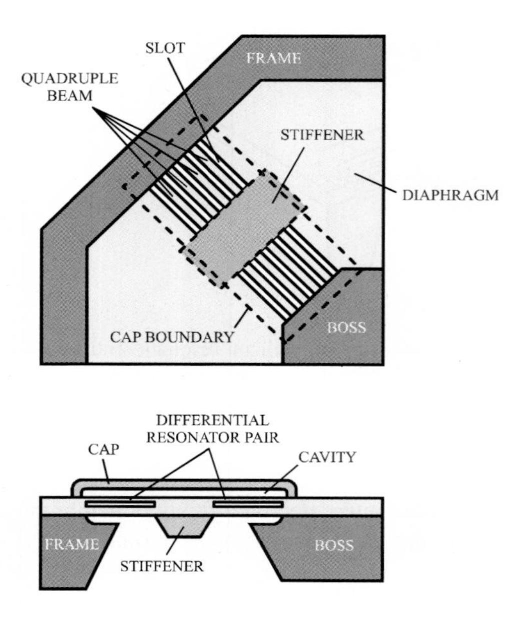

Fig. 6.36. Modified design of Fig. 3.35. To reduce undesired coupling between the resonators quadruple beams are used separated by a stiffener

6.2.5 Miniature Microphones

Microphones are in fact pressure sensors which respond to pressure changes in the audio frequency range. The basic structure of silicon miniature microphones is therefore similar to the structure of pressure sensors: they consist of a membrane which deflects as a result of a sound pressure. The deflection can be measured in several ways as indicated in Fig. 6.37. The first silicon microphone, which was presented by Royer at al. (Royer 1983) contained a piezoelectric ZnO layer on top of the membrane. Fig. 6.37(a) shows a schematic cross section of this device. The membrane had a thickness of 30 μm and a diameter of 3 mm. The ZnO layer had a thickness of 3-5 mm and was sandwiched between two SiO_2 layers containing the electrodes. Movements of the membrane cause stress in the piezoelectric layer, which generates an electrical voltage. A disadvantage of piezoelectric silicon microphones is the relatively high noise level (Scheeper 1994). A silicon microphone based on polysilicon piezoresistive strain gauges was presented by Schellin and Hess in 1992 (Schellin 1992). However, the sensitivity of their device was rather disappointing. At the moment, the majority of silicon microphones is based on capacitive detection.

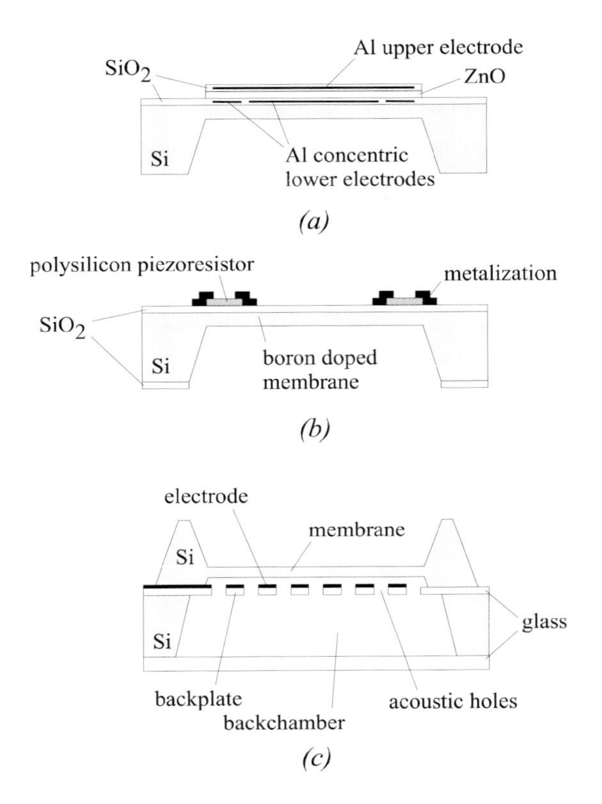

Fig. 6.37. Several read out techniques for micromachined silicon microphones: (a) piezoelectric (Royer 1983), (b) piezoresistive (Schellin 1992), and (c) capacitive (Bergqvist 1990)

Basically a capacitive microphone consists of a charged capacitor with one electrode formed by the deflecting membrane and another electrode positioned at a fixed distance of the membrane. The charge of the capacitor can be provided by a built-in charge (a so-called electret) or by an external voltage. The first electret microphones based on silicon technology used mylar foil for the membrane. An SiO_2 or Teflon layer was used as electret and was charged to a voltage of several hundreds of volts. Later designs use silicon or silicon nitride membranes. A problem with capacitive microphones is the viscous damping due to the air between the deflecting membrane and the static electrode beneath it. The first capacitive devices therefore used a rather large gap of 20 to 95 μm between the capacitor plates and one or more holes in the static electrode to allow for the air to flow. The microphones with silicon nitride and silicon diaphragms had a much smaller air gap (2…4 μm) and consequently a reduced bandwidth due to the increased air-streaming resistance. A solution was proposed by Bergqvist et al.

(Bergqvist 1991). They used a thin (10 μm) backplate with a very large number of acoustic holes (640...4000 holes per mm^2). The frequency response was flat between 2 and 20 kHz.

Another way to improve the frequency response is to use an electromechanical feedback system, which reduces the movements of the membrane as proposed by van der Donk (Donk 1992). Fig. 6.38 shows a top view and cross section of this device. Two interdigitated electrodes are used on the membrane, one is used as a normal sensing electrode and the other as the actuator electrode necessary to apply the feedback signal. The feedback system resulted in an increase of the cutoff frequency by a factor of 20.

Similar to capacitive pressure sensors, surface micromachining allows the fabrication of capacitive microphones in a single wafer, thus eliminating the need for wafer bonding or other techniques to attach the membrane to the microphone structure. Fig. 6.39 shows a cross sectional view of a single-wafer fabricated silicon condenser microphone (Scheeper 1992). Due to the narrow air gap this microphone can operate at a relatively low bias voltage and an electret is not necessary.

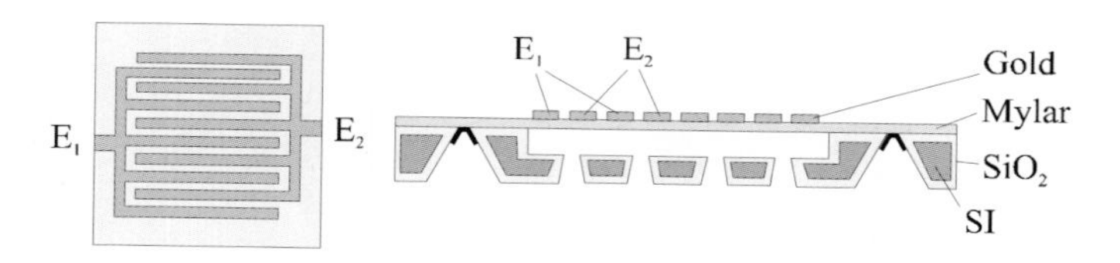

Fig. 6.38. Design of silicon condenser microphone which can be used for feedback. (a) top view showing the electrode configuration for actuating the membrane, (b) cross section of the microphone (Donk 1992)

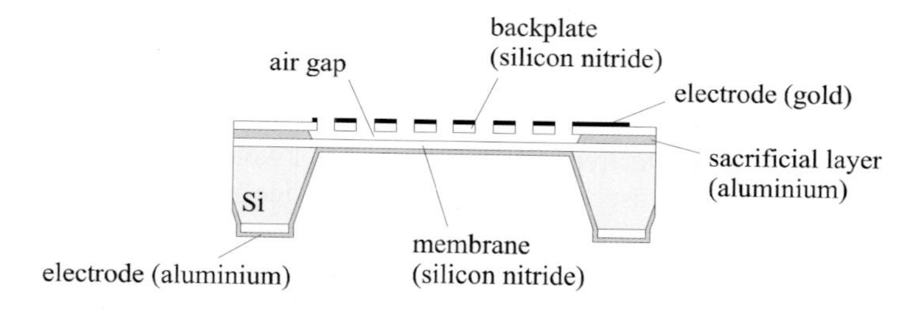

Fig. 6.39. Schematic cross section of a single wafer fabricated silicon condenser microphone (Scheeper 1992)

A similar structure can be made using plastic (polyimide) as the membrane and backplate material (see Fig. 6.40). This was done by Pedersen et al. (Pedersen 1998). In contrast with pressure sensors, for microphones hysteresis and creep in the membrane are not very important. The main advantage of using polyimide is the low temperature (<300°C) fabrication process. This allows the sensor to be realized on substrates which already contain the required interface electronics. Pedersen et al. demonstrated the successful integration of the microphone with a standard twin-well CMOS process. The CMOS process was used to integrate an amplifier and a dc-dc converter for generating the necessary bias voltage of 14 V from the supply voltage of 1.9 V.

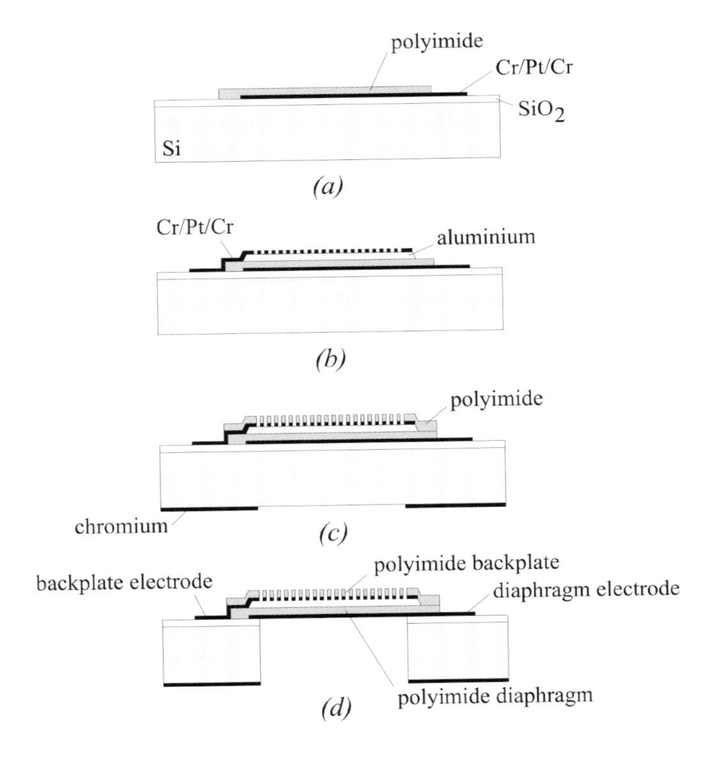

Fig. 6.40. Fabrication process of a polyimide condenser microphone (Pedersen 1998). (a) Deposition of Cr/Pt/Cr diaphragm electrode and polyimide diaphragm. (b) Deposition of Al sacrificial layer and Cr/Pt/Cr backplate electrode. (c) deposition of polyimide backplate and Cr etchmask on the back of the substrate. (d) Etching of sacrificial layer and substrate

6.2.6 Tactile Imaging Arrays

Tactile imaging arrays are arrays of force or pressure sensors. Applications for tactile imaging arrays are mainly in the field of robotics, where tactile imagers are needed to provide information on the distribution and amount of contact force between the workpiece and the robot gripper. Tactile imagers can be used for a variety of tasks, ranging from parts identification and orientation to surface texture measurements.

The requirements on tactile imagers are very application dependent. For relatively low-resolution applications an element spacing of 1 to 2 mm may be sufficient. In this case each element can be fabricated as an individual chip and and the elements can be assembled and wired into virtually any array size by simply mounting the chips side by side on a bare substrate surface. This has the significant advantage that it is possible to pre-test, characterize, and pre-sort the chips prior to the final array construction (Petersen 1985).
For higher resolutions the sensor elements have to be integrated on the same chip. A simple but effective structure was proposed by Chun and Wise (Chun 1985). Fig. 6.41 shows a cross section and top view of their device. The basic cell is formed between a selectively etched, boron-doped thin silicon membrane, which moves in response to the applied force, and a metalized pattern on an opposing glass substrate to which the silicon substrate is electrostatically bonded. Silicon dioxide is used to isolate the transducer plates on the silicon from the substrate and allow them to function as isolated row lines. In the layout, row conductors are run vertically across the silicon wafer in slots which are simple extensions of the capacitive gap recess. Metal column lines run horizontally on the glass under

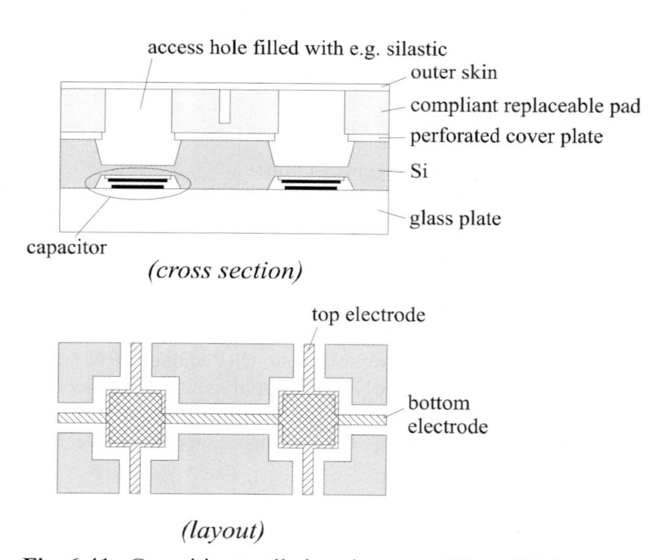

Fig. 6.41. Capacitive tactile imaging array (Chun 1985)

recesses in the silicon, expanding to form capacitor plates over the cell areas. A perforated cover plate is overlayed on the array for protection, and this plate is in turn covered by a compliant, replaceable pad and outer skin. The access holes coupling the force to the cell are filled with a substance (e.g. silastic) which acts as a force transmitter.

A high-resolution pressure imager was proposed by Sugiyama *et al.* (Sugiyama 1987). They realized a 32 x 32 (1k) element pressure imager with a spacing between the elements of only 250 μm. The pressure imager is in fact an array of miniature pressure sensors similar to the one we discussed in section 6.2.1, Fig. 6.17 (Sugiyama 1986). A silicon nitride membrane is used with polysilicon piezoresistive strain gauges. A small, shallow cavity is realized below the membrane by etching a sacrificial polysilicon layer through an etch hole. Fig. 6.45*(a)* shows a schematic cross section of the device. Sugiyama *et al.* succeeded in combining the fabrication process of the sensor elements with a 3 μm twin well CMOS process. The CMOS part contains the circuitry for selection and readout of the pressure elements. Only one pressure cell can be selected simultaneously and all other cells are turned of with no current flowing through the piezoresistive bridge. As a result, the total power consumption of the array is almost equivalent to the power consumption of one cell. The CMOS circuitry operates at a clock frequency of 4 MHz and the pressure cells are scanned at a frequency of about 60 kHz. Thus the entire array can be scanned in only 16 ms. Fig. 6.45*(b)* shows a possible package assembly for the pressure imager.

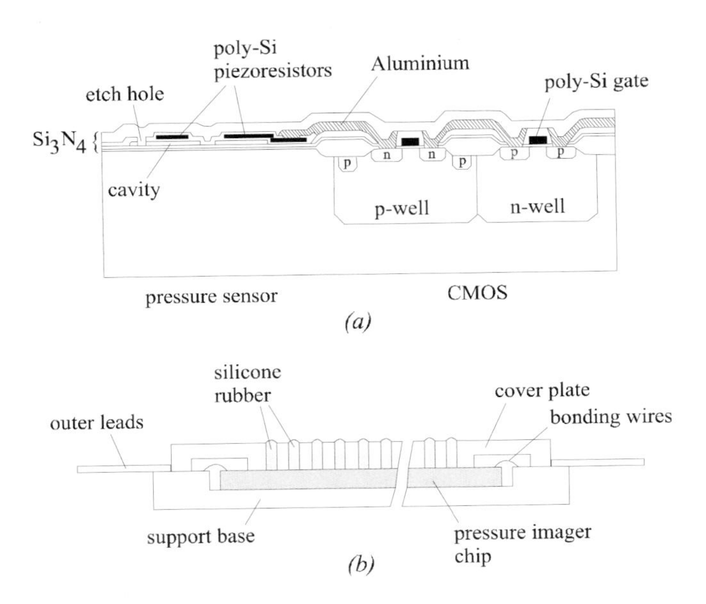

Fig. 6.42. Schematic cross-section of a high-resolution pressure image with integrated CMOS electronics. Polysilicon piezoresistors are used as readout elements

7. Acceleration and Angular Rate Sensors

Micromachined inertial sensors, consisting of acceleration and angular rate sensors are produced in large quantities mainly for automotive applications, where they are used to activate safety systems, including air bags, and to implement vehicle stability systems and electronic suspension. Besides these automotive applications accelerometers are used in many other applications where low cost and small size are important, e.g. in biomedical applications for activity monitoring and in consumer applications such as the active stabilization of camcorder pictures.

Inertial sensors can be divided in two categories: acceleration and angular rate sensors. Acceleration sensors generally consist of a proof mass which is suspended to a reference frame by a spring element. Accelerations cause a displacement of the proof mass, which is proportional to the acceleration. This displacement can be measured in several ways, e.g. capacitively by measuring a change in capacitance between the proof mass and an additional electrode or piezoresistively by integrating strain gauges in the spring element. To obtain large sensitivity and low noise a large proof mass is needed, which suggests the use of bulk micromachining techniques. For less demanding applications surface micromachined devices seem to be more attractive because of the easy integration with electronic circuits and the fact that bulk micromachining requires the use of a wafer bonding step. Recently, some designs have been presented which combine bulk and surface micromachining to realize a large proof mass in a single wafer process.

Angular rate sensors are based on vibrating structures with two modes of vibration. The structures are excited in one mode. An applied angular rate causes a vibration at the same frequency in the second mode due to the Coriolis effect. The amplitude of this vibration is proportional to the angular rate. The main problem is how to detect the sense amplitude because it is much smaller than the drive amplitude. A common technique is to increase the sense amplitude by matching the resonance frequencies of the drive and sense modes. This can be done either by electronic tuning of the spring constants or by using highly symmetrical structures.

In this chapter we will start with a discussion on acceleration sensors. Next, in Sect. 7.2, angular rate sensors will be discussed.

7.1 Acceleration Sensors

7.1.1 Introduction

As mentioned in the introduction, silicon accelerometers generally consist of a proof mass M which is attached to a fixed frame by one or more spring elements. Fig. 7.1 shows a simple lumped element model of such a structure. In this model, K is the effective spring constant of the spring element(s) and D is the damping factor. The operation of the device is based on Newton's second law of motion:

$$F = \frac{\mathrm{d}p}{\mathrm{d}t} = Ma \tag{7.1}$$

where F is the force acting on the mass, p is the impulse momentum and a is the acceleration of the mass. From (7.1) it is clear that an acceleration results in a force being exerted on the mass. This force results in a deformation of the spring element(s) and a displacement of the mass given by:

$$d_{\mathrm{static}} = \frac{F}{K} = \frac{Ma}{K} \tag{7.2}$$

The subscript *static* indicates that (7.2) is only valid for slow variations of the acceleration, i.e. well below the resonance frequency of the system.

The dynamic behaviour of the system can be analyzed by considering the differential equation:

$$M\frac{\mathrm{d}^2x}{\mathrm{d}t^2} + D\frac{\mathrm{d}x}{\mathrm{d}t} + Kx = F_{\mathrm{ext}} = Ma \tag{7.3}$$

where F_{ext} is the external force acting on the reference frame to which the proof mass is attached. Using Laplace transformation, the following second-order mechanical transfer function from an acceleration to a displacement of the mass is obtained:

$$H(s) = \frac{X(s)}{A(s)} = \frac{1}{s^2 + \frac{D}{M}s + \frac{K}{M}} = \frac{1}{s^2 + \frac{\omega_{\mathrm{r}}}{Q}s + \omega_{\mathrm{r}}^2} \tag{7.4}$$

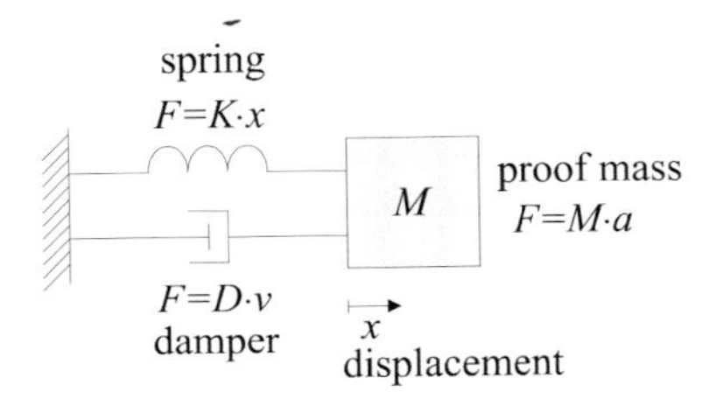

Fig. 7.1. Lumped element model of an accelerometer structure

where $\omega_r = \sqrt{\frac{K}{M}}$ is the resonance frequency and $Q = \frac{\omega_r M}{D}$ is the quality factor. From this equation we see that we can also write (7.2) in terms of the resonance frequency:

$$d_{\text{static}} = \frac{a}{\omega_r^2}, \tag{7.5}$$

which clearly illustrates the trade-off between bandwidth and sensitivity. For a high dc sensitivity we need a low resonance frequency. Feedback may be used to eliminate this trade-off as will be explained later.

The performance of accelerometers is limited by the thermal motion of the proof mass. According to the laws of thermodynamics, the thermal energy of a system in equilibrium is $k_B T/2$ for each energy storage mode, where k_B is the Boltzmann constant and T is the temperature. The small proof mass of micromachined accelerometers results in rather large movements. An equivalent acceleration spectral density, the so-called total noise equivalent acceleration (TNEA) can be calculated and is given by (Gabrielson 1993):

$$\text{TNEA} = \sqrt{\frac{\overline{a_n^2}}{\Delta f}} = \frac{\sqrt{4k_B TD}}{M} = \sqrt{\frac{4k_B T\omega_r}{QM}} \tag{7.6}$$

It is clear that for measuring low acceleration levels a large proof mass and high quality factor are required.

7.1.2 Bulk Micromachined Accelerometers

The fact that low-noise and high-sensitivity require a large proof mass suggests the use of bulk micromachining techniques in order to obtain a proof mass with a thickness equal to the wafer thickness. Fig. 7.2 shows a cross sectional drawing of the first micromachined accelerometer (Roylance 1979). It consists of a central silicon wafer containing the proof mass and a cantilever suspension beam. The wafer is bonded between two glass wafers to protect the structure and provide shock stop and damping. A piezoresistor is integrated in the suspension beam. When the support frame moves with respect to the proof mass the suspension beam bends, resulting in a change in resistance of the piezoresistor.

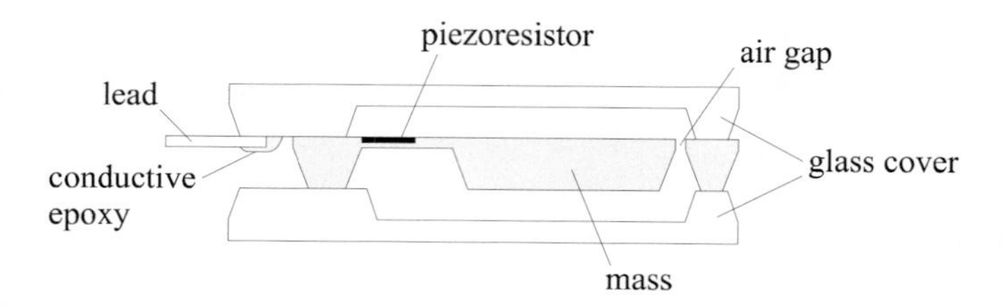

Fig. 7.2. Structure of piezoresistive accelerometer (Roylance 1979)

An important property of accelerometers is the off-axis sensitivity, that is the device should be sensitive to accelerations in one direction but insensitive to accelerations in other directions. In the case of a simple cantilever support the width of the beam is important as can be seen in Fig. 7.3(a). An acceleration in the y-direction causes a rotation of the mass around the x-axis, thereby torquing the suspension beam. Fig. 7.3 shows several other suspension structures that can be used in bulk-micromachined accelerometers. An extensive comparison between these structures was presented by Van Kampen et al. (Kampen 1998). The double cantilever support of Fig. 7.3(b) already has a much lower sensitivity for movements in the y-direction. Still, accelerations in the x-direction cause a bending of the suspension beams because the center of the mass is located beneath the plane of suspension. This bending cannot be distinguished from a bending due to an acceleration in the z-direction. This is different for the multiple-beam structures of Fig. 7.3(c–f). In these structures an acceleration in the z-direction causes a translation of the mass, while accelerations in the x or y direction cause a rotation of the mass. These effects can be detected separately and the off-axis sensitivity can be very low. A problem with the multiple beam suspensions may be residual strain in the beams. Furthermore, large deflections cause a tensile force in the beams resulting in a non-linear acceleration-deflection characteristic. Only the asymmetrical structure of Fig. 7.3(e) does not suffer from this effect as in this case a strain in the beams is relieved by a small rotation of the mass around the z-axis.

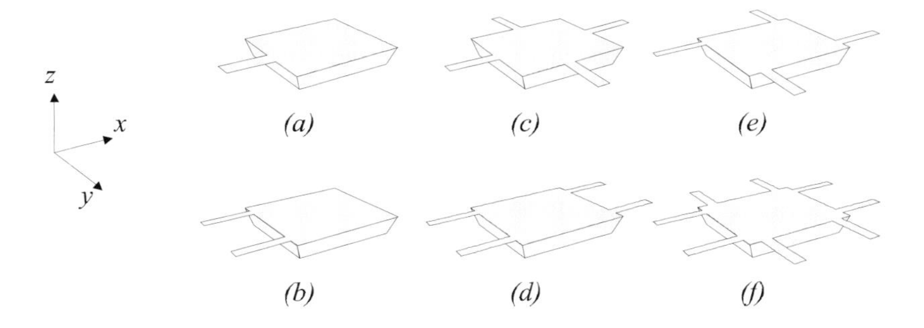

Fig. 7.3. Several suspension structures for bulk-micromachined accelerometers

As we have seen in the previous chapter, piezoresistive sensors have the advantage of a very simple fabrication process and also the electronic circuitry to detect the resistance changes is relatively simple. However, important disadvantages are the large temperature dependence and the low sensitivity compared to capacitive devices. To obtain a satisfactory sensitivity a relatively large proof mass is needed. Capacitive devices have several advantages: high sensitivity, good dc response and noise performance, low drift, low temperature sensitivity and low power dissipation. Furthermore, additional capacitors can be integrated and used in a closed-loop feedback configuration for electrostatic force-rebalancing or for self test purposes. The latter is especially important in critical applications like air bags. As with capacitive pressure sensors, a disadvantage is the increased complexity of the measurement electronics. Furthermore, the electronic interface circuit has to be placed in close proximity of the sensor to reduce parasitic capacitance and proper shielding is necessary to avoid electromagnetic interference.

The first micromachined capacitive accelerometers used bulk micromachining and wafer bonding to realize a thick, large proof mass resulting in high sensitivity. Fig. 7.4 shows the basic structure of such a device where the mass is suspended by one or two cantilever beams (e.g. (Rudolf 1990, Suzuki 1990)). Several other designs with basically the same topology but with more suspension beams have been reported, e.g. (Seidel 1990, Peeters 1992). The main advantage of these devices is the higher symmetry and, therefore, a reduced off axis sensitivity. On the other hand, an asymmetric suspension can be exploited to realize a three-axis accelerometer with a single proof mass (Mineta 1996, Puers 1998). Fig. 7.5 shows an example of such a structure (Puers 1998). In this structure, the proof mass contains 4 electrodes and has the shape of a four-leafed clover. It is suspended by four beams. The typical shape of the mass allows for relatively long beams and large capacitor plates in a small area. An acceleration in the z direction causes a translation of the mass in the same direction, resulting in the same capacitance change for all four capacitors (see Fig. 7.5(b)). An acceleration in the x or y direction causes a tilting of the mass (Fig. 7.5(c)). In this case two capacitors will increase and two capacitors will decrease. In first order approximation the three components of an arbitrary input acceleration vector can be independently derived from the values of the four capacitors.

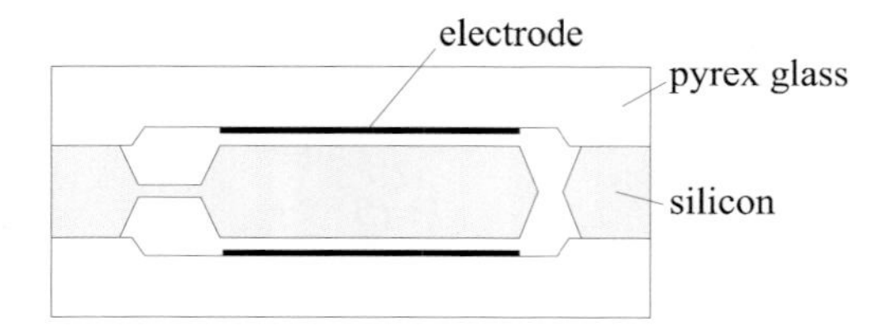

Fig. 7.4. Basic bulk micromachined capacitive accelerometer structure

Most single axis bulk micromachined accelerometers have the sensitive axis perpendicular to the wafer surface, so-called z-axis accelerometers. The reason for this is that in this case the mass can be realized relatively simple by wet etching. An accelerometer structure with a sensitive axis parallel to the wafer surface as indicated in Fig. 7.6(a) requires spring elements with a high vertical stiffness, i.e. the height of the springs should be significantly larger than the width. An important advantage of such a structure is the very low off-axis sensitivity because of the high symmetry and the relatively long distance between the springs (in z-axis accelerometers the vertical distance between the springs is limited to the thickness of the mass which is usually one or two times the silicon wafer thickness). One way of realizing the high aspect ratio springs is by deep reactive ion etching. A low-cost alternative was proposed by Schröpfer et al. (Schröpfer 1997). They used standard KOH wet etching on (100) wafers to etch the vertical (100) planes that are located at angles of +45 and -45 degrees with respect to the wafer flat. Fig. 7.6(b) illustrates the fabrication of a thin vertical spring using this process. Note that in this case the under etch of the oxide mask is at least equal to half the wafer thickness.

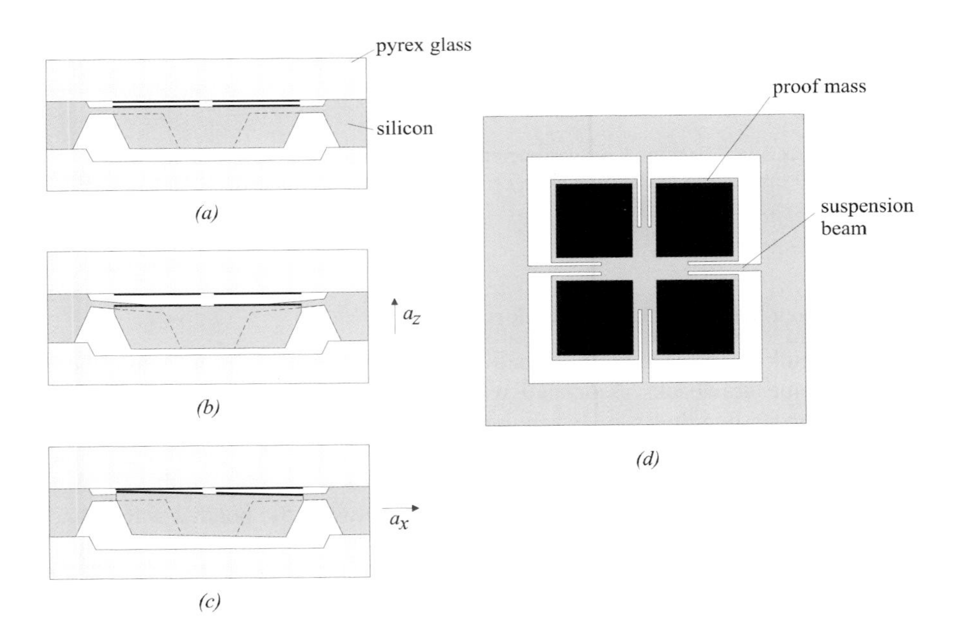

Fig. 7.5. Triaxial bulk-micromachined capacitive accelerometer: (a) rest position, (b) vertical displacement due to z-axis acceleration, (c) tilt due to x-axis acceleration, (d) top view of mass/suspension topology (Puers 1998)

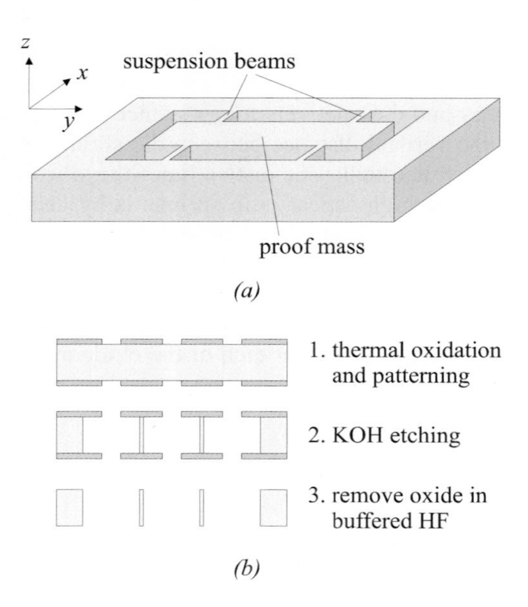

Fig. 7.6. Bulk micromachined accelerometer structure with sensitive axis in the wafer plane (a) and fabrication of the high aspect ratio suspension beams with a standard wet etching process (b) (Schröpfer 1997)

7.1.3 Surface Micromachined Accelerometers

Although bulk micromachining offers the advantage of a large proof mass, there are also some drawbacks associated with it. As we have seen in the previous section a wafer bonding step is required to realize an air gap that defines the amount of damping and allows for capacitive interfacing. As mentioned in the previous chapter, bonding of a silicon wafer to a glass wafer may result in a large temperature coefficient due to thermal mismatch between the bonded wafers and poor long-term stability due to slow relief of built-in mechanical stresses. Thus, for high accuracy devices only silicon wafers should be used. In the case of capacitive devices there is a trade-off between large capacitance values and low damping: the narrow gap required for a large capacitance also results in large damping. Therefore, bulk micromachined devices typically require packaging at a prescribed pressure to control the damping. Surface micromachining allows integration in a single wafer and it can be easily combined with integrated electronic interface circuits. Furthermore, the thin structures allow for the realization of arrays of holes to control the damping.

Just like bulk micromachined devices, surface micromachined accelerometers can be distinguished by their sensitive axis. In a so-called vertical structure the mass consists of a plate which moves in a direction perpendicular to the wafer surface and the mass forms a parallel plate capacitor with an electrode beneath it. In lateral structures, i.e. with the sensitive axis parallel to the wafer surface, a comb structure is attached to the mass and movements of the mass are sensed by measuring the capacitance between the fingers attached to the mass and fixed fingers attached to the substrate. The vertical structure has the advantage of a significantly larger capacitance value: up to 1 pF compared to less than 200 fF for lateral structures. However, the structure is asymmetric and a voltage between the capacitor plates will result in an electrostatic force pulling the mass towards the wafer surface.

Fig. 7.7 shows an example of a vertical structure as proposed by Lu et al (Lu 1995). It consists of a 400 x 400 μm^2 proof mass suspended above the substrate by four folded beams. The holes ensure complete etching of the sacrificial oxide layer between the proof mass and the silicon surface. Furthermore, the holes reduce the squeeze-film damping due to the air between the mass and the substrate. The sensitivity of the accelerometer is optimized by maximizing the inertial mass and minimizing the spring constant of the suspension in the *z*-direction. Polysilicon flatness imposes an upper limit on the dimensions of the mechanical structures and hence the achievable mass. A typical mass size for polysilicon surface micromachined accelerometers is 0.5 μgrams.

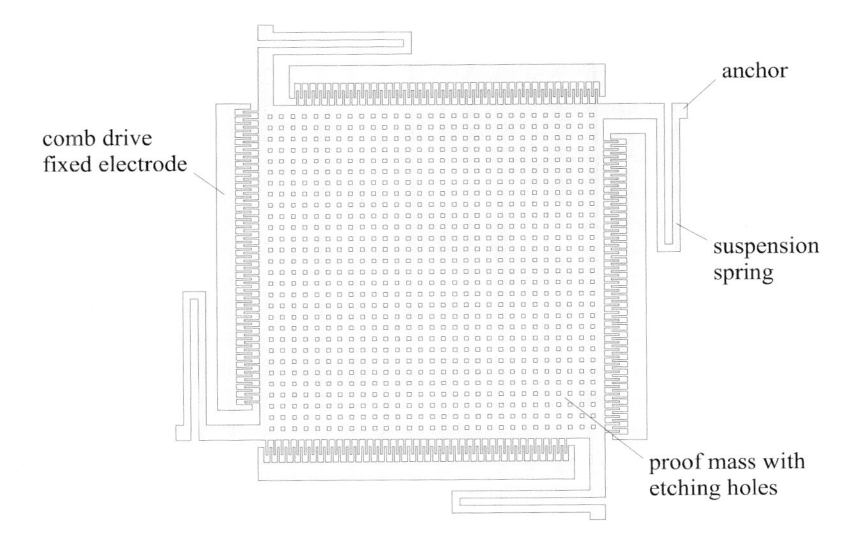

Fig. 7.7. Top view of surface micromachined *z*-axis accelerometer (Lu 1995). The proof mass is hanging free above an electrode on the surface. The capacitance between this electrode and the proof mass is a measure for the acceleration. Comb drives are located around the proof mass to provide an electrostatic pull-up force (see text and Fig. 7.8)

Because of the low inertial mass, very soft springs are needed to achieve a detectable displacement of the sensor. However, extremely weak springs are impractical because of sagging due to gravitation, susceptibility to self-resonance, and a tendency to stick to the substrate. Therefore, the folded beam suspension used by Lu et al. was dimensioned to obtain a vertical spring constant of 1.1 N/m. Thus, a 10 mg acceleration results in a displacement of only 0.05 nm and an extremely sensitive position sensing circuit is required.

As mentioned before, a problem with vertical structures is that the voltage used to detect capacitance changes also exerts an electrostatic pull-down force. Lu et al. solved this problem with the help of a comb structure around the periphery of the proof-mass. The comb fingers generate an opposing force to maintain the nominal position of the sensing element (Tang 1992). This is illustrated in Fig. 7.8. The silicon surface acts as a ground plane, resulting in an asymmetric electrical field distribution around the moveable finger. The result is a net pull-up force, which for small displacements is approximately proportional to the square of the applied bias voltage. An alternative solution is to use an additional electrode above the proof-mass, however this requires a more complex fabrication process.

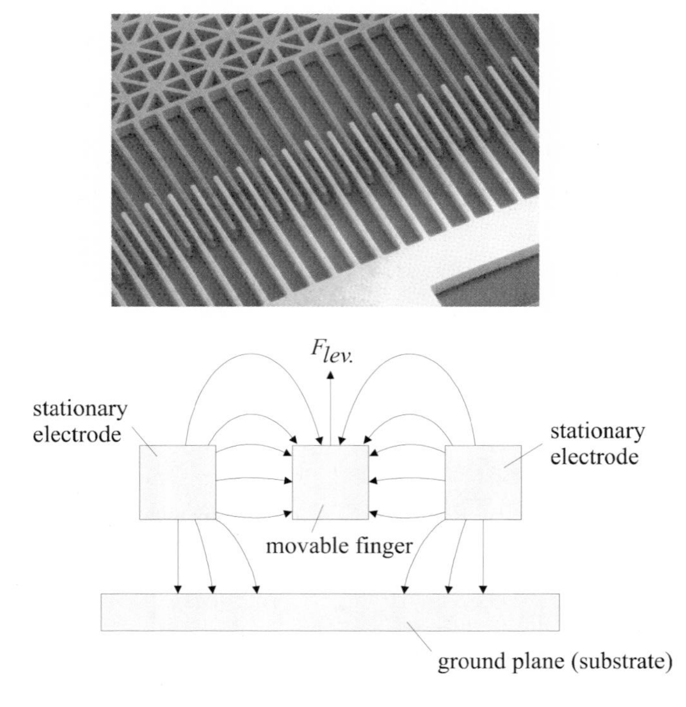

Fig. 7.8. Photograph of an electrostatic comb drive actuator (top) and illustration of the levitation effect due to the asymmetric electrical field resulting from the presence of a ground plane (the substrate) (bottom)

Fig. 7.9 shows the basic structure of a lateral accelerometer, i.e. with the sensitive axis in the wafer plane. This is also the structure used in the well-known accelerometers available from Analog Devices (Analog Devices 1995). The structure consists of a proof mass which is suspended to two anchors with U-shaped springs. Again, the mass contains a large number of holes to ensure complete etching of the sacrificial oxide layer underneath it. Contrary to the *z*-axis structure, the holes do not influence the damping as the mass now moves in the wafer plane. Comb structures are attached to the mass to both sense the position and exert electrostatic forces. The latter can be used for force feedback (see Sect. 7.1.4) and self test functions.

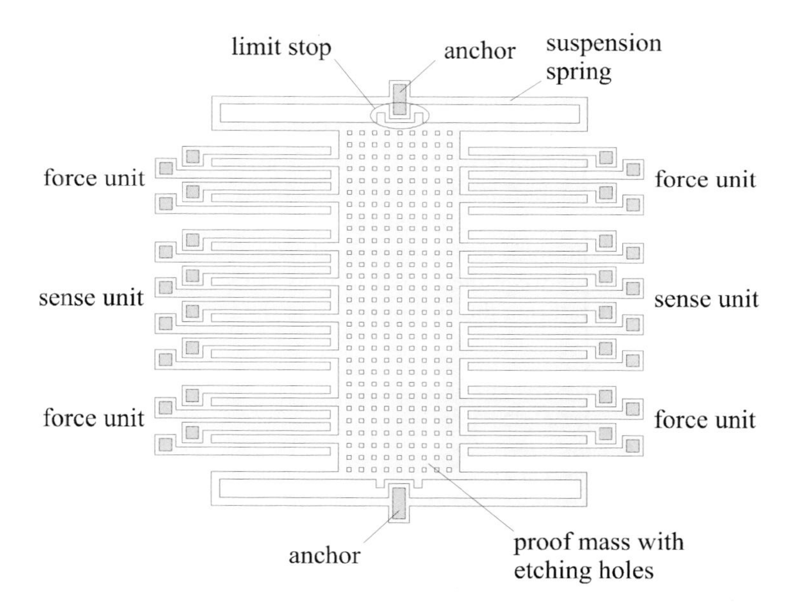

Fig. 7.9. Top view of a surface micromachined accelerometer with the sensitive axis in the wafer plane (Mukherjee 1999)

Of course, the realization of one vertical and two lateral accelerometers allows for the realization of a three axis accelerometer on a single chip. Such a chip was realized by Lemkin et al. (Lemkin 1997, 1999). They combined the surface micromachined accelerometer structures with a 2 μm CMOS process to realize the interface electronics. Fig. 7.10 shows a photograph of the complete accelerometer chip where the accelerometer structures are clearly visible in the center of the chip.

As mentioned before, surface micromached accelerometers have several advantages compared to bulk micromachined devices. However, one major disadvantage is the significantly smaller proof mass, which results in a much higher noise level. Because of their large proof mass, bulk-micromachined accelerometers obtain a high sensitivity with an equivalent noise level (eqn. (7.6)) below 1 $\mu g/\sqrt{Hz}$ over a bandwidth from dc up to 100 Hz. For a typical surface micromachined accelerometer the noise level is more than a hundred times larger. For many commercial applications (e.g. in the automotive sector) this is still acceptable, but inertial navigation and other precision applications such as tracking systems for head-mounted displays demand better performance (Boser 1996). A solution to this problem was proposed by Yazdi et al. (Yazdi 1999). They used a combination of bulk and surface micromachining to realize the accelerometer structure indicated in Fig. 7.11. Two polysilicon structural layers are used to realize a fixed electrode between two moving electrodes. The proof mass is suspended by a number of beams in the top polysilicon layer. The bottom poly layer forms an electrode between the proof mass and the top layer. This middle electrode is made rigid by embedding thick vertical stiffeners in it. These stiffeners are formed by thin film deposition and refilling of high aspect-ratio trenches in the proof mass (Selvakumar 1994). The top electrode is made rigid by making it short and wide, and supporting it through electrically-isolated standoffs on the proof mass. These standoffs are formed by the first poly layer and the dielectric layers on top and bottom of it. The sacrificial oxide layers between the proof mass and first poly and between the two electrodes are sealed by the poly layers and kept at the anchors to bring the anchor height to the level of the second poly.

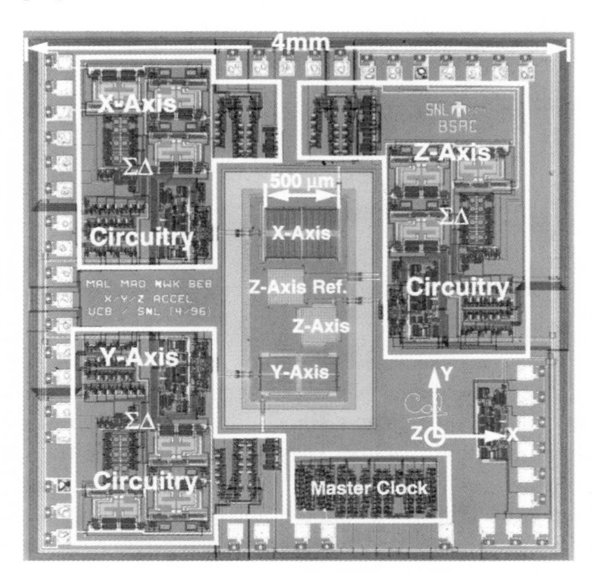

Fig. 7.10. Photograph of a monolithic three-axis polysilicon surface micromachined accelerometer with integrated sigma-delta readout (see chapter 10) and control circuitry (from: Lemkin 1999)

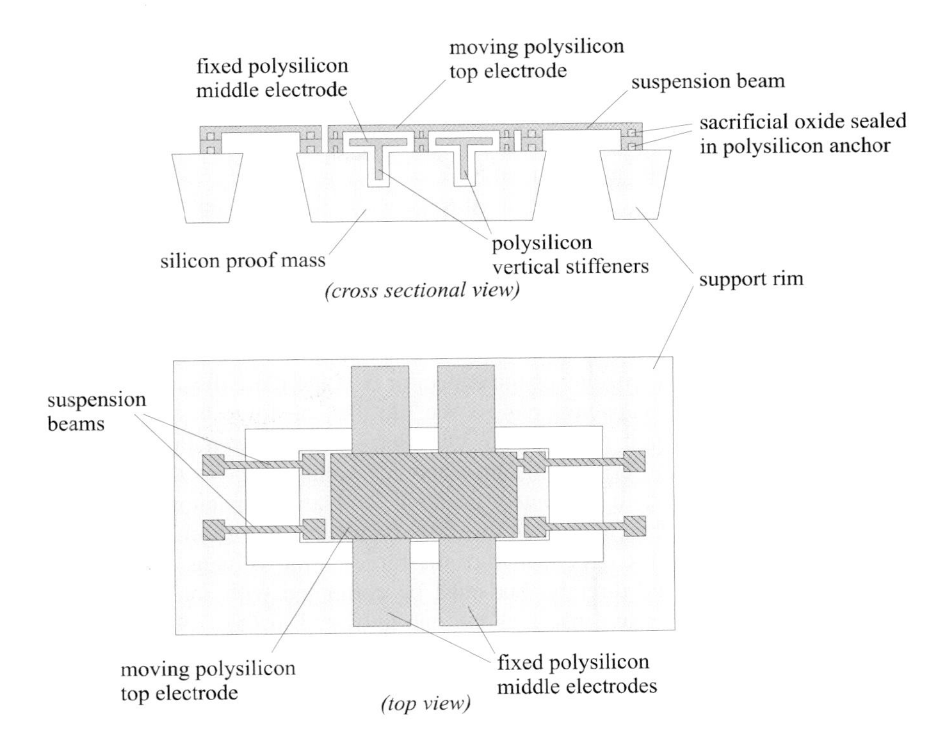

Fig. 7.11. Simplified structure of a combined bulk and surface micromachined accelerometer showing that a large wafer-thick proof mass can be realized without the need for a wafer bonding step (Yazdi 1999)

7.1.4 Force-feedback

As we have seen from equation (7.5) the sensitivity of an accelerometer can be improved by decreasing the resonant frequency. This trade-off between sensitivity and speed can be relieved significantly by using force-feedback (e.g. (Boser 1996)). Feedback increases the useful bandwidth by a factor equal to the loop gain, which, thanks to the added electronic circuitry, can be made large. As a result, the mechanical bandwidth of the accelerometer structure can be optimized for sensitivity, regardless of the desired bandwidth for the complete sensor. Furthermore, force-feedback may be used to improve dynamic range, linearity and drift. For low-damping, high-Q accelerometers force-feedback is important to limit the motion of the proof mass at the resonant frequency. Otherwise, these large motions would result in nonlinearity and, for capacitive devices, could easily exceed the small spacing between the electrodes of the sense capacitor.

Figure 7.12 shows an accelerometer embedded in a feedback loop. A compensator and a force-feedback transducer are added to the open loop sensor consisting of the proof mass and position measurement circuitry. The feedback force opposes displacements of the proof mass from its nominal position. The compensator is required for stability. Without the compensator the system will be unstable due to the 180 degree phase shift introduced by the accelerometer for frequencies above its resonance frequency.

Usually, electrostatic excitation is used to generate the feedback force. As we have seen in Chapter 5 (eqn. (5.19)) the electrodes of a parallel plate capacitor are attracted by an electrostatic force given by:

$$F_{\mathrm{el}} = \frac{1}{2}\frac{\varepsilon bl}{x^2}v^2 \tag{7.7}$$

where b and l are the dimensions of the capacitor, x is the plate distance and v is the voltage between the capacitor plates. We see that the force has a quadratic dependence on both x and v. In Chap. 5 we have already seen that for plate distances below a certain critical distance and for voltages above the so-called pull-in voltage the system becomes unstable and the capacitor collapses. Precautions should be taken to prevent this from damaging the sensor.

Because of the quadratic dependence of the force on the voltage, electrostatic actuators can not be used directly, but must be combined with some means of linearization. For symmetric sensors as the one indicated in Fig. 7.9 this can be easily accomplished. In this case each finger from the proof mass forms capacitors with two fixed fingers. By applying voltages $v_0+\Delta v$ and $v_0-\Delta v$ the effective force becomes linear with Δv:

$$F_{\mathrm{eff}} = \frac{1}{2}\frac{\varepsilon bl}{x^2}(v_0+\Delta v)^2 - \frac{1}{2}\frac{\varepsilon bl}{x^2}(v_0-\Delta v)^2 = \frac{\varepsilon bl}{x^2}v_0 \cdot \Delta v \tag{7.8}$$

assuming that the moving finger is located exactly in the middle between the fixed fingers. Of course, in practice this will not be the case resulting in some nonlinearity. Furthermore, this approach cannot be used for asymmetric structures like the vertical structure from Fig. 7.7.

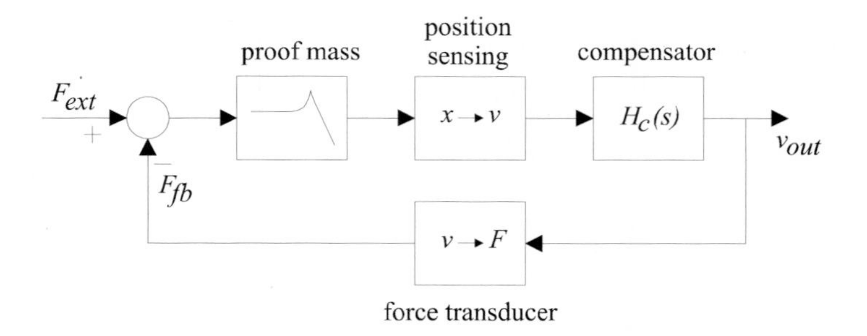

Fig. 7.12. Accelerometer with analog force-feedback loop

A more general and frequently used technique to obtain linearity consists in pulse modulating the feedback signal. In this case the feedback signal is quantized to only two levels and a comparator is used to select which force should be applied, as indicated in Fig. 7.13. The comparator switches only at prescribed moments, defined by the sample frequency f_s. In this way, all feedback pulses have the same duration (one period of the sample frequency) and an imbalance only results in an offset and/or gain error and does not cause distortion. The system is equivalent to a sigma-delta modulator as used in A/D conversion (Candy 1992), except that the noise shaping filter has been replaced by the mechanical sensor. The pulse-density of the one-bit digital output signal tracks the input acceleration, which is obtained by low-pass filtering and decimating the pulse-density code.

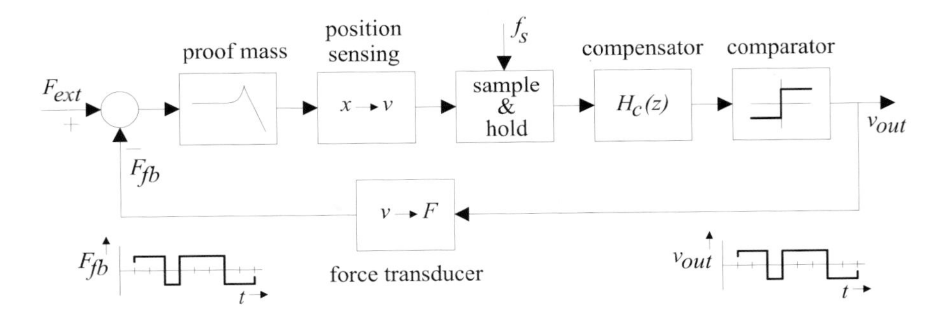

Fig. 7.13. Accelerometer with digital force-feedback loop

7.2. Angular Rate Sensors

Almost all micromachined angular rate sensors use vibrating elements to sense the rotation. The velocity of the vibrating element together with a rotation results in a Coriolis acceleration in a direction perpendicular to the direction of the vibration. As a result the vibrating element starts to vibrate in a second mode, perpendicular to the first mode. The amplitude of this secondary vibration is directly proportional to the rate of rotation.

The Coriolis effect can be most easily explained by imagining a ball moving over a rotating disk as indicated in Fig.7.14(a). Although the ball moves in a straight line from the center to the edge of the disk, it will pass through a curved trajectory on the disk. The degree of curvature is a measure for the rotation rate. In fact, an observer on the disk will see an apparent acceleration of the ball, i.e. the Coriolis acceleration, which is given by the cross product between the angular velocity vector Ω of the disk and the velocity vector v of the moving ball:

$$a_{\text{Corol}} = 2v \times \Omega \tag{7.9}$$

Thus, although no real force has been exerted on the ball, to an observer sitting on the rotating disk an apparent force has resulted proportional to the rate of rotation.

One well-known structure used in angular rate sensors is the tuning-fork structure depicted in Fig. 7.14(b). The tines are driven differentially and at a constant amplitude as indicated. A rotation around the stem of the tuning-fork results in a Coriolis acceleration of the tines in a direction perpendicular to the drive direction. The Coriolis force can be detected from the bending of the tines or from the torsional vibration of the tuning-fork stem.

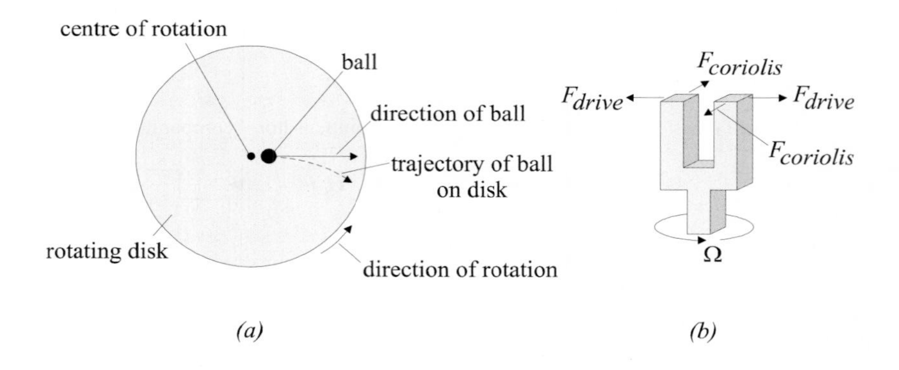

Fig. 7.14. (a) Illustration of the Coriolis effect: a ball moving from the center to the edge of a rotating disk moves along a curved trajectory on the disk, and (b) a tuning-fork angular rate sensor based the Coriolis effect.

A problem with angular rate sensors based on the Coriolis effect is the small amplitude of the secondary vibration. For example, for a sinusoidal drive vibration,

$$x(t) = x_0 \sin(\omega_{drve}\, t)\,, \tag{7.10}$$

with amplitude x_0 and frequency ω_{drive} the Coriolis acceleration is given by:

$$a_{Corol} = 2\dot{x}(t) \times \Omega = 2\Omega x_0 \omega_{drve} \cos(\omega_{drve}\, t) \tag{7.11}$$

Thus, for a drive amplitude x_0 of 1 μm, a drive frequency of 20 kHz and an angular rate Ω of 1 deg/s, the amplitude of the Coriolis acceleration is only 4.4 mm/s^2 (Clark 1996). Assuming a second-order spring mass system with a quality factor Q of 1 and a resonance frequency close to the drive frequency of 20 kHz this results in displacements in the order of 0.0003 nm (eqn. (7.5)). Two techniques are commonly used to increase the sense amplitude:

- Increase the drive amplitude. From (7.11) we see that the Coriolis acceleration is not only proportional to the angular rate Ω but also to the drive amplitude x_0. Thus, increasing the drive amplitude results in an increased sensitivity. Vibration amplitudes up to 50 μm have been achieved by using electromagnetic excitation (Lutz 1997). Although such a large amplitude can increase the output signal level, it increases the total power consumption and may cause fatigue problems over long-term operation (Yazdi 1998).

- Match the resonance frequency of the sense mode to the drive frequency in combination with a high quality factor. From (7.11) it appears that the Coriolis acceleration is an amplitude-modulated signal with carrier frequency ω_{drive}. Matching the resonant frequency of the sense mode to ω_{drive} results in a gain equal to Q. However, at the same time the bandwidth is limited to ω_{drive}/Q. Thus, for a typical Q (in vacuum) of 10 000 the bandwidth will be reduced to only a few Hertz. To obtain both an acceptable bandwidth and an increased sensitivity there should be a slight mismatch between the sense-mode resonance frequency and the drive frequency (Clark 1996).

To obtain a high drive amplitude electromagnetic excitation is the logical choice. Fig. 7.15 shows schematic drawings of two basic bulk micromachined tuning-fork structures based on this type of excitation (Hashimoto 1995, Choi 1996). Permanent magnets are mounted in the package to obtain the necessary static magnetic field perpendicular to the wafer surface. An alternating electrical current is fed through the tines of the tuning fork to generate the Lorenz drive forces. The silicon structure is bonded between two glass wafers. The vibration resulting from the Coriolis acceleration is sensed capacitively using electrodes on the glass wafers. The resonance frequencies of the drive and sense modes were closely matched with the help of a FEM analysis. Note that the single point suspension of Fig. 7.15(b) requires a metal pattern on the tuning fork to conduct the excitation current. However, the structure has several advantages compared to the two point suspension like a significantly higher Q factor because of the reduced losses into the suspension and a reduced influence of packaging stresses and temperature variations.

Note that the two masses in the tuning-fork structures of Fig. 7.15 are also subject to linear accelerations. This is why two masses are required. A linear acceleration will have the same effect on both masses, thus differential position sensing cancels the linear acceleration effect. Instead of using two masses one can also use a circular mass in a quasi-rotating mode of vibration (Juneau 1997) or four masses in a rotor like configuration as shown in Fig. 7.16 (An 1999). In this way it is possible to realize a dual-axis angular rate sensor. A rotation along one of the input axis causes a tilting of the structure which is sensed capacitively using electrodes on the substrate beneath the structure. Just like the structure in Fig. 7.15(b), this structure is only attached to the substrate in a single point which has the significant advantage of a reduced sensitivity to mechanical and thermal stress originating from the substrate and package.

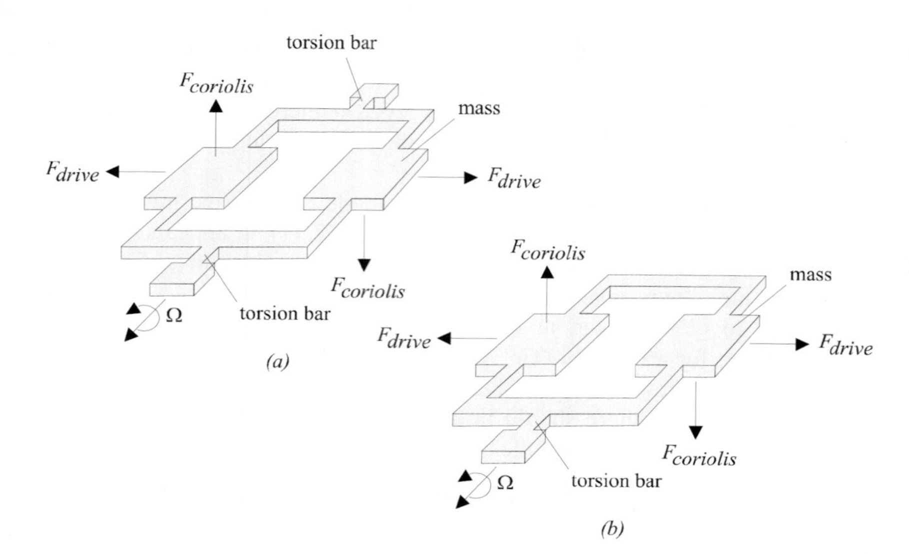

Fig. 7.15. Tuning fork angular rate sensors with (a) double torsion bar suspension (Hashimoto, 1995) and (b) single torsion bar suspension (Choi 1996)

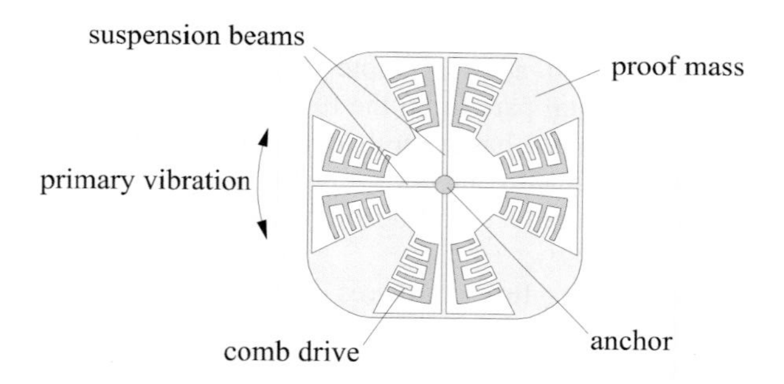

Fig. 7.16. Top view of a dual-axis angular rate sensor using 4 proof masses, 2 for each input-axis, and a rotary type of oscillation (An 1999)

Figure 7.17 shows the simplified fabrication process of the sensor. First, a thermal oxide and silicon nitride layer are deposited for electrical isolation from the substrate. The isolation layer is patterned to obtain a ground contact to the substrate. Next, a 300 nm polysilicon layer is deposited and patterned to serve as the capacitor electrodes. A 10% PSG layer is deposited as sacrificial layer, followed by the deposition of the 7 μm thick polysilicon structure layer. A second PSG layer is deposited on top of the structure layer to lower the resistivity of the polysilicon symmetrically by an annealing process. Next, the polysilicon layer is etched by a RIE process and the comb side is doped by a $POCl_3$ diffusion process. The sacrificial PSG layer is removed by buffered HF.

A problem with the dual-axis sensor is mechanical cross-talk between the two sense modes. As mentioned before, to increase the sense signal amplitude the resonance frequency of the sense mode should be close to the drive frequency. However, in the case of two sense modes, choosing the resonance frequencies close together will result in an undesired coupling. Therefore, An et al. chose to tune one sense mode to the downside of the drive mode and the other sense mode to the upside of the drive mode. In this way each sense mode is close to the drive mode and an optimal separation between the sense modes is obtained. The tuning of the sense modes is accomplished by controlling the bias voltages at the capacitor electrodes.

To detect the sense mode vibrations An et al. used a force feedback scheme as indicated in Fig. 7.18. This feedback is analogous to the feedback systems used for acceleration sensors as discussed in section 7.1.4, with the difference that it does not result in an increased bandwidth. This is because of the fact that the resonance peak of the sense modes is used to increase the sensitivity at the expense of a reduced bandwidth. The main advantage of closed loop operation is now the increased dynamic range and improved linearity. As shown in Fig. 7.18, differential sensing and driving is used. The feedback force is linearly dependent on the voltage Δv, see (7.8).

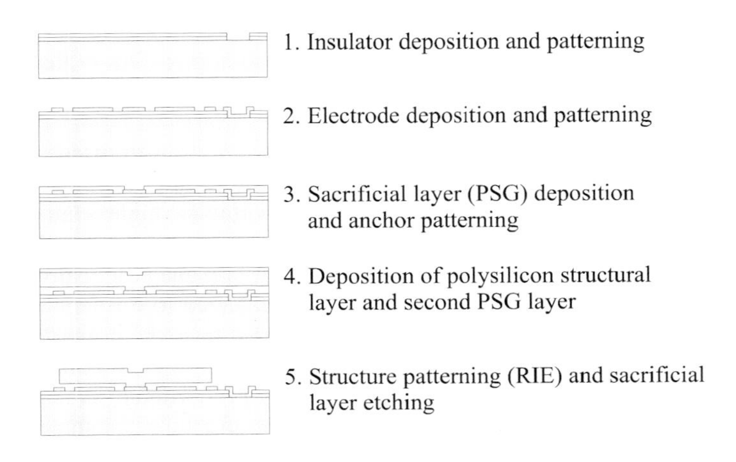

Fig. 7.17. Simplified fabrication process of the dual-axis angular rate sensor in Fig. 7.16

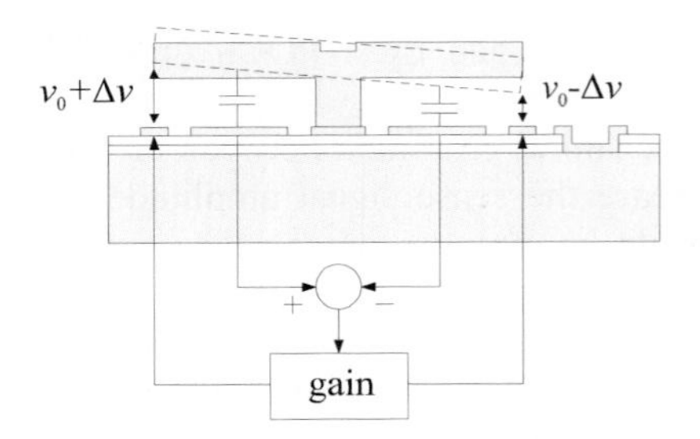

Fig. 7.18. Feedback scheme for dual axis angular rate sensor of Fig. 7.16

Figure 7.19 shows the structure of another angular rate sensor based on a rotary type of vibration (Geiger 1999). The device is called MARS-RR, which means *M*icromachined *A*ngular *R*ate *S*ensor with *two R*otary oscillation modes. A very good performance is obtained by decoupling the drive and sense modes. In Fig. 7.19, we see that the sensor structure consists of a circular part in the center and a rectangular part around it. The two parts are connected by means of two torsional springs. The entire structure is electrostatically driven in a rotary oscillation around the z-axis by comb drives which form the spokes of the inner wheel. Four comb drives are used for actuation and four combs are used for detection of the primary vibration. When the device is turned around its sensitive axis, the x-axis, the Coriolis forces cause a rotary oscillation around the y-axis. In this direction, the high stiffness of the beam suspension suppresses an oscillation of the inner wheel. Only the rectangular structure can follow the Coriolis forces, because it is decoupled from the inner wheel by the torsional springs. As in the previous example, the secondary vibration is detected capacitively using electrodes on the substrate.

The last structure of a micromachined angular rate sensor we want to discuss was developed by Putty and Najafi (Putty 1994). Fig. 7.20 shows the schematic structure of this device. It consists of a ring which is supported by eight semi-circular support springs. A number of drive and detection electrodes are located around the structure. The ring is electrostatically driven in an in-plane elliptically shaped flexural mode with a fixed amplitude. When it is subjected to a rotation around the z-axis, Coriolis forces cause energy to be transferred from the primary mode to the secondary flexural mode, which is located 45° apart from the primary mode. The amplitude of the secondary vibration is, again, proportional to the angular rate and is detected capacitively. The vibrating ring structure has some important features. The inherent symmetry of the structure makes it less sensitive to spurious vibrations. Furthermore, since the drive and sense mode are in principle identical, a good matching between the resonant frequencies is obtained resulting in a large sensitivity. Frequency mismatch due to mass or stiffness asymmetries occurring during the fabrication process can be electronically compensated by applying suitable bias voltages to the electrodes around the structure.

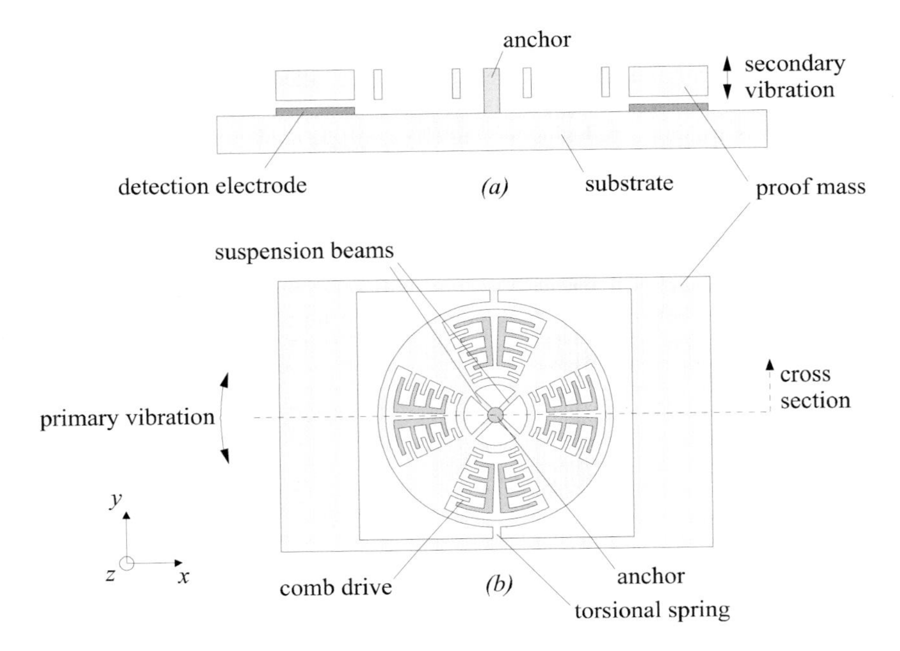

Fig. 7.19. Cross section (a) and top view (b) of angular rate sensor MARS-RR (Geiger 1999)

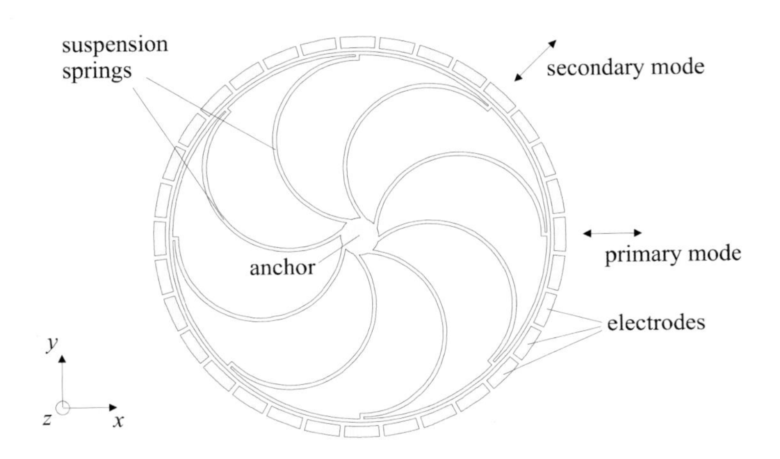

Fig. 7.20. Structure of a vibrating ring gyroscope. The ring is attached to an anchor in its center by semi-circular springs. The ring vibrates in a mode in which its shape changes into an ellipse and the direction of the main axis of the ellipse changes as a result of an angular rotation

The first version of the vibrating ring sensor was fabricated by electroforming nickel into a thick polyimide or photoresist mold on a silicon substrate containing the electronic control circuitry (Putty 1994, Sparks 1997). Recently, an improved version was made using a polysilicon ring (Ayazi 1998). This ring was realized by refilling deep dry-etched trenches with polysilicon deposited over a sacrificial oxide layer (Selvakumar 1994). Other designs have been proposed with a ring of monocrystalline silicon (Hopkin 1997, McNie 1999). Although good results have been obtained, a problem is the asymmetry introduced by the anisotropy of the mechanical parameters of monocrystalline silicon. This asymmetry causes significant coupling between the drive and detection mode and compensation is required, either electronically or mechanically by selectively adjusting the thickness of the ring.

8. Flow Sensors

Flow sensing is very complex. There are two distinct reasons for the complexity: First, flow is a science by itself and second, there are many phenomena that can be exploited for flow sensing. The designer of flow sensors must have a basic understanding of hydrodynamics. Micro flow sensors described so far are based on thermal principles or on the measurement of pressure distributions and shear stress. Optical and acoustical possibilities have not yet been explored.

Flow sensors based on a thermal principle add some heat locally to the streaming fluid, and measure the resulting temperature distribution close and far away from the heater, and/or the heat loss of the heater. The majority of designs are thermal flow sensors. The key to understand these flow sensors is a hydrodynamic phenomenon known as the boundary layer. Close to walls there is a region with large gradients in the flow velocity. This region is confined within a layer of a certain thickness, which is called the boundary layer. If there is a heater involved then in most fluids (liquid metals is the exception) also a thermal boundary layer builds up, which is submerged within the hydrodynamic boundary layer. For the design of flow sensors the boundary layer concept is very important. Extensions of sensors and obstructions must be compared with the extend of the boundary layer, and sensors operate completely different whether boundary layer is thick or thin compared to dimensions important in the sensor design. We therefore start with a description of the boundary layer. This chapter contains also a discussion of the relation of the heat loss of a heating element to skin friction. A chapter on flow sensing at very small flow rates follows, when the boundary layer is much thicker than the sensor dimensions. We then discuss thermal flow sensors, mechanical skin friction sensors, "dry fluid flow" and "wet fluid flow" sensors. With dry fluid flow we mean flow where viscous forces may be neglected, and wet fluid flow concerns flow dominated by viscous forces.

Electronics, interfacing and packaging will be treated in Chaps. 10 and 11.

8.1 The Laminar Boundary Layer

8.1.1. The Navier-Stokes Equations

A streaming liquid obeys the equations of Navier and Stokes, which read for an incompressible fluid

$$\rho\left(\frac{\partial \mathbf{v}}{\partial t}+(\mathbf{v}\nabla)\mathbf{v}\right)=-\mathrm{grad}p+\eta\nabla^2\mathbf{v} \tag{8.1}$$

with $\mathbf{v}(\mathbf{r},t)$ the velocity of the fluid at the point $\mathbf{r}$ at time t, $p(\mathbf{r},t)$ the pressure, ρ the density (mass per volume) and η the dynamic viscosity. The left hand represents the acceleration and the right hand side represents the forces (pressure and viscous forces, respectively).

A second equation is the continuity equation,

$$\frac{\partial \rho}{\partial t}=-\rho\mathrm{div}\mathbf{v}=0 \tag{8.2}$$

The last equality is for incompressible fluids. The assumption of incompressibility is a good approximation if the Mach number - the ratio of the characteristic flow velocity to the speed of sound - is small compared to 1.

The acceleration in eq. (8.1) is formulated in this particular way because of the following reasons. The acceleration of the fluid in a small volume at a point $(\mathbf{r},t)$ is needed, so we seek an expression for

$$\rho\frac{\mathrm{d}\mathbf{v}}{\mathrm{d}t}.$$

Following the formal rules for partial differentiation, we have for every component of v_i

$$\mathrm{d}v_i=\mathrm{d}t\frac{\partial v_i}{\partial t}+\mathrm{d}x\frac{\partial v_i}{\partial x}+\mathrm{d}y\frac{\partial v_i}{\partial y}+\mathrm{d}z\frac{\partial v_i}{\partial z}=\frac{\partial v_i}{\partial t}\mathrm{d}t+(\mathrm{d}\mathbf{r}\nabla)v_i$$

from which the form in (8.1) follows. Note that the acceleration is non zero also for cases in which the partial derivative $\partial\mathbf{v}/\partial t$ is zero. An example is the steady flow of water in a bent pipe: the flow is steady ($\partial\mathbf{v}/\partial t=0$) but obviously there is an acceleration.

For the purpose of clarity we write the x - component of (8.1) explicitly:

$$\frac{\partial v_x}{\partial t}+\left(v_x\frac{\partial}{\partial x}+v_y\frac{\partial}{\partial y}+v_z\frac{\partial}{\partial z}\right)v_x=-\frac{1}{\rho}\frac{\partial p}{\partial x}+\zeta\left(\frac{\partial^2}{\partial x^2}+\frac{\partial^2}{\partial y^2}+\frac{\partial^2}{\partial z^2}\right)v_x$$

with ζ the kinematic viscosity, $\zeta=\eta/\rho$.

We see that all components of the velocity are coupled by the terms in brackets on the left-hand side. This non-linear term makes the Navier-Stokes equation very complex. One of the consequences is that solutions may not be added like electromagnetic fields. A second consequence is that steady state solutions (where $\partial\mathbf{v}/\partial t=0$) may become unstable: an infinitely small perturbation to the steady state solution grows and the flow becomes non-steady: the flow is then turbulent.

For a sufficiently high flow velocity the viscosity can be neglected under certain circumstances, and only the Euler equation (the part of (8.1) without the term containing the Laplace operator) is left. Fluids, which obey the Euler equation, are sometimes called “ideal liquids”. Feynman calls his chapter on this equation the "flow of dry water", to remind that something essential has been left out, when neglecting the viscosity (Feynman 1964). For the Euler equation it can be shown that the circulation of the velocity field is conserved. That means that

$$\frac{\mathrm{d}}{\mathrm{d}t}\oint \mathbf{v}\mathrm{d}\mathbf{l} = 0$$

The vorticity, $\mathrm{rot}\mathbf{v}$, is carried along the streamlines.

Further the theorem of Bernoulli is valid for ideal fluids. The theorem says that

$$\frac{p}{\rho}+\frac{1}{2}v^2+\phi = \text{constant} \tag{8.3}$$

Here, ϕ denotes a potential energy of the fluid. (8.3) is valid along a streamline, and in the case that $\mathrm{rot}\mathbf{v} = 0$ everywhere, (8.3) holds everywhere. A very transparent derivation of (8.3) can be found in Feynman (1964). The Bernoulli equation is a consequence of the conservation of energy.

The phenomena related to the Bernoulli equation can be exploited for the measurement of the flow velocity. The pressure measured at a point where the fluid is streaming is smaller than the pressure at points where the fluid is at rest with respect to solid surfaces, for example at a stagnation point. Measurement devices that make use of this effect are called Prandtl flow sensors. A micro-version of a Prandtl flow sensor has been suggested by Berberig et al. (1997). Also the Pitot tube and the Recknagel disc (see e.g. Franke 1970) rely on the Bernoulli equation. An intriguing design using the pressure fall across an orifice due to the loss of kinetic energy has been proposed very recently by Richter et al. (1999).

Equation (8.1) can be written in dimensionless form, where a certain characteristic velocity U_0 and a characteristic length l fix the scales. As an example, for flow through a tube, these are the mean flow velocity and the radius of the tube, for flow along a cylinder these are the speed far away from the tube and the tube diameter. Equation (8.1) has similar solutions if the Reynolds number, defined by:

$$\mathrm{Re} = \frac{U_0 l}{\varsigma} \tag{8.4}$$

is equal: velocity and length must be scaled according to (8.4) including the kinematic viscosity.

The significance of the Reynolds number can be illustrated by the flow of a fluid past an infinitely long cylinder. We show the phenomena belonging to certain ranges of Re in Fig. 8.1. Fig. 8.2 shows the coefficient of the drag force the fluid exerts on the cylinder. The coefficient is defined as the ratio of the force per cross-section to the kinetic energy of the fluid per volume far away from the cylinder (with D and b the cylinder's diameter and length, respectively, and F the force):

$$C_D = \frac{F}{\frac{1}{2}\rho U_0^{\ 2} Db}$$

For Re > 100 the flow is turbulent. Close to Re = 10^6 the boundary layer separates which leads to the sudden drop of the friction coefficient.

We see that if Re is small enough, there is a one to one relation of the drag force coefficient with the flow velocity. Hence, for small Reynolds numbers the measurement of the drag force – not necessarily on a cylinder – is a way to measure flow velocities. This has been done in a microsensor by Gass et al. (1993).

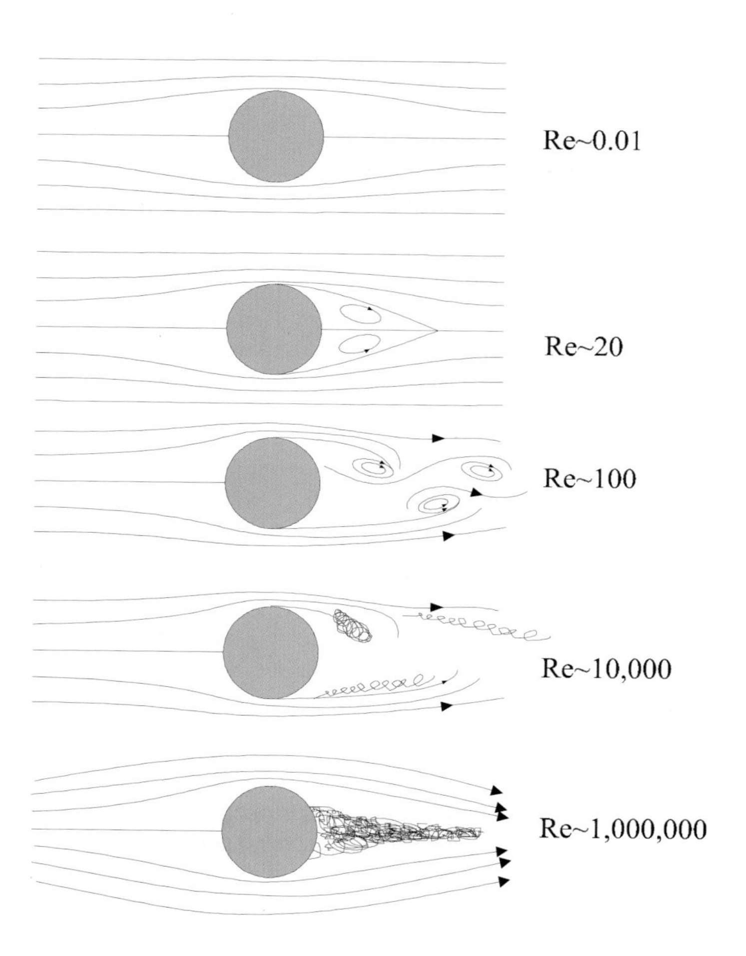

Fig. 8.1. Flow past an infinitely long cylinder at a range of Reynolds numbers.

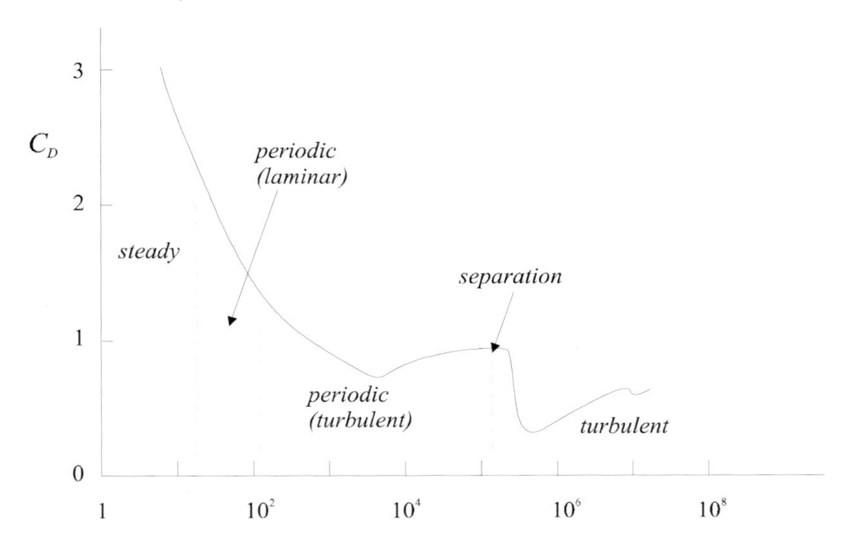

Fig. 8.2. The coefficient of drag force on a cylinder as a function of the Reynolds number

For small Reynolds numbers (<1600) there is a steady state solution of equation (8.1) for flow through pipes, which is the well-known law of Hagen-Poiseuille:

$$\dot{V} = \frac{\pi R^4}{8\zeta b}\Delta p \tag{8.5}$$

Here $\dot{V}$ is the volume flow (m^3/s), R and b the radius and length of a cylindrical pipe, respectively. Obviously, by measuring the pressure difference at points spaced by the distance b, the flow rate can be measured. The principle has been explored for microsensors several times (Cho 1992, Lui 1993, Boillat 1995, Oosterbroek 1997) and is discussed by Richter et al. (1999) for the limit of short channels. Schmidt et al. (1998) has presented another very recent approach.

8.1.2. Heat Transport

Many flow sensors are based on convective heat transport. These sensors comprise of one or more heat sources and temperature sensors. Sometimes both functions are integrated in a single element.

Landau (1974) gives the temperature distribution in a streaming incompressible fluid

$$\frac{\partial T}{\partial t} + \mathbf{v}\nabla T = D_T \nabla^2 T + \frac{\eta}{2\rho c_\mathrm{p}} \sum \left(\frac{\partial v_i}{\partial x_k} + \frac{\partial v_k}{\partial x_i} \right) \tag{8.6}$$

with T the temperature, D_T the thermal diffusion coefficient and c_p the specific heat at constant pressure. i and k represent the set of three Cartesian coordinates and the double sum runs over the three components of the Cartesian coordinate system (from 1 to 3). The last term describes the energy dissipation as a consequence of friction in the fluid. This term couples the heat distribution to the velocity field. In the following, this term will be ignored and we have

$$\frac{\partial T}{\partial t} + \mathbf{v}\nabla T = D_T \nabla^2 T \, . \tag{8.7}$$

In the case of a fluid in rest, (8.6) reduces to the diffusion equation,

$$\frac{\partial T}{\partial t} = D_T \nabla^2 T \tag{8.8}$$

The equations (8.1), (8.2) and (8.6) describe completely a homogeneous fluid in motion.

8.1.3 Hydrodynamic Boundary Layer

Walls in the fluid impose boundary conditions on the differential equations. Obviously, the velocity of a liquid normal to a wall must vanish at the wall. Also the tangential components vanish in the case of fluids and for gas ("no slip" condition, for large Knudsen numbers), and the shear force (the so-called skin friction) is given by

$$\tau = \eta \left. \frac{\partial v_x}{\partial y} \right|_{y=0} \tag{8.9}$$

A consequence is that close to walls there are large gradients in the velocity, so even when the Reynolds number is very large, or the viscosity is very small, the last term in (8.1) cannot be neglected, but dominates the flow.

Equation (8.9) gives the shear force on a surface in the $y = 0$ - plane due to a velocity in the x-direction.

In steady state so that $\partial \mathbf{v} / \partial t = 0$, a fluid streaming along a wall located in the $x - z$ plane (at $y = 0$) with the leading edge along the z-coordinate will have a large velocity gradient close to the wall and far away the velocity will be constant.

To calculate the velocity field close to a wall we assume the following geometry. The edge of the wall is along the z-direction and located at $x = 0$. At large negative x the velocity U is assumed to have only a constant component in the x-direction. In this situation no quantity can vary with z, and we may assume that the component of $\mathbf{v}$ in the x-direction is much larger than in the y-direction:

$$v_x >> v_y$$

We write out the x and y components of the Navier-Stokes equation,

$$\left(v_x \frac{\partial}{\partial x} + v_y \frac{\partial}{\partial y}\right) v_x = -\frac{1}{\rho}\frac{\partial p}{\partial x} + \zeta\left(\frac{\partial^2}{\partial x^2} + \frac{\partial^2}{\partial y^2}\right) v_x \tag{8.10}$$

$$\left(v_x \frac{\partial}{\partial x} + v_y \frac{\partial}{\partial y}\right) v_y = -\frac{1}{\rho}\frac{\partial p}{\partial y} + \zeta\left(\frac{\partial^2}{\partial x^2} + \frac{\partial^2}{\partial y^2}\right) v_y \tag{8.11}$$

where we omitted the derivative with respect to t (stationary flow). Also the components and derivatives which contain the z-coordinate are neglected. First of all we must note that all terms which contain v_y are much smaller than the ones with v_x. This means that the pressure derivative in x-direction is much larger than in y-direction. In the following, we assume therefore that

$$\frac{\partial p}{\partial y} = 0 \tag{8.12}$$

The pressure does not depend on y, along the coordinate normal to the wall, the pressure is nearly constant. The pressure change along the flow direction may be much larger.

Further we may assume that v_x changes faster in the direction normal to the wall. All derivatives with respect to y will be larger than the ones with respect to x. This means that additionally to (8.12), we have

$$v_x \frac{\partial v_x}{\partial x} + v_y \frac{\partial v_x}{\partial y} - \zeta \frac{\partial^2}{\partial y^2} v_x = -\frac{1}{\rho}\frac{\partial p}{\partial x} = U \frac{\mathrm{d}U}{\mathrm{d}x} \tag{8.13}$$

and the continuity equation

$$\frac{\partial v_x}{\partial x} + \frac{\partial v_y}{\partial y} = 0 \tag{8.14}$$

The second part of (8.13) is correct because of (8.12) as can be seen as follows: The partial derivatives of p with respect to y and z are equal to zero, which means that $\frac{1}{\rho}\frac{\partial p}{\partial x} = \frac{1}{\rho}\frac{\mathrm{d}p}{\mathrm{d}x}$. Far way from the wall, Bernoulli's equation can be used, which leads directly to the right hand side of (8.13). In the absence of an external potential, this means that

$$\frac{1}{\rho}\frac{\partial p}{\partial x} = \frac{1}{\rho}\frac{\mathrm{d}p}{\mathrm{d}x} = -U\frac{\mathrm{d}U}{\mathrm{d}x}$$

For the geometry we described here, the velocity U is constant, and consequently, (8.13) gets the simple form

$$v_x \frac{\partial v_x}{\partial x} + v_y \frac{\partial v_x}{\partial y} - \zeta \frac{\partial^2 v_x}{\partial y^2} = 0 \tag{8.15}$$

The boundary conditions are:

$$v_x = v_y = 0 \text{ at } x > 0, y = 0, v_x = U_0 \text{ for } y \to \infty$$

The equations (8.13, 8.14) can be made dimensionless if a characteristic length l is introduced. We define new quantities by

$$x = lx'; y = \frac{ly'}{\sqrt{\text{Re}}}; v_x = U_0 v_x'; v_y = \frac{U_0 v_y'}{\sqrt{\text{Re}}} \tag{8.16}$$

Here,

$$\text{Re} = \frac{U_0 l}{\zeta}$$

It can be easily seen that with this transformation the equations (8.13) and (8.15) do not contain any scale anymore, nor are they dependent on the Reynolds number. Inspecting the transformations (8.16), we see that v_x' must go from zero to 1 when y' increases and there will be a boundary layer which extends in the direction normal to the flow with a thickness in the order of

$$\delta = \frac{l}{\sqrt{\text{Re}}}$$

where the velocity changes appreciably. Further we can see that

$$v_y \approx \frac{U_0}{\sqrt{\text{Re}}}$$

The problem however does not contain any characteristic length. Consequently, the length l is related to a variable, namely x, the distance from the leading edge of the plate. We arrive at the conclusion that the thickness of the boundary layer increases with x as

$$\delta = \sqrt{\frac{\zeta x}{U_0}} \tag{8.17}$$

v_x (x,y) depends only on a single parameter ξ, which is a combination of x and y:

$$\xi = y\sqrt{\frac{U_0}{\zeta x}}$$

The equations (8.15) and (8.14) with the appropriate boundary conditions can be solved numerically. In Fig. 8.3 we show the function $v_x(\xi)/U_0$

The shear force, which the streaming fluid exerts on the wall, can be calculated as follows: Using the continuity equation (8.14) we may write (8.15) as

$$\frac{\partial}{\partial x} v_x^{\ 2} + \frac{\partial}{\partial y}(v_x v_y) - \zeta \frac{\partial^2 v_x}{\partial y^2} = 0$$

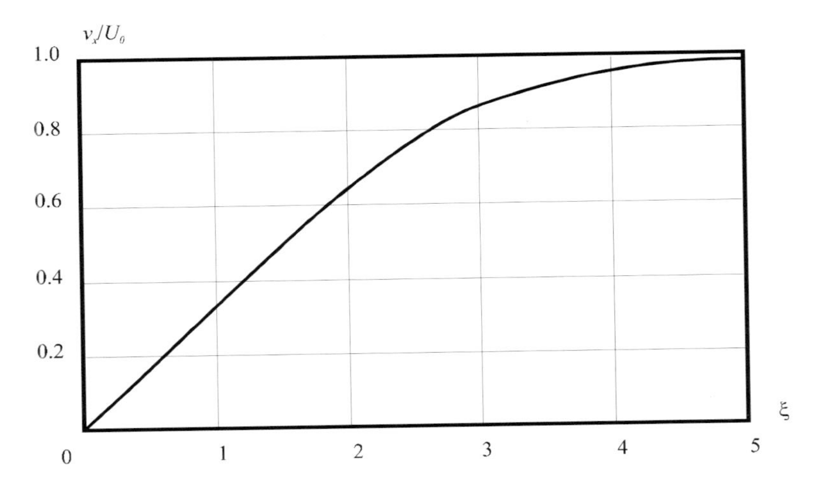

Fig. 8.3. The dependency of the stream wise velocity normalised on U of the parameter $\xi = y\sqrt{(U \;/\zeta x)}$ (Landau 1974)

We integrate over dy from zero to $y = \delta$ and note that v_x and v_y equals zero at $y = 0$ and $y = \delta$, respectively. The integral over the last term gives us the shear force at x divided by the density, and we are left with:

$$\frac{\partial}{\partial x}\int_0^\delta v_x{}^2 \mathrm{d}y = -\frac{\tau}{\rho}$$

Now we assume that

$$v_x = const.U_0 \frac{y}{\delta}$$

from which we obtain

$$\int_0^\delta v_x{}^2 \mathrm{d}y = const.U_0{}^2\delta = const.\frac{\eta U_0{}^3}{\tau}$$

here we have used

$$\tau = const.\eta\frac{U_0}{\delta}$$

Note that the numerical factor "*const.*" in the equations above has always a different value, and we absorbed the factor 1/3 after integration in *const.* We arrive at the following differential equation for τ:

$$\frac{\mathrm{d}\tau}{\mathrm{d}x} = const.\tau^3$$

The solution is (Landau 1974)

$$\tau = 0.332\sqrt{\frac{\eta\rho U_0^{\ 3}}{x}} \tag{8.18}$$

which is easily verified by differentiating τ with respect to x. The factor 0.332 is taken from the numerical solution shown in Fig. 8.3.

The friction force can be measured directly by flush-mounting a flexible element in the wall and determine its deflection. Pan (1995) has proposed micro sensors designed in order to exploit eq. (8.18) in this way.

The phenomenon of the boundary layer plays an important role if in fluid ducts there are changes in the geometry: sharp bends in ducts, changes of the diameter of the duct etc. The boundary layer extends into a duct until it reaches a thickness comparable to the radius of the duct. Within this region, the flow is boundary layer like, and further downstream there is the fully developed flow (for small enough Re the well known parabolic profile of Hagen-Poiseuille, eq. (8.5)). The order of magnitude of the entrance length l for a tube with radius a is

$$l \approx \frac{a^2 U}{\zeta} = a\,\mathrm{Re}\,. \tag{8.19}$$

Any characterisation of flow sensors must take care of this phenomenon: the flow set-up must be carefully designed in order to get reproducible results, which can be interpreted reliably.

Let us emphasise that the approximations made to derive the equations for the boundary layer are based on the assumption that Re >> 1. Re in this case is dependent on x, the distance from the leading edge. This means that the boundary layer gets its characteristic only far enough from the leading edge of that plate. Close to the leading edge v_x and v_y are of similar magnitude, and also $\partial p/\partial y$ cannot be neglected.

If the flow goes along a very long wall the flow becomes unstable and turbulent. This happens after a critical distance x_{cr} from the leading edge, which is given by (Eckert 1996)

$$R_{cr} = \frac{V x_{cr}}{\zeta} \approx 5\times10^5\,. \tag{8.20}$$

Schematically, the flow along a wall looks like in the one drawn in Fig. 8.4

Most flow microsensors operate below Re_{cr} so the turbulent boundary layer is not of much interest for most flow sensors. However for those interested in shear wall stress in turbulent flow, which is a field very interesting for the application of miniature flow sensors because of the small size and short response time, turbulence is of course the key. Here we will not dwell into turbulence, and we refer the interested reader to the literature (Stanton 1920, Winter 1977, Mendez 1985, Löfdahl 1992, Kälvesten 1996a, b, Padmanabhan 1997, Löfdahl 1999).

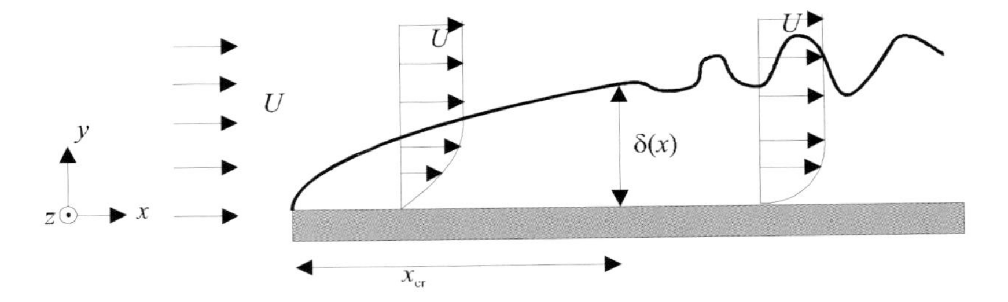

Fig. 8.4. Laminar and turbulent boundary layer

8.1.4. Thermal Boundary Layer

In the case that at a wall there is a heat source we get a temperature distribution in the fluid. The temperature is described according to (8.8) if the fluid is at rest, and to (8.7) if the fluid is moving at a speed small enough so that the heat produced by internal friction can be neglected. The velocity field is the one determined by (8.15). Close to wall, at large enough Re we again have the hydrodynamic boundary layer. At $\mathbf{v} = 0$ in steady state, the heat distribution is given by the Laplace equation, the solutions of which are completely analogous to electrostatic problems. If there is a heat source of strength s (the product $s\rho c_v$ is the heat generated per time and unit volume, with c_v the specific heat) the steady state equation is (ignoring a possible inhomogeneous medium with $D_T = D_T(\mathbf{r})$)

$$\nabla^2 T = -\frac{s}{D_T} \tag{8.21}$$

and the solution is[1]

$$T(\mathbf{r}) = \frac{1}{4\pi D_T} \int d\mathbf{r}' \frac{s(\mathbf{r}')}{|\mathbf{r} - \mathbf{r}'|} \tag{8.22}$$

The temperature distribution can be evaluated using this integral.

If the heat source is located in a wall along which there is a hydrodynamic boundary layer, a thermal boundary layer will build up if D_T is not too large. The idea is shown in Fig. 8.5. The relevant dimensionless quantity is the Prandtl number, defined as

$$\mathrm{Pr} = \frac{\mathrm{Pe}}{\mathrm{Re}} = \frac{\eta c_p}{\lambda} = \frac{\zeta}{D_T} \tag{8.23}$$

---

[1] Note that this solution is only valid if there are no additional boundaries which may disturb the temperature distribution

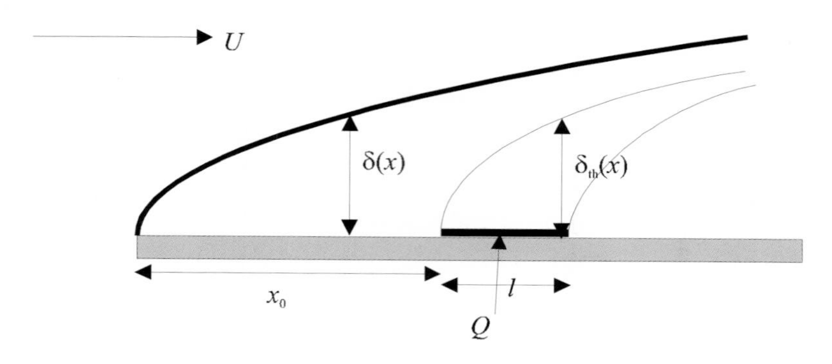

Fig. 8.5. Hydrodynamic and thermal boundary layer

where Pe is the Péclet number. Here c_p is the specific heat capacity of the fluid and λ is the heat conduction coefficient, related to D_T by

$$D_T = \frac{\lambda}{\rho c_p} \tag{8.24}$$

Pe is the ratio of the transport of heat by convection to that by diffusion

$$\text{Pe} = \frac{U_0 l}{D_T} \tag{8.25}$$

In the boundary layer, the continuity equation (8.14) and the Navier-Stokes equation in the approximation appropriate for the boundary layer (8.15) govern the flow. So the thickness of the boundary layer, the velocity and pressure distribution as well as the shear force are given by (8.17), Fig. 8.3, (8.18) and (8.12), respectively. In (8.7), we can leave out the z-dependence, in the Laplace operator, the derivative with respect to x is much smaller than the one with respect to y, and accordingly we neglect the former. The equation governing the heat transport in the boundary layer is then:

$$v_x \frac{\partial T}{\partial x} + v_y \frac{\partial T}{\partial y} - D_T \frac{\partial^2 T}{\partial y^2} = 0 \tag{8.26}$$

This equation has exactly the same form as (8.15) and will have similar solutions, if Pr is approximately 1 or larger. If Pr[2] is larger than 1 the thermal boundary layer will be submerged within the hydrodynamic one.

The heat flux density from the hot element is given by

---

[2] Examples for liquids: glycerol: 7250, water: 6.75, Mercury: 0,044. For gases, Pr~1 (air: Pr=0.733). All data from Landau (1974).

$$Q = -\lambda \frac{\partial T}{\partial z} \approx -\lambda \frac{T_0 - T_s}{\delta} = -\lambda \Delta T \sqrt{\frac{U_0}{x_0 \zeta}} \propto \sqrt{\text{Re}} \tag{8.27}$$

This equation is the basis of hot wire anemometers (King 1914). The heat transfer coefficient is proportional to the square root of the fluid velocity.

For sufficiently large Re and Pr, the dependency of Q on Pr can be estimated from (8.26). The thickness of the thermal boundary layer is denoted by δ_{th} and $\delta_{\text{th}} < \delta$. The temperature change is confined within a distance of δ_{th} counted from the wall. At $y = \delta_{\text{th}}$ the change of the velocity is U_0 ($\delta_{\text{th}} / \delta$). At $y \sim \delta_{\text{th}}$ the terms in (8.26) are of the order:

$$D_T \frac{\partial^2 T}{\partial y^2} \approx D_T \frac{\Delta T}{\delta_{\text{th}}{}^2}$$

and

$$v_x \frac{\partial T}{\partial x} \approx U_0 \frac{\delta_{\text{th}}}{\delta} \frac{\Delta T}{l}$$

The distance l must be counted downstream from the heating element onwards where the wall temperature is assumed to change abruptly. In the following we assume that the heating element is at the leading edge of the wall. In this case we have $l{=}x_0$. Neglecting the term proportional to v_y, the comparison of the two expressions gives

$$\delta_{\text{th}}{}^3 \approx D_T \frac{\delta l}{U_0}$$

or with $\delta = l/\sqrt{\text{Re}}$

$$\delta_{\text{th}} \approx \frac{l}{\text{Re}^{1/2}\,\text{Pr}^{1/3}} = \frac{\delta}{\text{Pr}^{1/3}} \tag{8.28}$$

Using this expression in (8.27) for the temperature gradient we get[3]

$$Q \approx \lambda \frac{\Delta T}{\delta_{\text{th}}} \propto \lambda\, \text{Pr}^{1/3}\, \text{Re}^{1/2}\, \Delta T \tag{8.29}$$

The heat transfer is often characterised by the Nusselt number Nu. The Nusselt number relates to the heat transport of a fluid to a solid wall. The heat flux density Q to a wall at a temperature T_0 from a fluid with temperature T is characterised by

$$Q = \alpha (T - T_0) = -\lambda \frac{\partial T}{\partial n} \tag{8.30}$$

---

[3] If the heat source lies a distance l_0 away from the leading edge, we find $\delta_{\text{th}} = \text{Pr}^{-1/3}\,\text{Re}^{-1/2}[(l_0 + l)^2 l]^{1/3}$.

(the derivative is taken in the direction normal to the surface). The Nusselt number is defined as

$$\mathrm{Nu} = \frac{\alpha l}{\lambda} \tag{8.31}$$

Comparison with (8.30) shows that

$$\mathrm{Nu} = const.\,\mathrm{Pr}^{1/3}\,\mathrm{Re}^{1/2} \tag{8.32}$$

8.1.5. Skin Friction and Heat Transfer

The skin friction has been given by equation (8.9), and in principle can be measured directly by a suitable micro sensor construction, which comprises a movable mass attached to springs. However, an alternative is to measure the heat transfer to the fluid from a hot element. The relation of the heat transfer to the skin friction will be treated here.

The theoretical problem has been treated at length recently by Mendez and Ramaprian (1985) and earlier by Bellhouse and Schulz (1966), with emphasis on turbulent (hence non-steady) flow. We follow the former treatment, however we restrict ourselves to steady laminar flow. The calculation is analogous to the one we used to derive the skin friction as a function of the streaming velocity far away from the boundary layer. The starting point is the assumption that there is a hydrodynamic boundary layer at a wall, and beginning at a heating element a thermal boundary has built up, which is submerged within the hydrodynamic boundary layer. Thus, the basic assumptions are; Re >> 1, Pr >> 1, and a sudden jump in the temperature at the heating element (we ignore thermal conductivity in the wall).

The starting equations are (8.14) and (8.26). Using the continuity equation (8.14), (8.26) can be written

$$\frac{\partial}{\partial x}(v_x T) + \frac{\partial}{\partial y}(v_y T) - D_T \frac{\partial^2 T}{\partial y^2} = 0 \tag{8.33}$$

We integrate (8.33) over y from the wall to the "edge" of the thermal boundary layer δ_{th}. This is the only region where the temperature changes.

$$\begin{aligned} &\frac{\partial}{\partial x}\int_0^{\delta_{\mathrm{th}}} (v_x(x,y)T(x,y))\mathrm{d}y + v_y(x, y=\delta_{th})T(x, y=\delta_{th}) \\ &\qquad - D_T\left(\left.\frac{\partial T}{\partial y}\right|_{\delta_{\mathrm{th}}} - \left.\frac{\partial T}{\partial y}\right|_0\right) = 0 \end{aligned} \tag{8.34}$$

The derivative $\partial T/\partial y$ at δ_{th} vanishes, and at $y = 0$, it is equal to the heat flux density transferred to the fluid Q divided by λ (see ((8.30)).

Next we integrate (8.14) over y. The integration gives

$$\frac{\partial}{\partial x}\int_0^{\delta_{\text{th}}} v_x \mathrm{d}y = -v_y^{\text{th}} \tag{8.35}$$

where v_y^{th} is the y-component of the velocity at the edge of the thermal boundary layer. Combining (8.34) and (8.35) we find

$$\frac{\partial}{\partial x}\int_0^{\delta_{\text{th}}} v_x(x,y)\Delta T(x,y)\mathrm{d}y = -\frac{1}{\rho c_{\text{p}}}Q \tag{8.36}$$

ΔT is the difference of temperature at the point (x,y) and far away from the wall (located at $(x,0)$). For the velocity $v_x(y)$ we assume that it depends linearly on y close to the wall. This is permissible if our assumptions are valid (the thermal boundary layer submerged in the hydrodynamic boundary layer). $v_x(y)$ is shown in Fig. 8.3).

$$v_x(y) \approx y\left.\frac{\partial v_x}{\partial y}\right|_{y=0} = y\frac{\tau}{\eta} \tag{8.37}$$

Using (8.36) and (8.37) we get

$$\frac{\partial}{\partial x}\frac{\tau}{\eta}\int_0^{\delta_{\text{th}}} y\Delta T\mathrm{d}y = -\frac{1}{\rho c_{\text{p}}}Q \tag{8.38}$$

The next assumption is that the temperature distribution in the thermal boundary layer depends on x and y in a similar way as the velocity $v_x(x)$. That means that

$$\Delta T(x,y) = \Delta T(\xi) \text{ with } \xi = \frac{y}{\delta_{\text{th}}} = y\frac{1}{\text{Pr}^{1/3}}\sqrt{\frac{\zeta x}{U_0}} \tag{8.39}$$

We now proceed in a way similar to that when we derived eq. (8.18). We assume that

$$\Delta T = const.\Delta T_0\frac{y}{\delta_{\text{th}}} \tag{8.40}$$

and we note that

$$\left.\frac{\partial T}{\partial y}\right|_{y=0} = -\frac{Q}{\lambda} \propto \frac{\Delta T_0}{\delta_{\text{th}}}$$

and we find

$$\int_0^{\delta_{\text{th}}} y\mathrm{d}y\Delta T(x,y) = \alpha\frac{\lambda^2 \Delta T_0{}^3}{Q^2} \tag{8.41}$$

with α a numerical constant. It has been shown by Curle (1962) by numerical integration that $\alpha = 0.22$. Menendez obtained $\alpha = 0.23$. Using this result and integrating (8.39) over the heating element (length L) we obtain

$$\tau = -\frac{\eta L}{\alpha\rho c_{\mathrm{p}}\lambda^2}\left(\frac{Q}{\Delta T_0}\right)^3 \tag{8.42}$$

Note that by combining (8.42) with (8.18), the heat transfer in the hydrodynamic boundary is proportional to $\sqrt{U_0}$. This is completely different if the heating element is flush mounted in the wall of a flow channel and the flow is settled according to a Hagen-Poiseuille distribution. Then, for a cylindrical tube, $\tau = \eta(v/\partial y)_{y=0} = \eta V_{\max}/4R$. Therefore in this case, the "heat conductance" $G = Q/\Delta T$ is proportional to $(V_{\max}/R)^{1/3}$ (R = radius of the tube), and dependent of the geometry of the tube. The dependency of the sensor signal on the duct geometry has been found experimentally by Mayer (1996).

8.2 Heat Transport in the Limit of very small Reynolds Numbers

In the last paragraph we treated the flow pattern and the heat transfer from a hot element in a wall for large Reynolds numbers. At small Re our assumptions are not valid and there will be no boundary layer.

Here we investigate the effects if the Reynolds number is very small. It is easy to see that a small Reynolds number means that the thickness of the boundary layer is much larger than the streamwise extension of the flow sensor. If we have

$$\delta = \sqrt{\frac{\varsigma l}{U_0}} >> l$$

we see immediately that this is equivalent to

$$\sqrt{\frac{U_0 l}{\varsigma}} = \mathrm{Re} << 1 .$$

In order to see what the dominant effects are at Re→0, we assume that the flow disturbs the temperature distribution around a heater very little. The temperature distribution is given by

$$v_x \frac{\partial T}{\partial x} - D_T \nabla^2 T = 0 \tag{8.43}$$

We assume a flow in the x-direction. If the flow gives rise for a very small change of the temperature distribution we may write

$$T(\mathbf{r}) = T_0(\mathbf{r}) + T_1(\mathbf{r}) \tag{8.44}$$

with $T_1 << T$. T is the solution of (8.43) with $v_x = 0$. Inserting (8.44) in (8.43), and noting that $\nabla^2 T_0 = 0$ we get

$$v_x \frac{\partial T_0}{\partial x} - D_T \nabla^2 T_1 = 0 \tag{8.45}$$

where we neglected a term of the order of $v_x\ T_1$.

If T is known, which in principle is a question of integrating (8.22), the first term in (8.45) is a known function of **r** and we can solve (8.45) in principle by a straightforward (numerical) integration over $v_x \partial T_0 / \partial x$:

$$T_1(\mathbf{r}) = -\frac{1}{4\pi D_T} \int \mathrm{d}\mathbf{r}' \frac{v_x(\mathbf{r}') T_0'(\mathbf{r}')}{|\mathbf{r} - \mathbf{r}'|} \tag{8.46}$$

with $T_0' = \partial T_0 / \partial x$. d**r** denotes the volume element.

Similar to (8.22) this equation is only valid under the condition that no walls disturb the temperature distribution. Independent of additional boundaries, the correction to the temperature distribution will be *linear* in v_x. This means that if a temperature sensor is placed closely to a heater, the sensor will measure a temperature raise proportional to v_x.

The first term in (8.45) – which is the same as the numerator in (8.46) – plays the role of an effective heat source just like s in (8.21) and (8.22). This term has a simple physical meaning:

- The temperature gradient is proportional to the heat flux, so the temperature field is analogous to the electrical potential of a charge distribution, which looks like the dot product of the velocity and the heat flux. The dot product of the velocity with the heat flux looks very much like an electrostatic dipole. The heat flux from a heating element is a function, which is directed outwards. The product with the x- component with the velocity will lead to a term, which contains the cosine of the angle between the heat flux and the velocity. So downstream with respect to the heat source we get a negative heat source and downstream a positive one, of equal size for a symmetric arrangement. If the heat flux is radial (outward from a point like heat source) we get a positive and a negative term centred around the heat source on the axis parallel to the flow, equivalent to two charge distributions of opposite sign. If the heat source has cylinder symmetry, we get a term under the integral, which is equivalent to two wires of opposite charge, located on the axis formed by the velocity through the heat source. So, the equivalent heat source $v_x\ \partial T_0 / \partial x = \mathbf{v} \cdot \mathrm{grad} T_0$ is in fact one heat source and one heat sink for the temperature distribution T .
- The reduction of the temperature distribution to a problem of the Poisson equation allows an important conclusion: In regions without an equivalent heat source there is no maximum of the temperature.

Far away from the heat source, we may develop the numerator under the integral in (8.46) in a Taylor expansion:

$$\frac{1}{|\mathbf{r}'-\mathbf{r}|}=\frac{1}{\sqrt{(\mathbf{r}^2-2\mathbf{r}\mathbf{r}'-\mathbf{r}'^2)}}\approx\frac{1}{r}\left(1+\frac{\mathbf{r}\mathbf{r}'}{r^2}\right)+\text{higher order terms} \tag{8.47}$$

Combining (8.47) and (8.46) leads to

$$T_1(\mathbf{r})=-\frac{1}{4\pi D_T}\frac{1}{r^3}\mathbf{r}\cdot\int d\mathbf{r}'\mathbf{r}'v_x(\mathbf{r}')T_0'(\mathbf{r}') \tag{8.48}$$

where we assumed that the temperature distribution is symmetric, so that

$$\int d\mathbf{r}'v_x(\mathbf{r}')T_0'(\mathbf{r}')=0$$

The integral in (8.48) is equivalent to the electric dipole moment.

If the temperature decreases abruptly at the edges of the heater the temperature gradient is a δ-function at the edges. With a uniform velocity within a cylinder of radius R (what we would have in a tube close to the entrance), we find then

$$T_1(x)\approx\frac{U_0T_0}{2D_T}b\frac{x}{R}\left(\frac{R}{x}\right)^3 \tag{8.49}$$

where the former approximation holds for $x >> b$ and $x >> R$. Note that the expression $\frac{U_0}{D_T}\frac{b}{R}$ has the dimension of a reciprocal length. This length defines the amplitude to the temperature elevation by the flow around a heater. T_1 is an antisymmetric function around the heater. T_1 is negative upstream and positive downstream.

To illustrate in more detail what this term does we make the following more realistic assumption. We have a heater located in a tube, which gives rise to a local temperature elevation at the heater, with a linear drop of the temperature at the edges, as shown in Fig. 8.6(a). In Fig. 8.6(b) the x-component of the temperature gradient is given, which is constant because of the linear decrease of the temperature. The temperature T_1 is equivalent to the electrostatic potential of a charge distribution like the one shown in Fig. 8.6(b). The resulting functions (equivalent to the potentials of the individual sources) are shown in Fig. 8.6(c) as dashed lines, and the full line is the final temperature correction due to the flow. We see that maximum temperature changes occur just outside the regions with a temperature gradient. A faster decay of the temperature would shift this maximum closer to the heater.

In agreement with (8.48, 8.49), the signal is proportional to the flow rate at small velocities. In sensors which operate in the range we discuss here (Johnson 1987, Jouwsma 88, Komiya 1988, Branebjerg 1991, Lammerink 1993, Boer 95, Mayer 1996, Ashauer 1998), the temperature is measured a short distance upstream and downstream with respect to the heater, and the temperature difference ΔT is taken as a signal, see Fig. 8.7. From our discussion we see that the optimum position of the temperature sensors is close to the outer edge of the largest x-component of temperature gradients.

When increasing the velocity, the temperature difference increase becomes slower, and decreases finally. The turn over occurs when the thickness of the thermal boundary layer is comparable to the distance between the temperature sensors and the heater. This will be discussed in more detail below (section 8.3.2.). At large Re, the heat transport from the fluid to the temperature sensor is dominated by the boundary layer, and Kings law (King 1914) applies. One must note in this connection that a hot element has two jumps in the temperature (see Fig. 8.5): one down stream and one upstream. The temperature sensor downstream "sees" a fluid at the wall temperature with a distance δ_{th} away a hot layer. The temperature gradient is mainly directed towards this warm layer, and the heat transport is proportional to $1/\delta_{th}$. In this case the upstream sensor only has the function of measuring the temperature of the fluid, the heater is meant to be invisible to this sensor. Examples are papers by van Putten (1974), van Riet (1976), Huising (1980), Rehn (1980), van Putten (1983), van Oudhuisden (1990a, b), Stephan (1991), van der Wiel (1993).

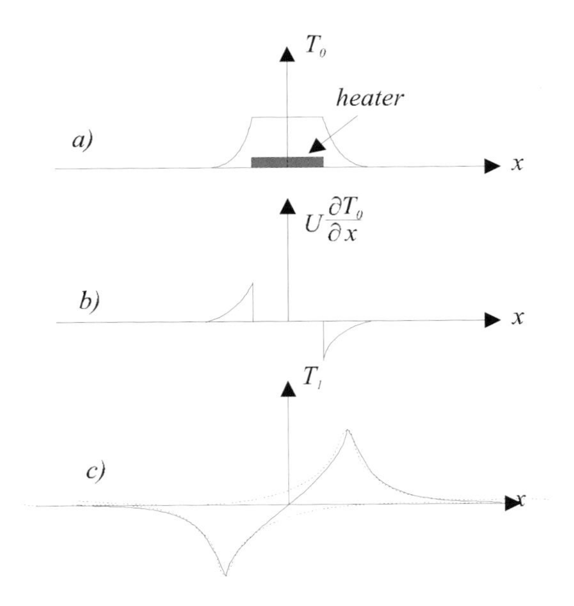

Fig. 8.6. Schematic temperature distribution for a tube with a heater. a) is the assumed temperature distribution at U=0, b) shows the resulting temperature gradient in the x-direction, c) shows the resulting corrections to the temperature distribution according to equation (8.46); here the dashed lines are the temperature distributions resulting from the individual temperature gradients upstream and downstream, the solid line is the sum

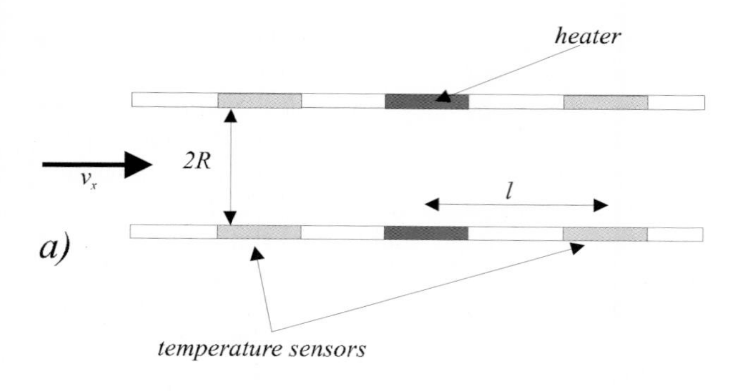

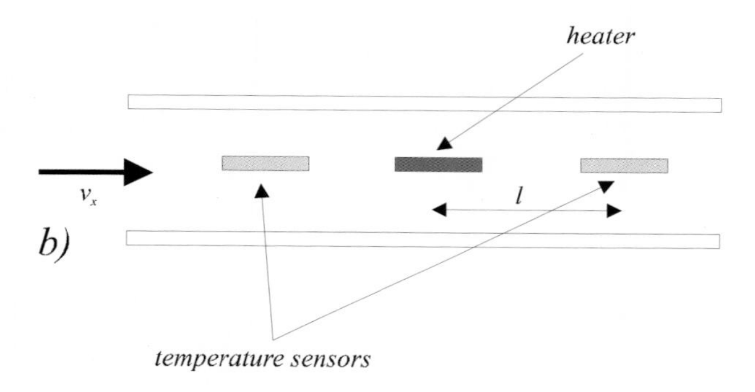

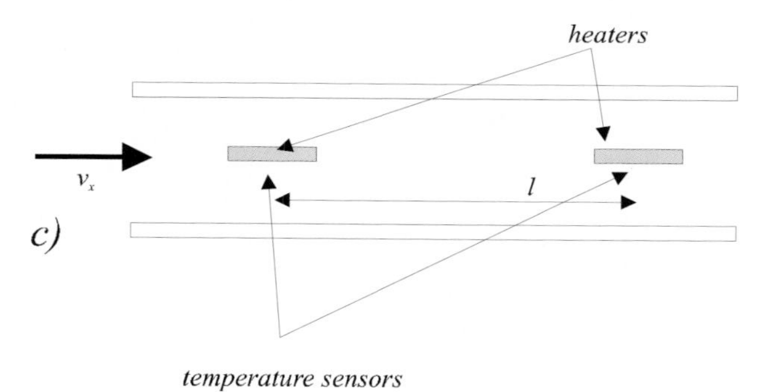

Fig. 8.7. Three configurations of thermal flow sensors. a) Heater and temperature sensors machined in the wall of the tube. b) Heater and temperature sensors mounted in the centre of the tube. c) Like b, but heater and temperature sensors integrated on single elements, and reduction to only two elements

Sensors of the types shown in Fig. 8.7(b) and (c) may have an offset velocity (Kohl 1994, Wachutka 1991). These sensors do not react on flow if the velocity of the fluid is too small. The reason is that the heat transport by flow competes with the heat diffusion from the hot element to the close wall. Heat will be carried away from the heater only if there is a thermal boundary layer over the heater. However, if the distance to the wall is much less than the thermal boundary layer, heat diffusion to the wall dominates, and there is practically no heat transfer to the streaming fluid. That means that if $\delta_{th} >> R$ the sensitivity of the flow sensor drops to zero. This condition can be formulated as

$$U_{cr} = (b/R^2)\, \lambda^{2/3} \zeta^{1/3} \tag{8.50}$$

where U_{cr} is approximately the velocity where the sensitivity to flow starts. This agrees well with experimental data published by Kohl (1994).

8.3 Thermal Flow Sensors

We can classify thermal flow sensors in three basic categories:

- Anemometers
- Calorimetric flow sensors
- Time of flight sensors

Generally, anemometers comprise of a single element which is heated and the temperature of which is measured. Since the electric resistivity of most materials depends strongly on the temperature and the element is heated by an electrical current, the measurement of the resistivity of the element is a natural choice. Any flow will enhance the rate at which heat is transported away from the element. The anemometer can be operated in two modes: constant power and constant temperature. The simplest mode is the mode where the anemometer is fed with constant power, and the temperature of the hot element is measured. The response time in this mode is given by the heat capacity of the hot element and the rate at which heat is transferred to the medium (RC-time). When the temperature is held constant - this requires a feed back loop - the power needed to keep the temperature constant is measured. In this mode the anemometer is considerably faster than in the constant power mode.

The signal of the anemometer is proportional to the square root of the flow velocity as a consequence of the boundary layer as outlined in chapter 8.1.

Calorimetric flow sensors and time of flight sensors require two or more elements. A standard configuration consists of a heater surrounded by temperature sensitive elements arranged symmetrically downstream and upstream. Flow will carry away heat in the direction of the flow and accordingly will cool down the heater and change the temperature distribution around it. The temperature difference upstream and downstream is measured. The signal (proportional to ΔT) is proportional to the flow velocity close to $U = 0$, but saturates and even decreases at higher flow velocity. Calorimetric flow sensors operate at very low flow velocities, a case we have analysed in Sect. 8.2.

8.3.1 Anemometer Type Flow Sensors

In the following we describe and discuss briefly issues of design, fabrication, and performance of a few examples for micro anemometers.

The traditional anemometer technology is based on the hot wire. Here a wire, typically with a length in the mm range and a thickness below 10 μm, is suspended between two holders. In order to make the sensor fast, the wire must be thin to minimise the heat capacity, and to make it sensitive, the heat conduction to the support must be as small as possible.

We treated aspects of heat transfer at length in chapters 8.1 and 8.2. Here we restrict ourselves to a few remarks:

Response Time

The response time of a hot wire is given by (see. e.g. Lomas 1986)

$$f_c = \frac{P(U)R\alpha}{2\pi\rho cAl} \frac{R}{\Delta R} \tag{8.51}$$

with $P(U)$ the velocity dependent power fed to the hot wire. The other symbols mean: R: electrical resistance, α: temperature coefficient of the resistance (TCR), ρ: mass density, c: specific heat capacity, A: cross-senction of the wire, l: its lenth. Typical response times are in the range of ms. When driving the anemometer with constant temperature, the response time can be smaller by orders of magnitude (Fingerson and Freymuth 1996).

In micro technology, the wire is replaced by a thin film deposited on a silicon substrate. In fact, first designs were fabricated in this way (van Riet 1976, van Putten 1974, Huysing 1980, Yang 1992). This has a disadvantage, which is immediately apparent: the heat capacity is the defined by the chip, and the heat conduction to the support is large due to the high thermal conductivity of silicon. Three different roads were went to solve these problems:

- The micromachined element carrying the resistor is thermally isolated from the rest of the structure using a material with low heat conductivity. An example of this approach has been presented by e.g. Stemme (1986), Löfdahl (1992), see Fig. 8.8.
- The thin film resistor is placed on a membrane. This has the effect that heat diffusion is restricted to lateral diffusion in the thin membrane, and that the substrate supporting the membrane is more or less at the temperature of the fluid (the surrounding). Examples for this techniques (see Fig. 8.9) are provided by van der Wiel (1993), Jiang (1996), Ashauer (1998) and Kälvesten (1996).
- The thin film resistor is placed on top of micro bridges crossing a flow channel. This technique restricts the heat conduction to the substrate within a one-dimensional structure. Examples for this technique can be found in Urban (1990), Ebefors (1998), Kohl (1994), Lyons (1998), Stephan (1991), Lammerink (1993), van Kuik (1995).

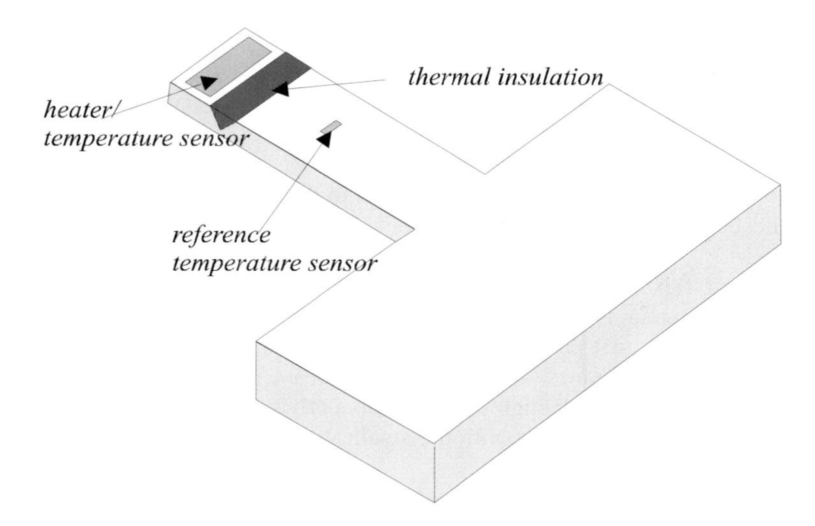

Fig. 8.8. In Stemme's micro-anemometer design (Stemme 1986) polyimide is used to isolate the hot part of the sensor from the support. Temperature sensors and heater elements are diodes

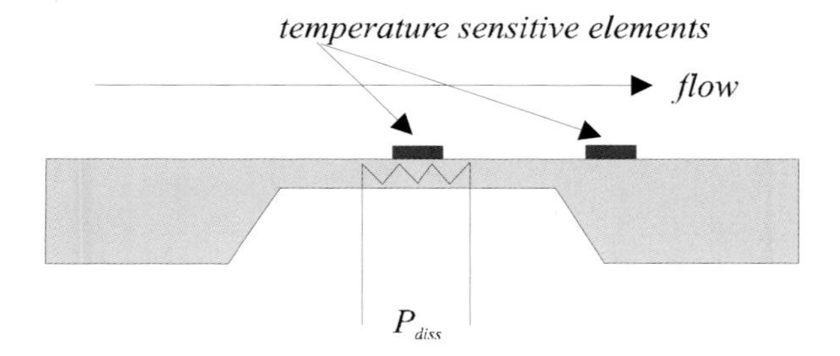

Fig. 8.9. Heater element on top of a membrane

The MEMS group at ETH Zürich used a design somewhere between a membrane and a bridge (Fig. 8.10): The structure of their flow sensors is a wide and thin cantilever beam (e.g. Funk (1993) Moser (1991), Mayer (1996), Wachutka (1991). Swart (1991) proposes a membrane suspended by four beams.

Micromachined flow sensors are not much faster than the classical hot wire anemometer. Ebefors reports cooling and heating times of 0.1 and 0.3 ms, respectively. The fastest sensors of de Bree et al. (de Bree 1995, 1996, 1997, 1998) have cut-off frequencies a little bit above 1 kHz. Also Lyons reports response times of not shorter than 2 ms.

Sensitivity
The anemometers loose their heat to the fluid by forced convection and by heat diffusion to the support. The total heat flux therefore can be expressed as

$$Q = (\alpha + \beta\sqrt{\mathrm{Re}})\Delta T \tag{8.52}$$

α represents the conductive heat flux to support and to the fluid, and $\beta\sqrt{\mathrm{Re}}$ to the streaming fluid due to convection. Anemometers must be designed so that α is minimised.

Conductive heat flux to the support is minimised by minimising the cross section of the thin film and its mechanical carrier and by choosing a material with low heat conduction. The first point shows that beams as support are expected to perform superior over membranes, and that a narrow beam is better than a wide beam. Silicon is a material with high thermal conductivity (100W/mK), while silicon nitride is one of the materials with the smallest thermal conductivity (see e.g. Sanchez et al. (1996)).

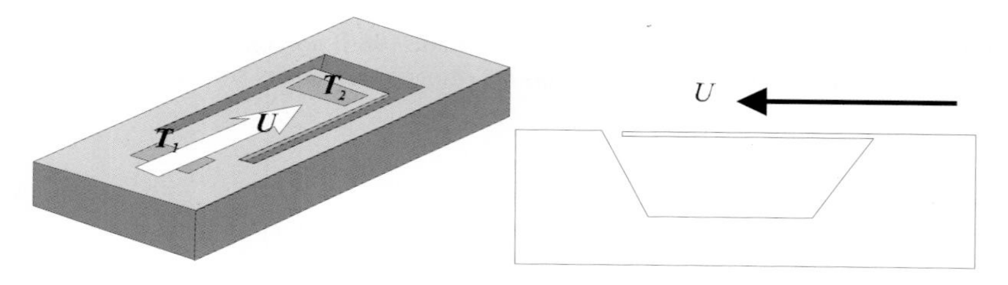

Fig. 8.10. Flow sensor with thermal elements placed on a wide cantilever beam. Left: perspective view, right: cross section (Moser 1991)

Fig.11. Microbridge type anemometer with Ge thermistors (Kohl et al. 1994)

Conductive heat flux via the fluid to the walls seems to be important if the sensor element is close to walls. As an example consider Fig. 8.12. Fingerson and Freymuth (1996) mention that for hot wire anemometers, the distance to the wall must be 500 times larger than the diameter of the wire in order to eliminate the effect of heat conduction to the wall completely. Van der Wiel gives an estimate for the heat conduction from the sensor to the duct, by simplifying the geometry of conductive heat flow, as indicated in Fig. 8.12. Van der Wiel's result (van der Wiel 1993) is (we correct by a factor of 2 to account for the heat flux through the rear side of the membrane)

$$Q = 4\lambda\sqrt{A}(1+\sqrt{A}/H)\Delta T \tag{8.53}$$

with $2H$ the cross-section of the tube (the chip is located in its centre) and A is the area of the heater. For the heat loss to the substrate a similar geometry is assumed, which leads to

$$Q = 8\lambda_{\text{substrate}} d/\ln(L/b)\Delta T\,. \tag{8.54}$$

The meaning of the symbols is given in Fig. 8.12. Note that the hydrodynamic equation are not linear, therefore solutions for the case that $U = 0$ cannot be superimposed to solution with $U \neq 0$. In particular, if the convection is large enough no conductive heat flux will reach the wall opposite to the sensor element.

In van der Wiel's paper (1993) no figures for the size of the chip is given. The chip is placed on a wing-shaped ceramic in the middle of a tube of 20 mm diameter, to detect a flow velocity from 0.005 to 2.65 m/s (Re_{ch} = 100–53000[4]). That means that there may be always vortices and turbulence in the largest range of the measurement results, and they see these phenomena as a very high noise level. Unfortunately, no details of the noise level are given in the paper. Tests of the device show good agreement of the model with experimental results for $Re<1000$ and large deviations at $Re>10{,}000$. The authors also observed that a tilt of the sensor with respect to the flow influences the performance of the sensor. Their interpretation is that a small tilt angle leads to a stabilisation of the boundary layer.

Kohl (1994) notes that the sensitivity to flow becomes zero if the flow is too small. In Kohl's set up, there is a bridge (similar to the one in Fig. 8.11) crossing a channel. The distance from the sensor to the channel wall is in the range of 80 - 150 μm, with a heater width of 50 μm it is clear that the conductive heat transport is very important. Contrary to the claims of Fingerson and Freymuth (1996), the sensitivity to the flow drops to zero if the flow velocity becomes smaller than a critical value. Kohl's experimental set up was such that entrance effects may play an important role. The length of the duct used by them was only 9 mm, while the entrance length ~2mm×Re_{ch}, with Re_{ch} up to 10^4; 2mm is the wetted perimeter of the channel. This might be the reason why they find distinct deviations form

[4] Re_{ch} is the Reynolds number based on the dimensions of the channel, not of the sensor.

King's law. The threshold velocity was discussed in chapter 8.2. and attributed to a competition of heat conduction and convective heat transport; the latter looses if there is no boundary layer.

Wachutka et al. (1991) describe two types of flow sensors: An anemometer, one of which is shown schematically in Fig. 8.10, the second one operates with a bridge suspended over an anisotropically etched recess, carrying a heater and two temperature sensors up- and down stream, respectively. The other sensors have a similar geometry, the basic difference is that the second one is a bridge in place of a cantilever beam. The device shown in Fig. 8.10 has a thermopile (polysilicon/aluminium) on top with the hot end at the end of the bridge and the cold one on the substrate. The bridges have been made from a sandwich of field oxide, CVD oxide and silicon nitride with a total thickness of 2 μm. The length and width are in the range of a few 100 μm, with an aspect ratio in the order of 2 - 3. The fabrication technology is described in detail in Moser (1991).

The sensors suffer substantially from heat conduction to the substrate and do not compare in performance to the designs of Stemme et al (1986), Fig. 8.8, who integrated a thermal insulation into a cantilever beam.

In studying thermal shear stress sensors, Jiang et al. (1996) compared three geometries to find the optimum one for the sensitivity. The designs are shown in Fig. 8.13 and 14.

The result is that the vacuum cavity increases the sensitivity by a factor of 4 compared to the bottom one without cavity (Fig. 8.13), while the two sensors with cavity show very similar results. In Jiangs work, the sensors are used for shear stress imaging in turbulent flow. For this purpose, the sensors are quite small (membranes in the size of 150 μm, width of the wire of 3 μm) and they were placed on rows of 25 sensors. This way a shear stress image can be obtained.

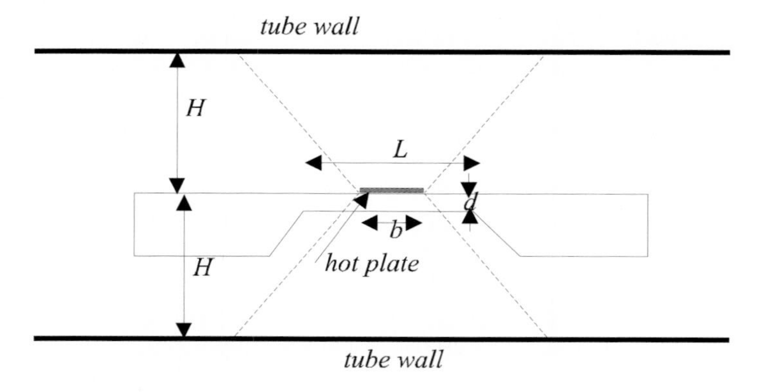

Fig. 8.12. Geometry of conductive heat transfer from a sensor chip to tube walls. The dashed lines indicate the cone through which heat flows according to the model

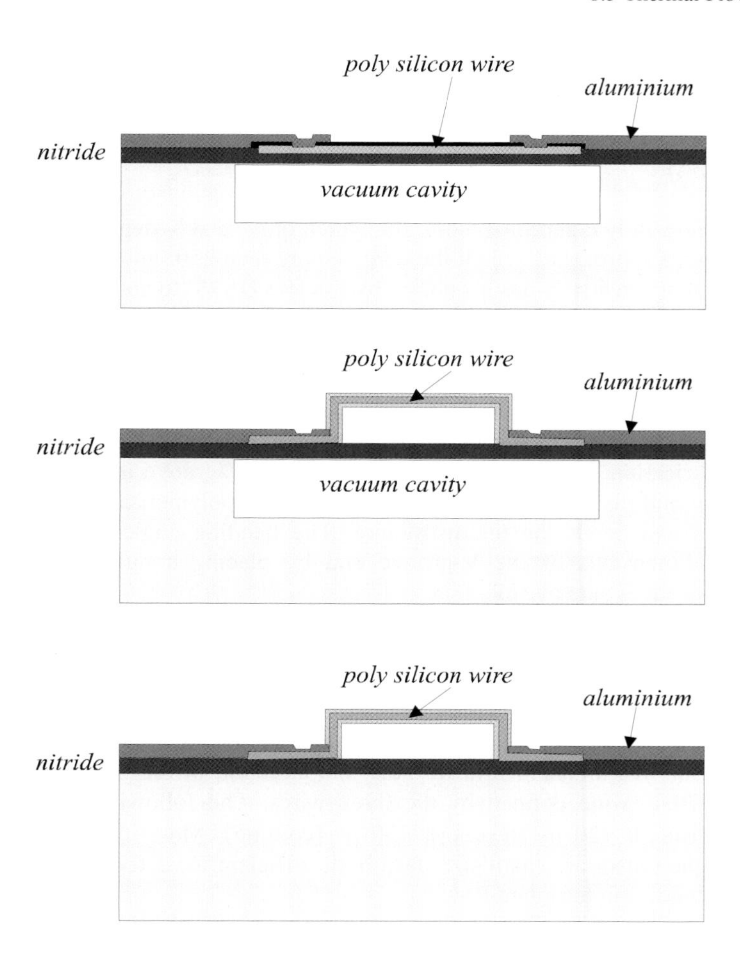

Fig. 8.13. Designs of three types of flow sensors: Top: A polysilicon wire machined on top of a silicon nitride membrane over a sealed cavity held at low pressure. The depth of the cavity is 2 μm. Middle: similar to top, but with the polysilicon wire elevated over the membrane. Bottom: similar to middle, but without a vacuum cavity

Angular sensitivity
The sensitivity of a hot wire anemometer depends on the orientation of the velocity vector with respect to the direction of the wire. For infinitely long wires, a wind parallel to the wire cannot carry away any heat, therefore in this case, we have

$$Q = (\alpha + \beta|\cos\theta|\sqrt{\mathrm{Re}})\Delta T \tag{8.55}$$

The sensitivity is proportional to $|\cos\theta|$, with θ the angle between the normal on the wire and the flow direction. There is no sensitivity at all with respect to the side from which the flow comes. In the limit of a wire which is as long as thick, the angular sensitivity of course must vanish. Therefore, hot wire anemometers with a finite length have a sensitivity more like (Fingerson 1996)

$$Q = (\alpha + \beta\sqrt{\cos^2\theta + f\sin^2\theta}\sqrt{\mathrm{Re}})\Delta T \tag{8.56}$$

with f a phenomenologically determined constant, which only weakly depends on the Reynolds number. f approaches zero if the wire is more than 250 times longer than thick. This angular sensitivity has been used by Ebefors (1998) to construct a microsensor for turbulence studies in which the vector flow velocity was the target. The smallest eddies in turbulent flow are in the range of 100 µm, with a lifetime of 100 µs (Löfdahl 1992). The idea was to fabricate three hot wires in *x*,y,*z*-direction, to get real three-dimensional information on flow. Having used polyimide for thermal isolation earlier (Stemme 1986), they had the availability of the following fabrication trick (Fig. 8.15): after etching a V-groove in silicon (KOH-etching) polyimide is deposited and patterned. Due to polymerisation the polyimide contracts and bends the microstructure. The bending angle can be controlled by the dimensions of the V-groove and by placing a number of polyimide-filled V-grooves side by side.

In Fig. 8.16 we show the design of the resulting three-axis hot wire micro-anemometer.

The sensor was inserted in a flow pipe 16x16mm^2 cross-section, length 30 mm, and checked with a reference flow sensor. Measurements were performed at constant power supply (5.77 mW). The wire temperature is then ca. 200 °C. The three axis flow sensor was mounted in a way that the direction of the flow was inside a cone of 70.4° wide defined by the three wires. The following time constants were measured: 120 µs (heating), 330 µs (cooling). Most important features of this unique sensor are small size, fast, both sufficient for detecting the smallest and fastest eddies in turbulent flow.

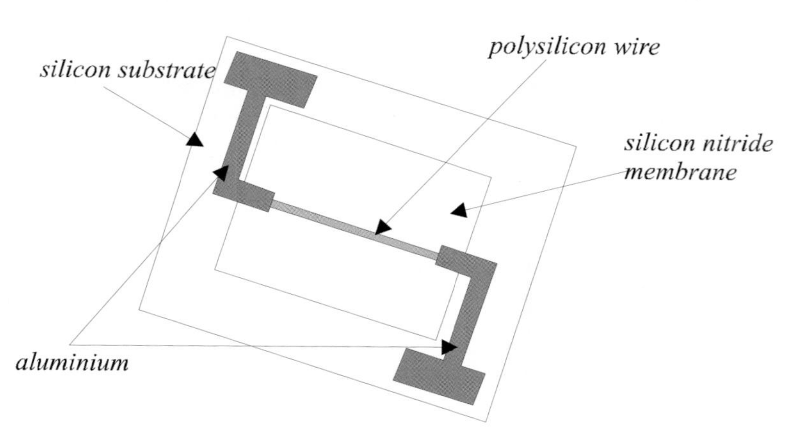

Fig. 8.14. Drawing of a flow sensor with a polysilicon wire carried by a silicon nitride membrane (perspective view of the sensor shown in Fig. 8.13, top)

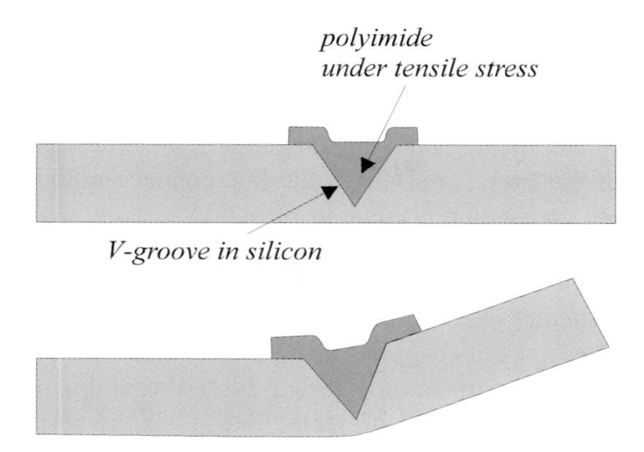

Fig. 8.15. Bending of a silicon member by polyimide (Ebefors 1998)

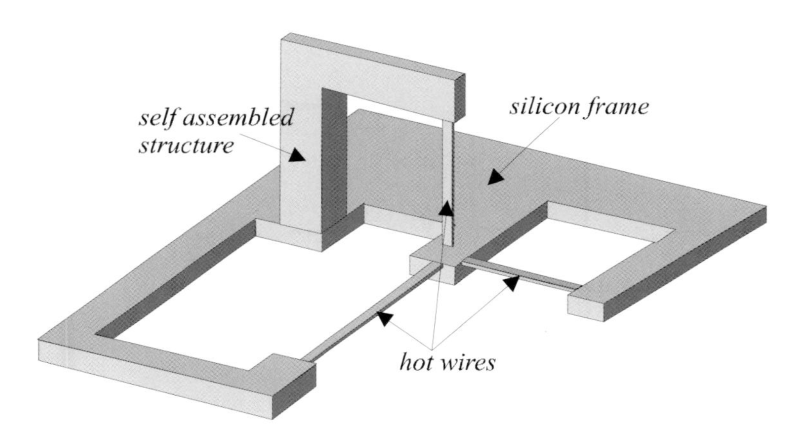

Fig. 8.16. Design of a three-dimensional hot wire anemometer. The dimensions of the hot wires are: 500 μm×5 μm×2 μm, the total chip size is 3.5 mm×3mm×0.5mm (Ebefors 1998)

8.3.2. Two-wire Anemometers

Early work on flow sensors with two temperature sensors on the chip can be found in Huising (1980), van Putten (1983), van Riet (1976), Rehn (1980), van Oudhuisden (1990 a and b) and Stephan (1991). The main purpose to use two elements is to get a differential signal to eliminate common disturbances, such as temperature drifts. It is clear that the ambient temperature changes the resistance of a resistor, so the change in ambient temperature gives a signal without flow.

A second purpose is the following. Two wires can be used to measure the temperature difference between two points in the streaming liquid, when one of the points or both are heated. In this case there is a sensitivity to the direction of the flow, contrary to the single element anemometer. The signal changes the sign if the flow direction reverses and if the flow is normal to the line connecting the temperature sensors the signal due to convection is equal to zero.

If it is assumed that the convective heat transport from the two points still is dominated by the thermal boundary layer - the boundary layers are thinner closer to the leading edge of the sensor structure, see fig. 8.17 and 8.18, than at the far end, the temperature difference between the two points still will depend on $\sqrt{U}$: For the case in Fig. 8.17, the upstream temperature sensor is at the ambient temperature, so the downstream sensor functions basically like a single wire anemometer. For the case shown in Fig. 8.17, we would have at the upstream side:

$$Q = (\alpha + \beta / \delta_1)(T_1 - T_0) \tag{8.57}$$

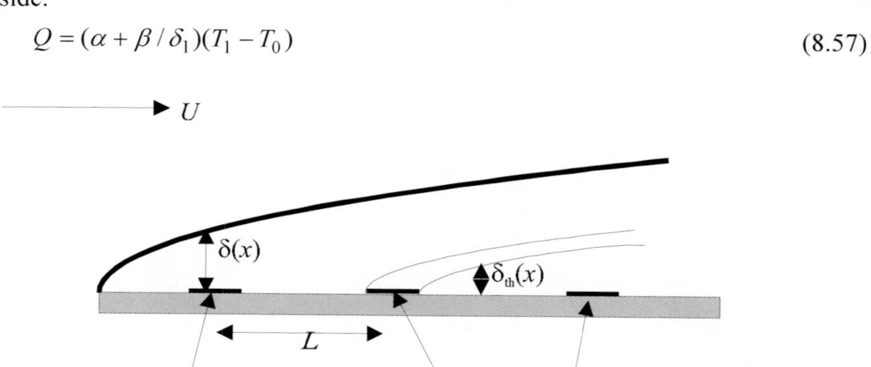

Fig. 8.17. Two-wire anemometer. In this version, the heaters and the temperature sensors are separate elements

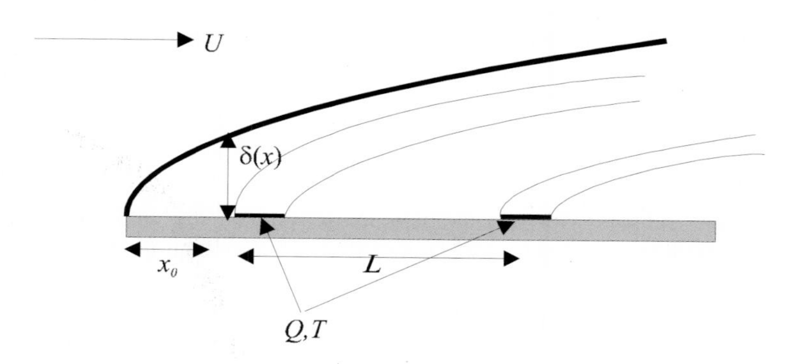

Fig. 8.18. A two-wire anemometer. In this version the heater and temperature sensors are physically the same structures

and down stream:

$$Q = (\alpha + \beta / \delta_2)(T_2 - T_0) \tag{8.58}$$

with T_0 the temperature of the liquid. If the heat dissipation is equal at both points, we have

$$\Delta T = (T_1 - T_2) = \frac{1}{\alpha' + \beta' \sqrt{U}} \tag{8.59}$$

So the signal will have the characteristic square root U dependence just as the single hot wire anemometer. This has been confirmed experimentally (van Riet 1976, van Putten 1974, van Putten 1983).

The boundary layer is dominant if its thickness is much smaller than the distance between the elements of the microstructure. If this is true, the heat loss is dominated by the heat diffusion over the boundary layer, and much less by heat diffusion directly between the elements. In the sensors described by van Riet (1976), the distance between the two elements is l = 0.2 cm, and approximately equal to the streamwise length of the chip. The maximum thickness of the boundary layer is $\delta_{\max} = \sqrt{\zeta l / U_{\min}}$ = 500 µm for ζ_{air} =0.15 cm^2/s (Landau 1974), $U_{\min}$ V_{min} ~ 10 cm/s, considerably smaller than l. Analysis of the geometry used by van Oudhuisden (1990 a, b) and van Putten (1974, 1983) gives similar results: within the ranges studied experimentally we have always $\delta << l$.

8.3.2 Calorimetric Type Flow Sensors

Calorimetric flow sensors are constructed similar to the arrangements of heaters and temperature sensors as shown in Figs. 7, 17 and 18. In most designs, calorimetric flow sensors comprise a heater with temperature sensors arranged symmetrically upstream and downstream with respect to the heater. The output signal is then the temperature difference at the sensors: the downstream sensor is heated, and the upstream sensor is cooled. Similar to the two element anemometers the direction of the flow can be determined by calorimetric flow sensors. A schematic of the temperature distribution is shown in Fig. 8.19. Commercial sensors have these elements mounted on the wall of the tube, most microsensors insert a chip carrying these elements into the flow or fabricate bridges across a micromachined flow channel. The former has the advantage of robustness and the disadvantage of slow speed and low sensitivity.

Basically, calorimetric flow sensors are mass flow sensors (see e.g. Hohenstatt 1990). The principal idea is to transfer some heat to a fluid, which is carried away by convection. So it is the amount of heat, which is intended to measure, which is proportional to the mass flow. Note that the relation of the transported heat is dependent on the type of fluid: the specific heat is the important parameter.

Calorimetric flow sensors have an output, which is characterised by a linear dependency on the flow rate at small flow, a maximum and a decline (Johnson

1987, Komiya 1988, Lammerink 1993, Lyons 1998[5], de Bree 1995, 1997, 1998, Ashauer 1998). A typical example is shown in Fig. 8.20. These data are obtained with Lammerink's flow sensors (Lammerink 1993), for water flow. Geometric details are given in the figure caption. The temperature is measured via the electrical resistance of the a chromium-gold thin film on top of a silicon nitride carrier. The chromium is necessary for the adhesion of gold to the silicon nitride, but has the disturbing property that it forms an alloy with the gold which has a much smaller temperature dependence of the resistivity than gold itself. The slow diffusion process gives additionally rise to a drift in the resistivity of the material. Sensors using this type of material must be "burned in" (de Bree 1997).

We found earlier than anemometers have a sensor characteristic, which contains the square root of the flow velocity. The physical reason is the boundary layer that builds up on walls and heated elements. In calorimetric flow sensors the temperature sensing elements are very close to each other, and the heat transport from the heater to the temperature sensors is not dominated by the boundary layer. The condition for the $\sqrt{U}$ -characteristic is that $\delta \ll l$, in agreement with the anemometers described in the literature. Calorimetric flow sensors are designed such that the distance between the temperature sensors and the heater is very short, ranging from 1 mm (Lammerink 1993) down to 10 μm (Ashauer 1998). Analysis shows that the linear regime is characterised by $\delta \gg l$ (Elwenspoek 1999). As an example, Ashauer et al. (1998) place heater and temperature sensors at a distance in a range of l = 10 - 150 μm, on a chip with a streamwise length of at least 1 mm (they only give the size of the membranes but not of the chip in the flow). They are using water, the plate length is 1 mm at least, and within a range of ±4 mm/s for the flow velocity, we find $\delta_{min} \sim 500\mu m \gg l$.

Thus, in the regime of linear dependency of the temperature difference on the flow the Reynolds number is much smaller than 1 and the theory in Sect. 8.2 is valid, with a temperature distribution around the heater given by (8.46) and in a rough approximation by (8.48).

The maximum of the curve is roughly given by the condition $\delta_{th} \sim l$ with δ_{th} the thermal boundary layer and l the distance between heater and temperature sensor. We have approximately

$$\delta_{th} \approx \frac{\delta}{Pr^{1/3}} \tag{8.60}$$

or with (8.17),

$$Re^{m} \approx Pr^{2/3} \tag{8.61}$$

[5] Unfortunately, Lyons et al. are very reluctant to give any details of their geometry. Judging from their SEM-images, the temperature sensors are at a distance of less than 100 mm, with chip size probably in the mm range

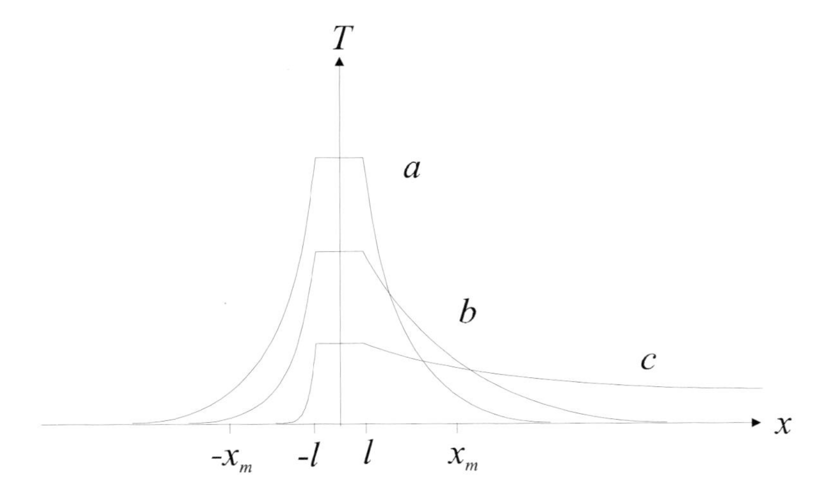

Fig. 8.19. Temperature distributions in the middle of the channel as a function of the position x. The heater extends from x=-l to x=l. x_m denote positions of temperature sensors. The curves characterise three typical flow velocities: a: $U = 0$. b: U is small enough to allow heat to diffuse to the upstream temperature sensor. c: U is too large to allow heat to diffuse to the upstream temperature sensor (Elwenspoek 1994)

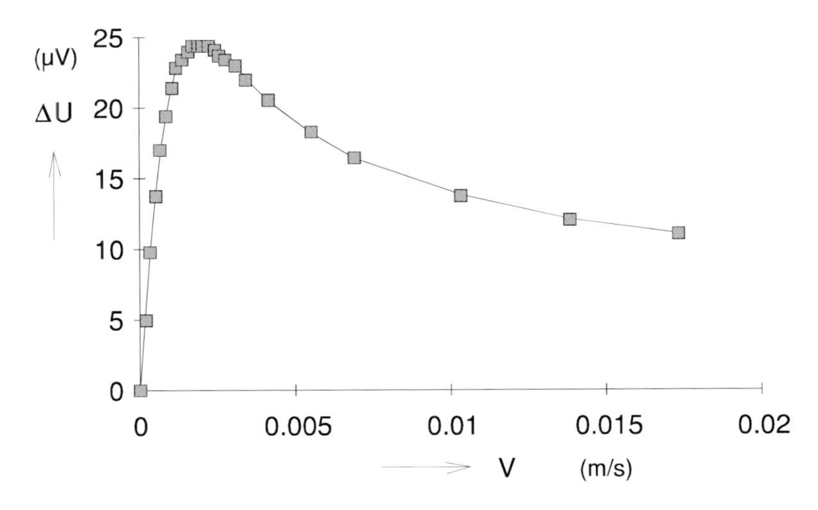

Fig. 8.20. Flow-sensor output with water as liquid. With a flow channel of 1000 μm x 500 μm, a flow velocity v = 20 mm/s corresponds to a volume flow of 600 μl/min. For water: κ = 0.6 W/mK, D = 1.4 10^{-7} m^2/s. The distance of the temperature sensors to the heater was 1mm (Lammerink 1993)

at the maximum. Interestingly a very similar equation can be derived (Elwenspoek 1994). If we assume that the maximum occurs in the case that the flow is too large that no heat can reach the upstream temperature sensor by diffusion, we may compare the diffusion time and the flow time:

$$\frac{l^2}{2D_T} \approx \frac{l}{U_{\max}}$$

which leads to

$$\frac{U_{\max} l}{2D_T} \approx 1 \tag{8.62}$$

Here $U_{\max}$ is the flow velocity, which corresponds to the maximum signal. (8.62) is essentially the same as (8.61) if Pr = 1 (except the factor 2, which we grant for the estimates of order of magnitude). Since normally for gases Pr ≈ 1, (8.62) and (8.61) can only be distinguished for fluids with Pr far from 1 (e.g. glycerol or mercury).

Equation (8.61) corresponds quite satisfactorily with the experimental results published so far. We see a somewhat better correspondence with (8.61) than with (8.62).

Lammerink (1993) has proposed a model to describe the temperature distribution around a heater in the presence of fluid flow. This model is essentially the same as the one given by Komiya (1988). The idea is to describe the heat transport from the hot element to the fluid and to the wall of the duct by lumped elements. The crucial assumption is that from the centre of the duct upstream and downstream heat flows to the walls as if there were a heater at temperature T. The resulting differential equation is of the form (Komiya 1988, Hohenstatt 1990, Lammerink 1993),

$$D_T \frac{\mathrm{d}^2 T}{\mathrm{d}x^2} + U_x \frac{\mathrm{d}T}{\mathrm{d}x} - \beta T + s = 0 \tag{8.63}$$

with s representing the strength of the heat source (see (8.21)) and β a parameter depending on the geometry of the duct. Lammerink gives a solution of (8.63), which is

$$\text{upstream: } T(x) = T_0 \exp\{\gamma_2 (x - L)\} \tag{8.64a}$$

$$\text{down stream: } T(x) = T_0 \exp\{\gamma_1 (x + L)\} \tag{8.64b}$$

with

$$\gamma_{1,2} = \frac{1}{2D_T}\{U \pm \sqrt{U^2 + 4D_T{}^2 / h^2}\} \tag{8.64c}$$

T_0 is the temperature rise of the wire, L = width of the heater and G=$2w/h$ (w and h represent width and half height of the channel, respectively). It is assumed that the heater and the temperature sensors are located in the centre of the duct, as in Fig. 8.7b. The temperature difference between positions symmetrically downstream and upstream of the heater indeed has a qualitative dependency as the experimental results shown in Fig. 8.20, but a quantitative comparison cannot be made. In this model, the maximum of the curve is given by the geometry of the duct and does not correspond to experimental results.

A new configuration of combined heater/sensor elements has been proposed recently by de Bree et al. (1998), shown in Fig. 8.21, with a SEM image in Fig. 8.22. This instrument can be used as both: a calorimetric flow sensor, where the temperature difference of the two elements is measured at small flow rates and as an anemometer at larger flow, where the sum of the temperature of both elements is measured. The calorimetric sensor is much more sensitive than the anemometer at small velocities, but at larger velocity the anemometer becomes more advantageous.

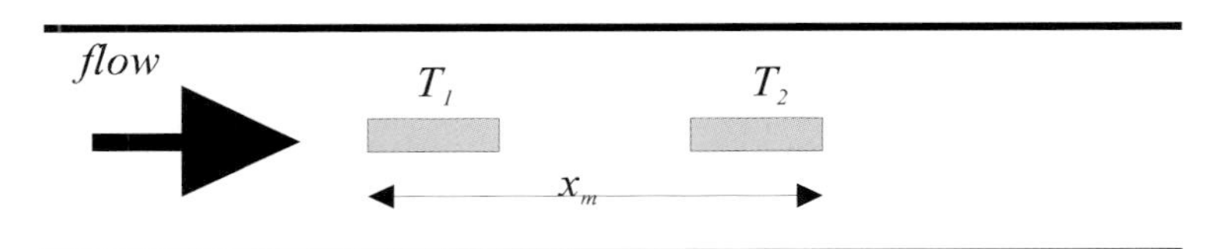

Fig. 8.21. Two element calorimetric flow sensor

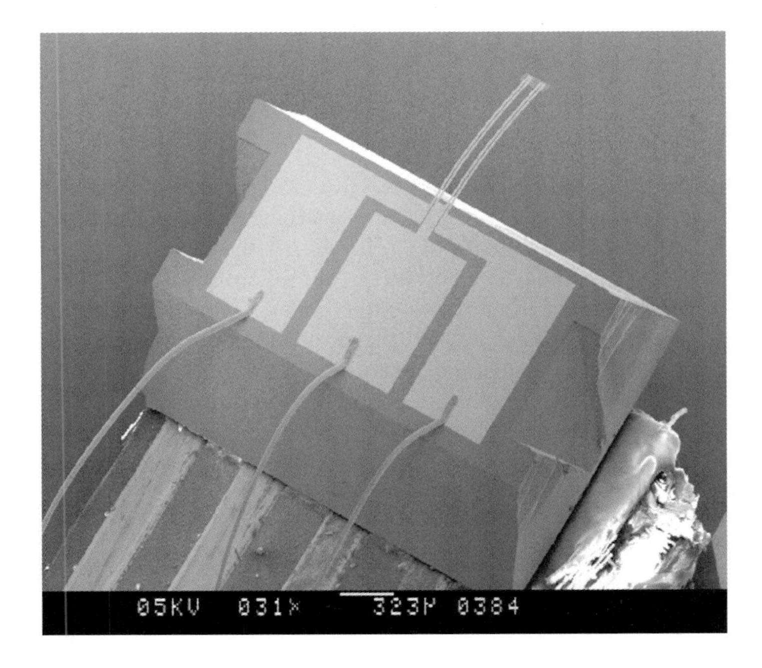

Fig. 8.22. The μ-flown, used as a bi-directional flow sensor

Experimental results are shown in Fig. 8.23.

Packaging issues of thermal flow sensors are addressed by Masana (1998), with the conclusion that flow sensors must be designed together with the package.

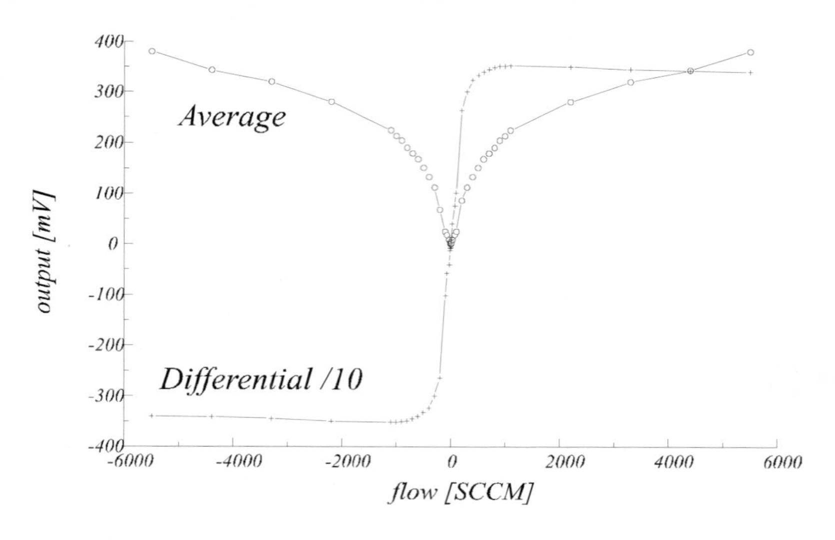

Fig. 8.23. Differential and common output signal of the sensor. Solid line: sum of signals, dashed line difference (de Bree 1998)

8.3.3. Sound Intensity Sensors – The Microflown

The sensor shown in Fig. 8.22 can also be used as a low-frequency microphone, a so-called microflown (de Bree 1996, 1997). These instruments are sensitive to sound with a horizontal response function up to a cut-off frequency given by (8.51), which is in practice limited to a few kHz. The microflown has a few characteristics very much different from pressure microphone. Most important of course, they are sensitive to a particle flow in place to a pressure, they do not have a low frequency cut-off, the sensitivity must be measured in V/(cm/s), in place of V/Pa and they are sensitive to the direction of the sound wave.

Particle velocity and pressure are the two conjugate quantities in acoustics. There is no unique relation between particle flow and pressure, except under well defined conditions, such as in plane waves or spherical waves. The sound intensity, which is the product of the flow and the pressure – hence a vector quantity – cannot be measured by a single pressure microphone. Normally, two well phase matched microphones are mounted a certain distance apart, and the sum gives the pressure, the difference is a signal proportional to the particle flow. The demand for phase matching is very rigid and accordingly sound intensity measurement instrumentation is pretty expensive. The combination of a microflown with a pressure microphone could be a good solution.

The power flux of a sound wave through a surface is given by the product of the sound pressure p and the particle velocity $\mathbf{u}$. The "particle velocity" is the velocity of air in a small volume. Both quantities refer to the local quantity averaged over a volume containing many molecules (a sufficiently large volume is 1 μm^3). p and $\boldsymbol{u}$ are related by Newton's equation and the equation for mass conservation:

$$\rho \frac{\partial \mathbf{u}}{\partial t} = -\mathrm{grad}\, p \tag{8.65a}$$

and (eq. (8.2) for the case that the fluid is compressible)

$$\frac{\partial \rho}{\partial t} = -\rho \left(\frac{\partial u_x}{\partial x} + \frac{\partial u_y}{\partial y} + \frac{\partial u_z}{\partial z} \right) = -\rho \mathrm{div}\mathbf{u} \,. \tag{8.65b}$$

The term $(\mathbf{u}\nabla)\mathbf{u}$ has been neglected in (8.64) due to the very small particle velocity in sound waves at normal intensity. Combination of these equations using the assumption that the density and the pressure are coupled by the adiabatic ideal gas equation (in linear approximation) leads to the wave equation:

$$\frac{\partial^2 \rho}{\partial t^2} = c^2 \nabla^2 \rho \tag{8.66}$$

which must be solved with the appropriate boundary conditions.

The sound intensity (powerflux, in W/m^2) at a given point and time is given by

$$\mathbf{I} = p\mathbf{u} \tag{8.67}$$

Eqs. (8.64) and (8.65) give relations between pressure and flow, however the relation depends on the boundary conditions. Only when the boundary conditions are such that one gets plane waves[6], the ratio of the pressure to the flow – the acoustic impedance – simply is:

$$Z = \frac{p}{u} = \rho c \tag{8.68}$$

and the intensity is given by:

$$|\mathbf{I}| = \frac{p^2}{2\rho c} \tag{8.69}$$

In general circumstances the waves are not planar, therefore there is no general simple relation (no general impedance) between pressure and flow. This makes it impossible to determine sound intensity by using a single microphone (a dynamic pressure sensor) only: one needs to measure both: pressure and flow (particle velocity).

---

[6] one gets plane waves far from sources and in the absence of reflecting surfaces

In principle there are three ways to do this:

(i) One can measure pressure and flow using two appropriate sensors within a very small volume.

(ii) One can measure the pressure and the pressure gradient with a pair of pressure sensors, and by using eq. (8.64) the flow can be calculated.

(iii) One can measure flow and flow divergence using a three-dimensional array of flow sensors such that the derivatives of the particle velocity in the three directions (see eq. (8.65)) are measured and calculate the pressure using eq. (8.64).

All three methods are sensitive to phase shifts generated in the separate sensors, the least sensitive to phase shift errors being method (i). Measuring a gradient or a divergence therefore requires sensor pairs with identical phase shifts.

An instrument based on method (i) is on the market (Norwegian electronics). The flow is determined by measuring a Doppler-shift using two pairs of ultrasonic transducers. A photograph of the instrument is shown in Fig. 8.24. The dimensions of this instrument are rather large (order 10 cm). The size limits the reliable performance to below a few 1000 Hz.

Instruments based on method (ii) are fabricated by e.g. Brüel and Kjær. An example is shown in Fig. 8.25. A three-D sound intensity probe has been described very recently by de Bree (1999).

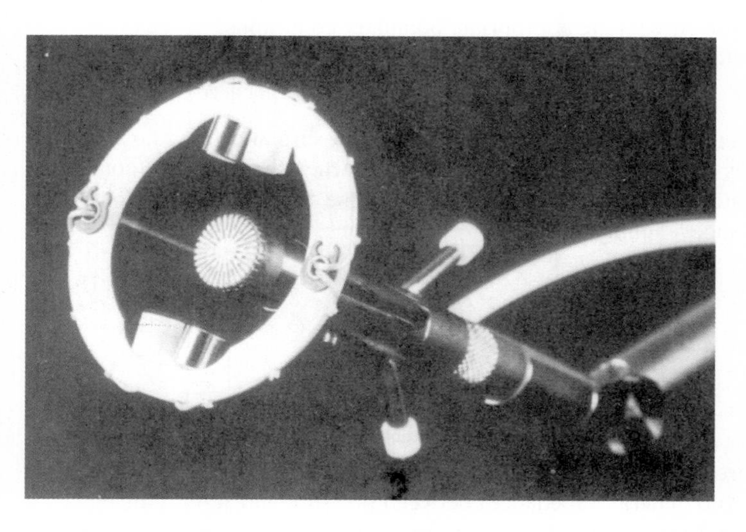

Fig. 8.24. Type 216 p-u intensity probe (Norwegian electronics) [product information Norwegian Electronics of the intensity probes on the market]

A scanning electron microscopic image of a microflown can be seen in Fig. 8.26.

The microstructures are remarkably robust. They can be touched, and mounted in a device they stand shock tests when falling from a height of several meters.

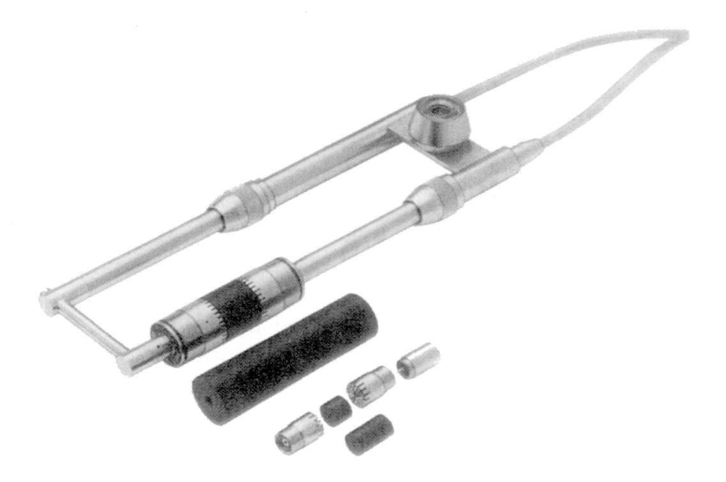

Fig. 8.25. A Brüel and Kjær sound intensity probe [product information B&K of the intensity probes on the market]

Fig. 8.26. An example of a micro flow sensor. The cantilever beams have a length of 0.8 mm. The thickness of the structures is 1 µm. This example consists of a central heater and two temperature sensors (Courtesy de Bree)

A schematic signal conditioning circuit is shown in Fig. 8.27. The probe comprises two distinct sensors, a conventional microphone and a dynamic flow sensor. The sensors give electrical signals proportional to flow and pressure, respectively, are amplified, the signals are multiplied and averaged to obtain the root mean square value.

This method was tested with encouraging results in a 40 meters long pipe (to control the acoustic waveform). A conventional microphone (B&K4133) and a flow sensor similar to the one shown in Fig. 8.26 were used.

The results of the measurements are shown in Fig. 8.29(a).

If the sound wave is a plane wave the particle velocity only has a component normal to the wave front. In this simple case the divergence equals the change of flow over a short distance parallel to the particle velocity. Hence, the differential signal from a pair of flow sensors with a spacing much smaller than the wavelength integrated over time gives a signal proportional to the pressure. The appropriate signal conditioning circuitry is shown schematically in Fig. 8.28.

Results of methods (i) and method (iii) are compared in Fig. 29. A similar experiment has been performed in a closed tube with standing waves. The results are shown in Fig. 8.29(b). The flow sensors arrangement for measuring dynamic pressure is shown in Fig. 8.30.

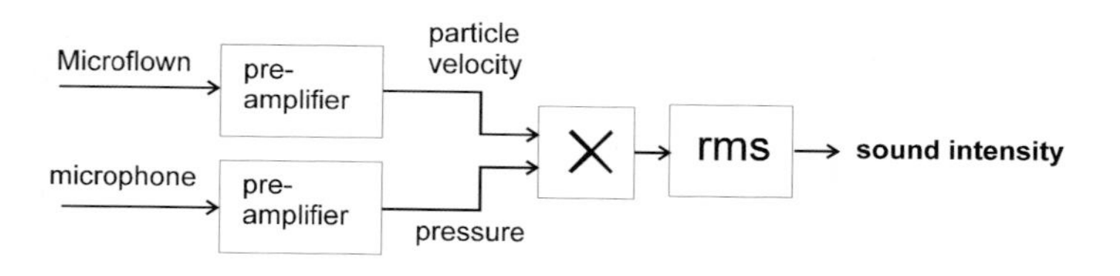

Fig. 8.27. Possible implementation of the product method (i)

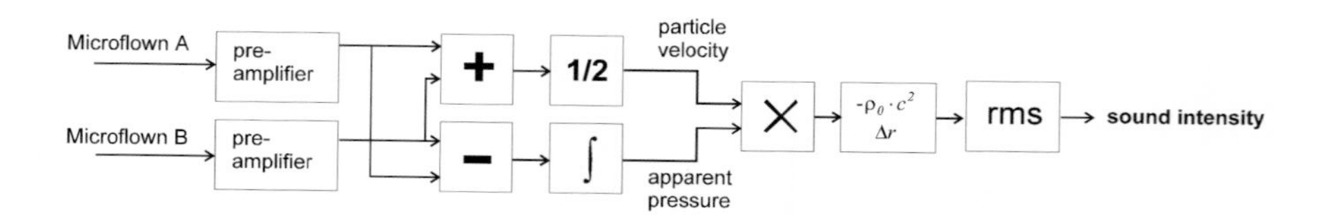

Fig. 8.28. A block diagram of the differential flow method (iii)

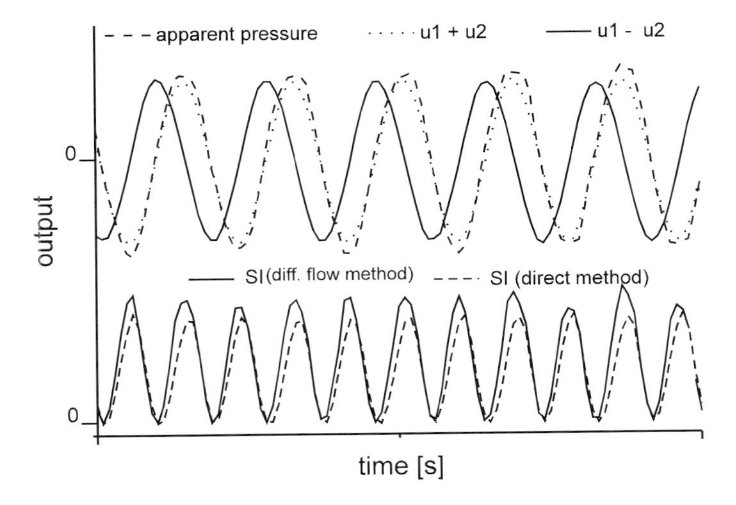

Fig. 8.29a. Pressure measurement and flow measurement in an acoustic burst in a long pipe. The upper part describes the differences and sum of the flow sensor output (dotted line and full line), and the broken line is obtained after integrating the difference signal. This line is proportional to the pressure. The lower part shows the result of intensity measurement using method (i) (broken line, using a pressure microphone and a flow sensor) and method (iii) (using two flow sensors)

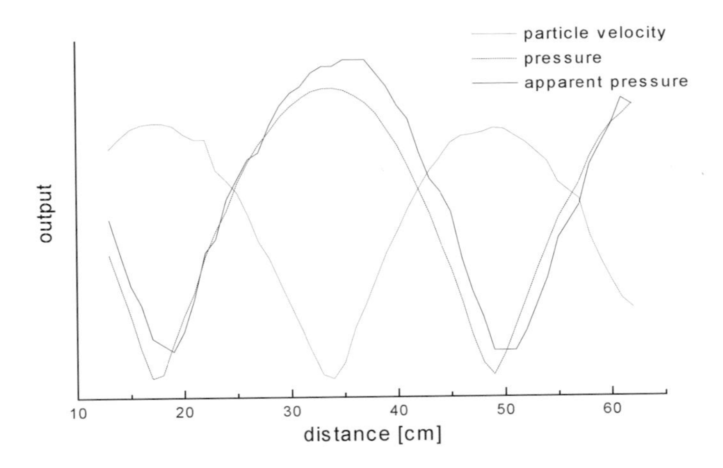

Fig. 8.29b. The minima and maxima of the particle velocity (flow sensor signal), pressure (pressure microphone signal) and apparent pressure (integrated differential flow sensors signal) in a closed tube (standing waves)

Fig. 8.30. Two closely spaced microflowns for pressure measurement. The distance between the microflown is app. 5 cm (De Bree 1997)

8.3.4 Time of Flight Sensors

Calorimetric flow sensors can be operated in a dynamic mode. A heat pulse from the heating element will be transported by the convection to the downstream sensor. The pulse will deform by the velocity profile (Fig. 8.31) and broaden at the same time by heat diffusion (Fig. 8.32). Operated in this way, a pure time signal is measured, consequently only the velocity of the streaming fluid is measured, independent of the type of fluid. Since the heat transport by diffusion is important for the calorimetric flow sensor, in a static mode the sensor works in a different way for different fluids. The distortion of the signal by diffusion is serious only at small flow velocities. Roughly we can say that if the dispersion of the heat pulse is of the order of the distance between sensor and heater, the measurement of the time of flight becomes inaccurate. This means for flow velocities smaller than $U_c = D_T / l$ the signal tends to be too broad to be useful. But this relation approximately determines the maximum in the ΔT-U curves, see (8.62). This means that calorimetric flow sensors operate optimal at velocities below U_c and in the time of flight mode, there the optimal regime of operation starts. A discussion of this point can be found in Yang (1992) and in Ashauer (1998)

If both are measured simultaneously, more information than just the flow can be determined because materials properties of the fluid and the flow velocity influence heat transfer and time of flight in a different way. This point has been discussed by Lammerink (1995) and van Kuijk (1995). Feeding a time varying signal to the heater leads to a mean temperature elevation of the heater plus a time varying component. The result is a temperature difference between the sensors located upstream and downstream, which depends on the time. Lammerink (1995) and van Kuijk (1995) demonstrated that it is possible to extract parameters from the time dependent signal. They determined the flow of a gas and its composition, provided the components in the gas mixture are known. To recognise the composition they used a neural network (Lammerink 1995).

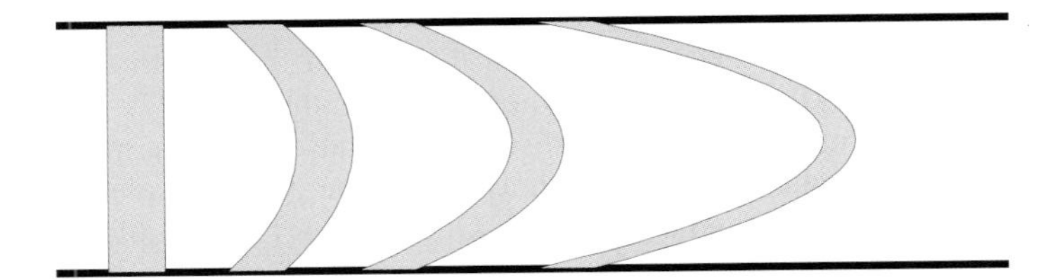

Fig. 8.31. Schematic of a heat pulse distortion in a flow with parabolic flow profile

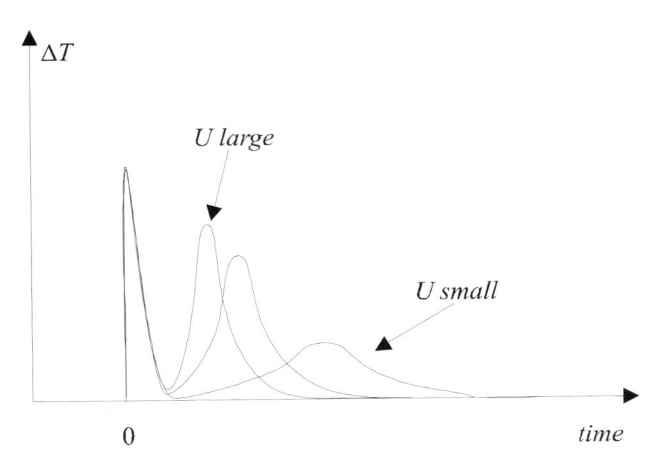

Fig. 8.32. Schematic of a pulse distortion by dispersion in a time of flight flow sensor. A heat pulse is fed to the fluid at $t = 0$. For fast flowing fluids the pulse comes quickly to the temperature sensor located downstream, and has no time to disperse. Small flow rates lead to broad, less intense and late pulses

8.4 Skin Friction Sensors

Many friction sensors are developed for the study of turbulence. Turbulence is still a phenomenon, which is subject to intense basic research, and sophisticated instrumentation is needed to further the field. Here microtechnology, MEMS, may play an important role. The elements of microsensors can be made small and fast enough to resolve phenomena in turbulent flow. This is why in the past ten years a rather large number of papers appeared addressing the problem of microsensors especially designed for turbulence studies (Yoon 1992, Liu 1994, Kälvesten 1994a, Kälvesten 1994b, Pan 1995, Jiang 1995, van Oudhuisden 1995, Padmanabhan 1995 Jiang 1996, Kälvesten 1996, Schmidt 1987, Padmanabhan 1997, Ebefors 1998, Schmidt 1988). Furthermore, due to the small size of the sensors, many sensors may be placed close to each other and the friction field (and the pressure and the velocity field) can be measured. Successful attempts to produce space and time resolved images of the shear stress in the presence of turbulence have been made by Jiang et al (1996) and Padmanabhan (1997). We referred to the former work in Sect. 8.3.1. Microsensors can also be made much more sensitive because there is an arsenal of techniques for the thermal isolation of the heater (van Putten 1983, Stemme 1986, van Oudhuisden 1990a, Löfdahl 1992, Jiang 1996). This sensitivity is sufficient to detect the heat flow to a turbulent gas flow. As said earlier, the requirement in space and time resolution is app. 100μm and 100μs.

Skin friction can be measured in a number of ways. An overview on instrumentation fabricated by traditional machining can be found in a paper by Winter (1977). In micromachined versions there are only two basic types described: Thermal skin friction sensors and floating element skin friction sensors. The theory of skin friction in relation to heat transfer has been described in Sect. 8.1.5. The main result is that the skin friction in a developed boundary layer is proportional to the heat flux to the fluid to the third power.

A thermal measurement is indirect, so some researchers have preference for floating element sensors. This sensor consists of a movable element attached to springs to a substrate, flush mounted into the wall on which the skin friction is to be measured. The disadvantage of these devices are the strict requirement for flush mounting and the fact that inevitably there are openings in the wall surrounding the floating element to allow a small movement. It is the latter reason that researchers often complain over the poor reproducibility when using floating elements (e.g. Winter 1977, Hanratti 1996, Löfdahl 1992). Micromachining allows however the fabrication of floating elements with very narrow gaps (2 μm is standard), and surface micromachining ensures that the element is neatly in plane with the substrate. With the integration of an actuator, a force feedback is possible which enables one to keep the gap at constant width (Pan 1995). So, in our view, at the moment the question what the best type of skin friction microsensor is, seems to be open.

A floating element consists of a movable plate attached to springs with well defined spring constants. The deflection of the plate is measured, the amount of which gives us information on the force acting on the plate. Most of the floating elements are fabricated using surface micromachining.

Ng et al. (1991) proposed a design of plates attached to springs as shown in Fig. 8.33. They used a fabrication technology involving wafer bonding and etch-back instead of surface micromachining. The element shown in Fig. 8.33 was positioned so that the tethers were parallel to the flow. The dimensions of the

floating element were 120×140×5 μm^3. The tethers are then loaded such that the upstream ones are in tensile, the downstream ones in compressive stress. Their idea was to use the tethers as piezoresistive elements. The structural layer was lightly doped to get suitable electrical conductivity, and when the springs are electrically connected in a half Wheatstone bridge, one gets a differential signal proportional to the deflection. The authors tested their device in a liquid (Cannon N2700000) and were able to detect shear stresses in the range of a few kPa. This corresponds to a force resolution in the range of 0.1 nN.
Ng's design is quite similar to earlier work by M.A. Schmidt (1987, 1988), who designed a sensor with a plate size in the 1 mm^2 range made from polyimide. Polyimide is not the most appropriate material for this purpose: it swells by absorbing water in a humid environment and it is subject to plastic deformation. Silicon is a perfect material regarding its mechanical properties for this type of sensors. It should be noted that the construction in Ng's design is mechanically overdetermined. Compressive stress in the structure may lead to buckling with the result of complete different mechanical properties. Another disadvantage of this design is that the tethers are rather stiff, which results in a relatively poor force resolution. The deflection of the plate would be much larger if the flow were directed normal to the springs, but this deformation would make the piezoresistive detection in this way impossible (piezoresistors on top of the tethers would be an outcome, see Burger 1998).

An alternative transduction mechanism is to measure the change of a capacity due to the motion of the plate. This has been done by Pan (1995) in two variants, in collaboration with Analog Devices. The designs are shown in Fig. 8.34 and 35.

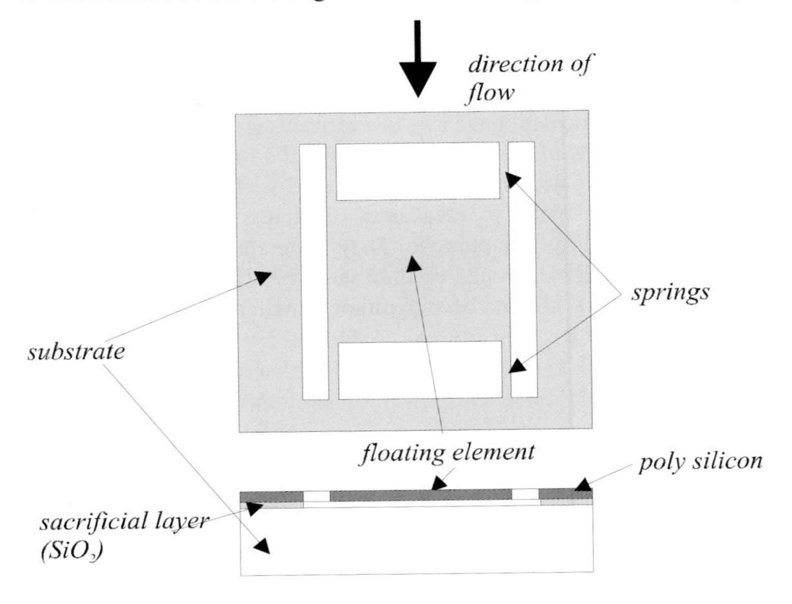

Fig. 8.33. Basic construction of a floating element

The first design is shown in Fig. 8.34, which in fact is a structure originally designed for a lateral resonator. The dimensions of the plate are app. 100 by 100 by 2.2 μm^3. A load in horizontal direction will give rise to a differential change of the capacities formed by the floating plate and the two sense electrodes beneath the plate. The comb drive structures serve as actuators to keep the plate in the unloaded position; the voltage needed to do this is then taken as a signal. This voltage is directly proportional to the force acting on the plate. The sensors were calibrated in a flow channel fed by nitrogen at flows up to 25% of the speed of sound. Pressure sensors in the tube were used to determine the pressure gradient in the tube. The shear stress is related to the pressure gradient by

$$\tau = -\frac{b}{2}\nabla p(1 - \frac{c_p}{c_v}\mathrm{M}^2) \qquad (8.70)$$

with M the Mach number, b the height of the channel, c_p and c_v the specific heat at constant pressure and volume, respectively.

The sensitivity depends on the design of the folded beams, (the stiffness is limited to a lower limit by sticking problems), and a resolution of app. 2 Pa has been reached. This corresponds to a force resolution in the range 0.2 pN. Contrary to Pan's claim, the spring construction is still so that the whole structure is mechanically overdetermined, and similar problems with buckling due to compressive stress must be expected. In fact, it is even worse because the springs are weaker due to their longer length.

The sensor developed in collaboration with Analog Devices is based on their AD integrated accelerometer. The principle of the design is shown in Fig. 8.35. The size is similar to the sensor of Fig. 8.34, however the shear stress resolution is larger by a factor of 10. The sensors was packaged in a way that Pan et al. had difficulties to align the sensor carefully in the wind tunnel. An inclination of the sensors with respect to the tube wall gave rise to enormous errors: a deviation of a few degrees leads to variations of the sensor sensitivity of a factor of 10.

The sensitivity of the skin friction sensors described so far is limited by the way the deflection of the plate is measured. Padmanabhan (1995, 1997) introduced an optical method for this purpose. They integrated photodiodes close to the edges (1995) and under openings of the plate (1997) to achieve a stress resolution of 0.0014 Pa, two orders of magnitude smaller that Pan's sensor. Dimensions of Padmanabhan's sensor is again in the 100 by 100 μm^2 range, with a resonant frequency of 100 kHz, sufficient for the characterisation turbulent flow. The shear stress resolution corresponds to an impressive resolution of 1.4×10^{-13} N.

The high resolution of the mechanical skin friction sensors is not reached by thermal skin friction sensors: Kälvesten (1996) reports a resolution in the range of 0.2 Pa, a figure similar to Jiang's results (Jiang 1996).

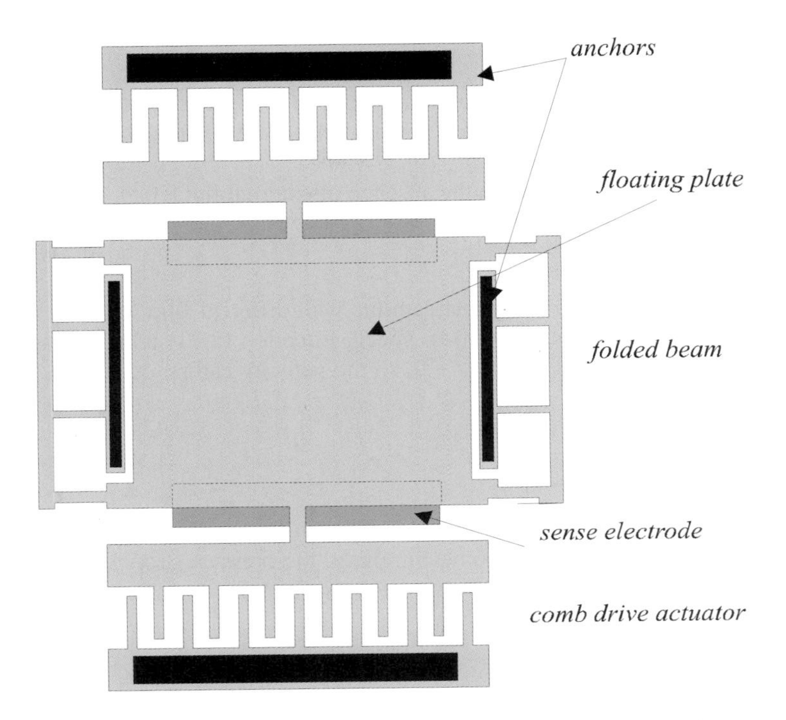

Fig. 8.34. Electrostatic skin friction sensor with sense electrodes and actuators for force feedback (Pan 1995)

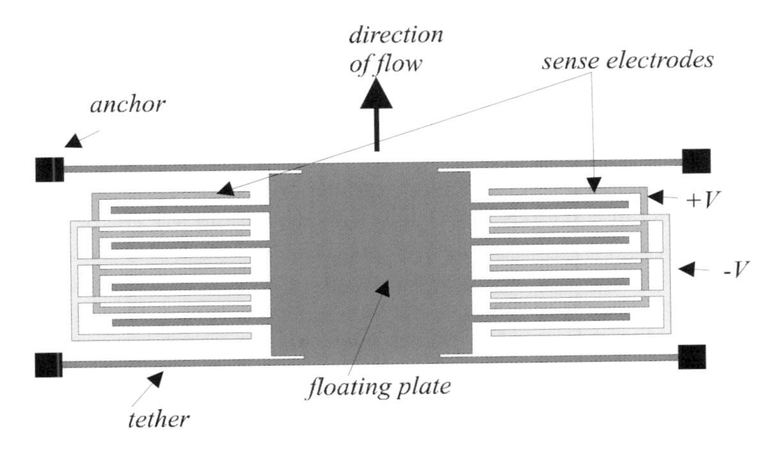

Fig. 8.35. Electrostatic skin friction sensor (Pan 1995; Analog Devices design)

8.5 "Dry Fluid Flow" Sensors

Under the head of dry flow sensors we treat all types of flow sensors where the kinetic energy of the fluid plays a role, and viscosity not. A fluid flow is "dry" if the Reynolds number is large.

In ideal liquids along a streamline (in the absence of external fields)

$$\frac{p}{\rho}+\frac{1}{2}U^2 = \text{constant} . \tag{8.71}$$

The pressure difference at points along a streamline with different flow velocities relates directly to the square root of the flow speed. Equation (8.71) tells us that the pressure is maximal at points where $U = 0$, so the velocity can be determined by

$$U = \sqrt{\frac{2\Delta p}{\rho}} \tag{8.72}$$

at a point far away from the stagnation point. Δp is then the pressure difference between this point and the stagnation point. Similarly, the pressure drop Δp over an orifice varies at large Re with the flow rate in a quadratic way, known as Toricelli's law,

$$\Phi \propto A\sqrt{\frac{2\Delta P}{\rho}} . \tag{8.73}$$

Φ is the volume flow and A the area of the orifice. Equation (8.73) is very similar to (8.72), the main difference between (8.72) and (8.73) being the proportionality factor. The expression (8.73) is just a consequence of the Bernoulli equation; a discussion can be found in Feynman (1965). Toricelli's law applies if the Reynolds number is large (here the characteristic length is the length of the channel forming the orifice) if the length of the orifice is much smaller than its cross section. The hydrodynamic boundary layer is then much thinner than the cross section, which means that viscous losses are confined to a very small section of the orifice, close to the wall, where there is the velocity gradient. The proportionality constant depends on the geometry of the orifice.

The interesting aspect in equations (8.72) and (8.73) is twofold: (i) The design of a flow sensor is reduced to the design of pressure sensors and a suitable geometry to find the points where to measure the pressure. (ii) The only temperature dependent quantity defining the relation of the flow rate and the pressure difference is the density. The density depends much weaker on the temperature than the viscosity. The main draw back of the sensors discussed in this section is their insensitivity to flow at small velocities.

In the following we describe a few micro sensor designs that rely on equations (8.72) and (8.73).

Schmidt et al. (1998) manufactured a plate attached to a flexible cantilever beam in an orifice. The plate reduces the cross-section of the opening in the orifice and a force on the plate results. The deformation of the cantilever beams was determined by piezoresistive elements. At very high Reynolds numbers the pressure drop across the plate is due to kinetic energy loss and eq. (8.73) applies. A schematic drawing of the plate with some qualitative stream lines is shown in Fig. 8.36. The dimensions of the plate were between 500×500 and 700×700 μm, a flow speed in the range of m/s, leading to a pressure drop over the orifice of a few thousand Pa.

A similar principle has been used by Richter et al. (1999) who machined an orifice in the centre of a membrane, also with piezoresistive elements to detect its deflection. In fact, they simply used a commercial pressure sensor and drilled a hole through the membrane. The drilling can be done by laser machining, reactive ion etching or powder blasting. The membrane size was in the range 1–2 mm^2, with a thickness of 20μm and holes had a diameter of 30–400 μm, with a conical shape due to the manufacturing process. Clearly, for these relatively thick membranes the hole must be rather large. For the small hole diameter, viscous losses become important. In Fig. 8.37 we show the design and results of measurements for the largest hole investigated by Richter.

If these sensors are used for fluid flow bubbles are of concern since they can clog the orifice. However, as shown by Richter (1999) the pressure needed to overcome the pressure associated with larger curvature is below the pressure range this design operates.

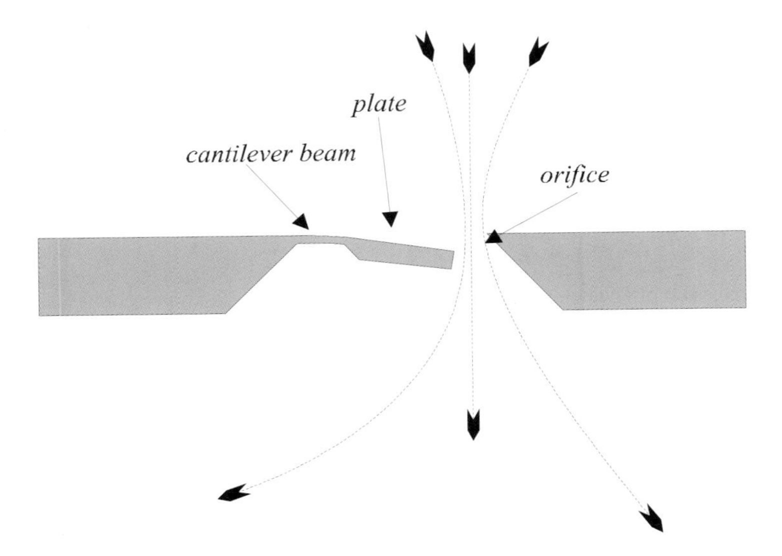

Fig. 8.36. A plate attached to a substrate by an elastic cantilever beam bends under a flow through an orifice due to kinetic pressure loss. Design by Schmidt (1998)

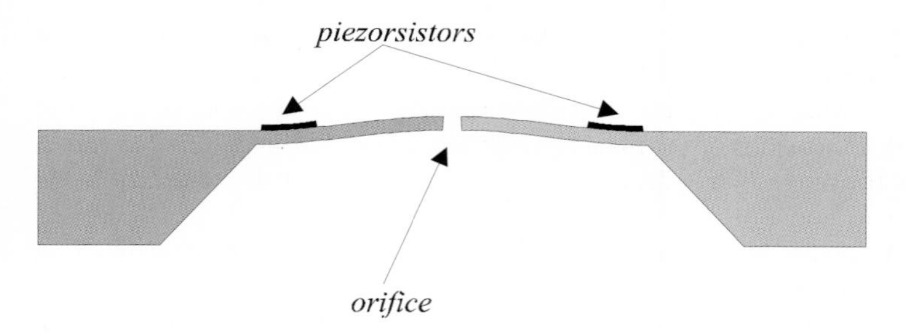

Fig. 8.37a. Richter's design for a flow sensors based on eq. (8.71). The sensor is a commercial pressure sensors with a hole drilled through the membrane (Richter 1999)

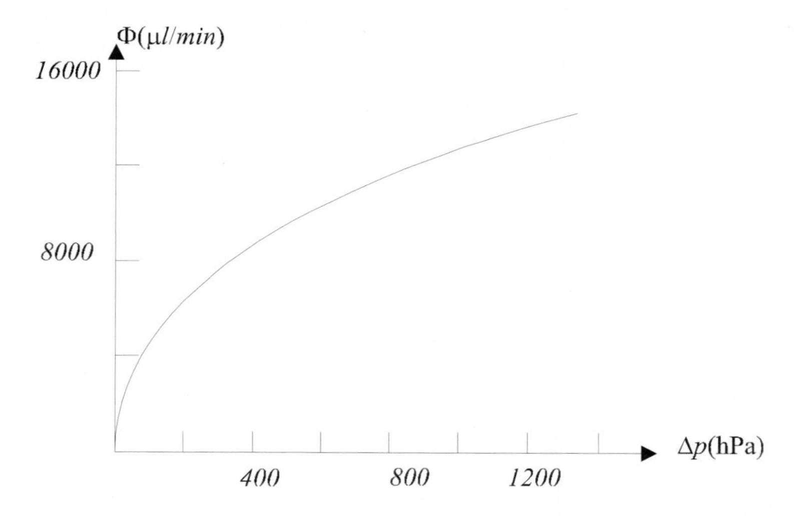

Fig. 8.37b. Characteristic of Richter's sensor for a Honeywell 24PC (30psi) pressure sensor with an orifice of 160 μm.

In Prandtl type flow meters, the pressure of the fluid is compared at a stagnation point where $U = 0$ and at a point far enough away from the stagnation point. According to Franke (1970) classical Prandtl flowmeters require a tube with double walls. The inner tube must have a diameter 1/3 of the outer one. The end opening of the inner tube is directed towards the flow, and there is a tiny opening (size: 1/10 of the outer tube diameter) at a distance 3 × outer tube diameter connecting the outer tube to the pressure at the opening. Any supports must be a distance at least 10 × the dimensions of the support away from the end.

The inner and outer tubes are connected to barometers from which the pressure difference can be determined.

Berberig suggested for a micro version of a Prandlt flow sensor the design shown in Fig. 8.38(a) An opening of a capacitive pressure sensor connects the outside to the pressure sensor cavity. The sensor is arranged in the flow so that the opening is located at the stagnation point as shown in Fig. 8.38(b), at the inside of the cavity the pressure will be higher than at the other side of the membrane, which experiences the dynamic pressure. If the deflection is not too large, the change in the capacity will be proportional to Δp, which in turn is proportional to the square of the velocity. Overall dimensions of Berberig's sensor is $5 \times 5 \times 1.4$ mm^3. The pressure sensor is designed to operate in linear range for air flow up to 75 m/s. Then a maximum deflection of 1 μm is reached (the gap spacing is 7 μm). Although there is some scatter in the experimental data the sensor functions according to the simple theory.

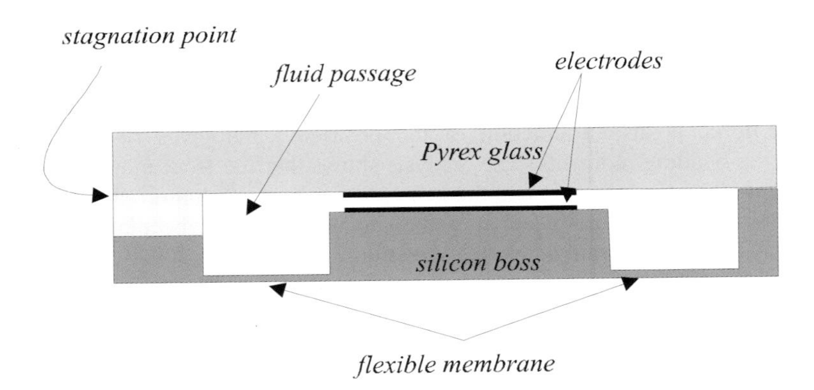

Fig. 8.38a. Design of a capacitive flow sensor using Prandtl's principle (redrawn from Berberig 1997)

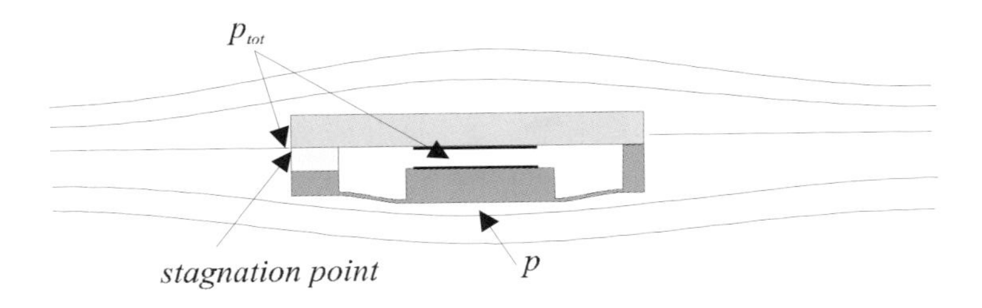

Fig. 8.38b. Deflection of the Prandtl micro flow sensor when subject to a flow (Berberig 1997)

One must note that in this design the capacitor gap is connected to the medium in which the flow must be determined. So the measurement will be possibly adulterated due to corrosion and contamination. It is not necessary at all (see Oosterbroek, as discussed below, Fig. 8.43) to connect the flow medium to the capacitor gap, but this requires a different design and probably a more involved fabrication process: The pressure sensor must be located inside the cavity which is connected to the stagnation point. We must also mention that when the medium is a liquid, one must expect problems with filling the device with the medium. Further, the medium has a dielectric constant (not important, of course, if the medium is a gas), which might be temperature dependent and composition dependent as in aqueous solutions and water.

The last example we give is due to work by Svedin et al. (1997). A plate with an inclination with respect to a flow field experiences a lift force, which varies over the plate. The lift force acts on walls perpendicular to the flow direction and the variation of the lift force gives rise to a deformation of the plate.

The flow velocity on both sides of the plates differs, so there is a total force acting on the plates. In the middle the plates are fixed, so there is a bending moment. The force is also a function of the position. This gives rise to a difference of the bending moments. The analysis shows that the front moment is ca. 8 times larger than the rear moment, the front plate is bent much more than the rear plate. Therefore it is possible to measure the deflection of the plate in a differential way by placing strain gauges as shown in Fig. 8.39.

The output of a Wheatstone bridge measuring the differential deflection turns out to be proportional to the square of the flow velocity, in agreement with theory. The output is also dependent on the angle of the airfoil with respect to the flow. If the angle becomes too large, the viscous drag force dominates with the result that the deflection of both wings becomes symmetric. The maximum sensitivity is at an angle 20°, weekly dependent on the flow velocity and dependent on the geometry of the air foil.

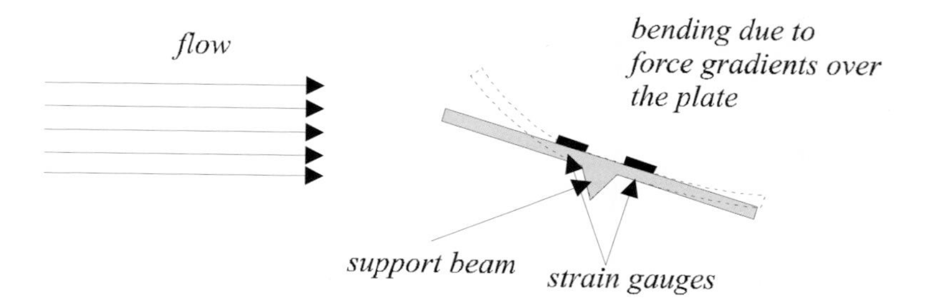

Fig. 8.39a. Design idea of a gas flow sensor based on a lift force (redrawn from Svedin 1997)

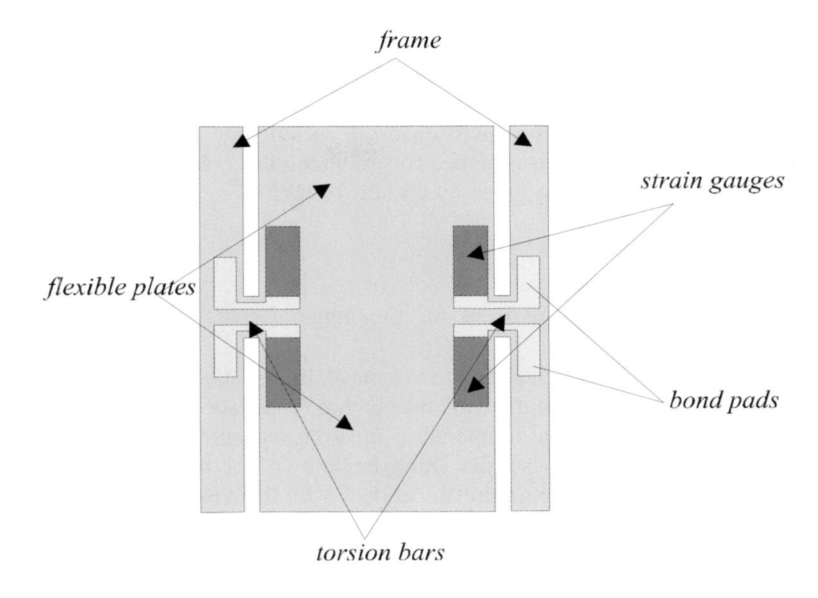

Fig. 8.39b. Layout of the lift force sensor (redrawn from Svedin 1997)

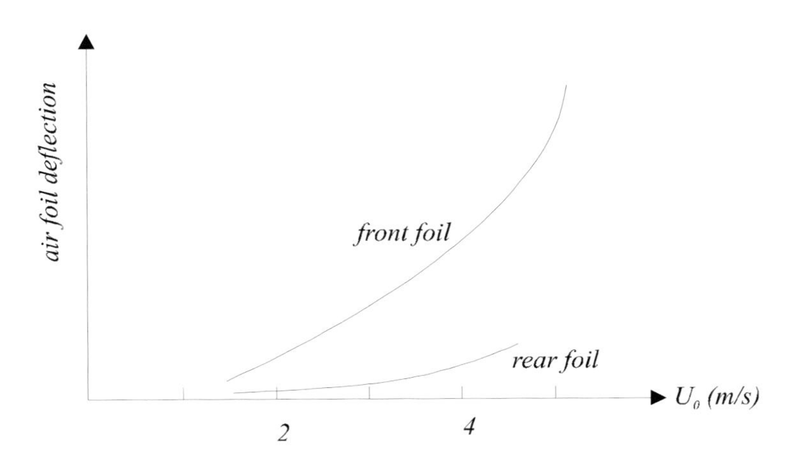

Fig. 8.40. Deflection of the air foils upstream and downstream (Svedin 1997)

8.6. "Wet fluid flow" sensors

If a viscous fluid flows through a duct there is pressure difference needed to keep the fluid flowing. Generally, the volume flow rate is proportional to the pressure difference, quite analogous to Ohm's law for electric conductors[7]. For a cylindrical duct the relation (5) applies. The resistance is a function of geometry. In Gravesen (1993) an overview can be found on the flow resistance of ducts of various types of cross-sections. For microsensors a rectangular cross-section is the most important one. The flow resistance for a rectangular cross-section with side length a and b and a length l is given by (White 1994)

$$R = \frac{\eta l}{512} {f_H}^2 \frac{(a+b)^2}{a^3 b^3}. \tag{8.74}$$

The numerical factor f_H varies from 56.91 for a rectangular cross section ($a = b$) to 96 for $b/a \to 0$.

In any case, the resistance depends on the viscosity. In liquids, the viscosity is strongly dependent on the temperature. This means that for sensors measuring the viscous pressure drop in ducts, the temperature must be measured, too, and the temperature dependence of the viscosity must be known for the temperature compensation. In gases, the temperature dependence of the viscosity is much smaller than in liquids.

There are two basic designs for wet fluid flow sensors. One can measure the pressure difference between the ends of a duct, either by carrying out a differential pressure measurement using a single pressure sensor, or by two absolute pressure sensors. We begin with the first option. Here the two sides of the membrane of the pressure sensor is connected to the medium via an obstructing channel. A remark we made concerning Berberig's sensor applies also here when a capacitive pressure sensor is used: The medium has access to the gap of the capacitor, with possible problems.

Cho et al. (1992) use a capacitive pressure sensor made by the dissolved wafer process developed at the University of Michigan (see Chau 1988). This process involves shallow and deep doping of silicon by high boron concentration, anodic bonding of the doped side of the silicon wafer to a Pyrex wafer and dissolving the undoped silicon in KOH or EDP. The result is a silicon structure of a maximum height of 20 μm on top of a glass wafer. In their design for the flow sensor Cho et al. connected the cavity of the pressure sensor to the outlet of the duct and to a narrow micromachined duct over which the pressure falls. The outlet of the narrow duct is connected to the outer side of the membrane of the pressure sensor. A schematic drawing is given in Fig. 8.41. The actual channel (with a cross section of 60×3.5 μm^2) for flow obstruction has been machined in the same process steps as the pressure sensor, the channel meanders over the Pyrex. The thickness and size of the diaphragm is 2,9 μm and 2×2 mm^2, electrode separation 3.5 μm. Pressure resolution with such a pressure sensor is in

---

[7] Of course the analogy breaks down when the resistance is considered: the electric resistance is proportional to $1/A$, the hydraulic one to $1/A^2$. This is a result of the parabolic flow profile.

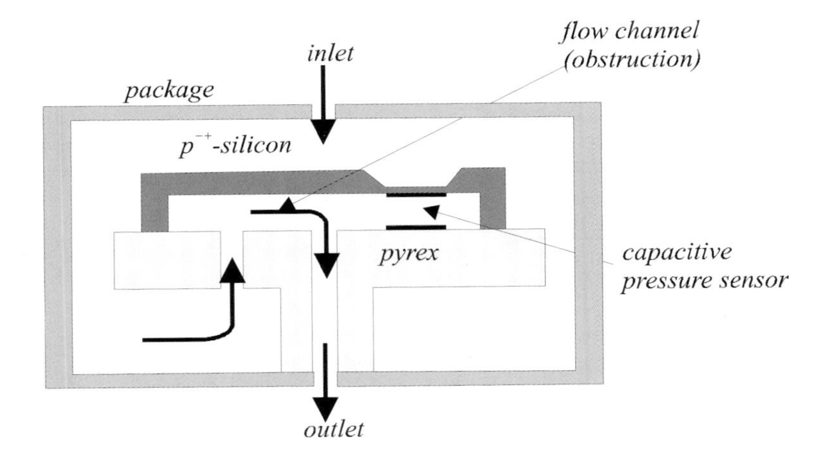

Fig. 8.41. Schematic of a flow sensor with a differential pressure sensor (Cho et al. 1992)

the range of 10 Pa, and the flow resolution with the obstruction leads to 10^{-8} SCCM[8] with a full range of 1 mSCCM.

Flow sensors involving two pressure sensors (principle shown in Fig. 8.42) have been proposed by Boillat et al. (1995) and by Oosterbroek et al. (1997) for the measurement of liquid flow. Boillat used piezoresistive pressure sensors for flow sensors with measurement range from 1 to 300 μl/min with a pressure drop in the range of 50 hPa. A strong temperature dependence of the flow signal was observed, which turned out to be dominated by the temperature dependence of the viscosity of water. Oosterbroek machined capacitive pressure sensors (Fig. 8.43) in silicon sandwiched between Pyrex wafers. The device was designed for flow rates in the range up to 3 ml/min. Flow resolution in both designs is a few % full scale. Note a detail: In Oosterbroek's design, Pyrex has been micromachined, too. In his paper a full account is given for the models used for the design of the sensor (including membrane design for the pressure sensors, channel design for the pressure drop and considerations regarding the size of the pressure sensors to ensure that no effects of kinetic pressure affect the measurements (Oosterbroek 1999). Both designs avoid problems with contamination and corrosion of sensitive parts of the pressure sensors.

Gass et al (Gass 1993a, 1993b) machined a flow sensor very similar to the one shown in Fig. 8.36, but for operation in the range where viscous losses over the opening dominate. He demonstrated that liquid flow can be measured with a resolution of nl/min.

---

[8] SCCM = standard cubic centilitre/minute

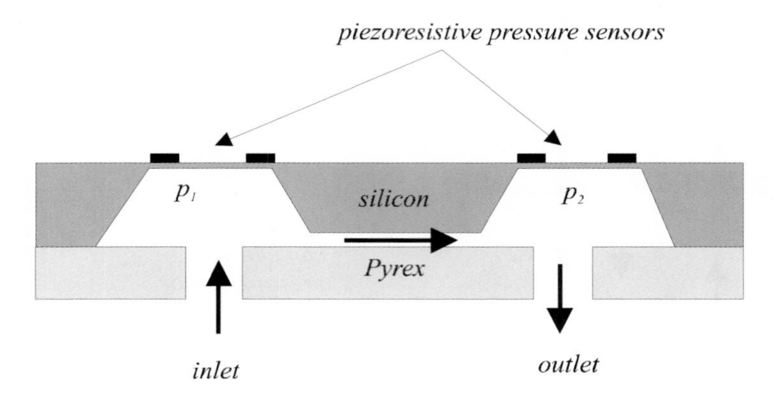

Fig. 8.42. Flow sensor based on two piezoresistive pressure sensors connected by a narrow channel (Boillat 1995)

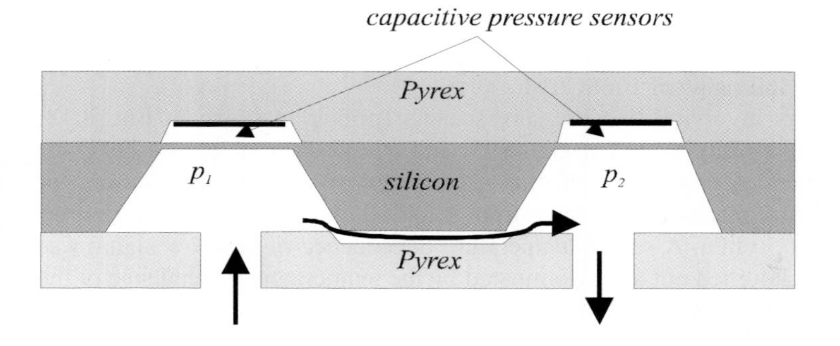

Fig. 8.43. Flow sensor based on two capacitive pressure sensors connected by a narrow channel (Oosterbroek 1997)

9. Resonant Sensors

This chapter differs from the previous chapters in the sense that we will now discuss a class of sensors based on the operating principle of the sensor, i.e. resonant, and not on the measurand. In the previous chapters, resonant sensors have already been mentioned several times. This is not surprising as virtually every measurand can somehow be measured with a resonant sensor. In this chapter, we will present an overview of the operation and performance of resonating microsensors. A more complete discussion on the theory and applications of resonant sensors can be found in (Prak 1993) and (Wagner 1995). Other valuable papers discussing the operation of resonant sensors are (Tilmans 1992, Stemme 1991, Guckel 1990, Howe 1987). Resonant sensors are frequency output sensors and as such have several advantages:

- The (digital) output signal can be directly connected to digital signal processing electronics. Analog to digital conversion is not necessary. Furthermore, a frequency signal can be transported over long distances with no loss of accuracy.
- Large dynamic range. Time is by far the best measurand. Furthermore, the upper bound of the dynamic range is limited by the measurement time: increasing the measurement time automatically results in a larger dynamic range.

However, it should be emphasised that these advantages do not include a high accuracy, good repeatability, low hysteresis, etc.. These properties are dependent on the specific sensor design; not on the type of output signal.

In Sect. 9.1 we will start with a discussion on the basic operation and physics of resonant sensors. Next, in Sect. 9.2 several mechanisms for excitation and detection of the vibration will be discussed. Finally, in Sect. 9.3 some examples and commercially available devices will be discussed.

9.1 Basic Principles and Physics

9.1.1 Introduction

The vibrating element in a resonant sensor can have various shapes, such as beams, diaphragms or membranes, the "butterfly" structure, the "H" structure, etc., as shown in Fig. 9.1. Furthermore, each of these shapes can have several types of vibrations, e.g. transversal, longitudinal, torsional, and lateral. Since the structures generally have an infinite number of degrees of freedom (they have a distribute mass and spring), an infinite number of resonances or vibrational modes is possible for every type of vibration. Usually, a resonant sensor is designed such that one of these modes dominates. The resonance frequency of this mode is made dependent on the measurand by letting the measurand change the stress, mass or shape of the resonator.

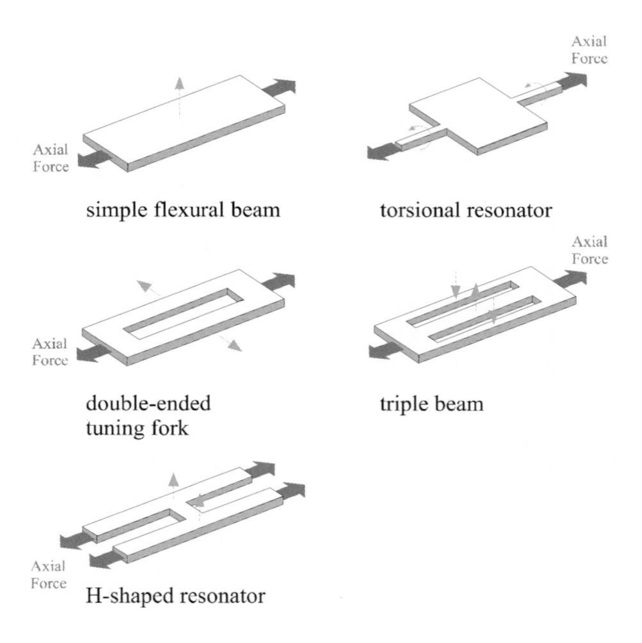

Fig. 9.1. Various structures used in resonant microsensors

Since a large percentage of all resonant sensors comprise transversally vibrating prismatic beams, we focus our attention on such structures (prismatic beams are beams with a flexural stiffness which is independent of the length coordinate). A further restriction concerns the boundary conditions of such a beam. We concentrate on a clamped-clamped microbridge, thus boundary conditions are $v(x)=0$, $v'(x)=0$, $v(l)=0$ and $v'(l)=0$, where v is the deflection of the beam as shown in Fig. 9.2. The boundary conditions of a clamped-free beam, for example, would be $v(x)=0$, $v'(x)=0$, $v''(l)=0$ and $v'''(l)=0$.

It should be emphasized that the solving methods outlined in this section are applicable to other types of vibrations and other structures with only minor changes. The transversal vibrations are very illustrative for a more general case.

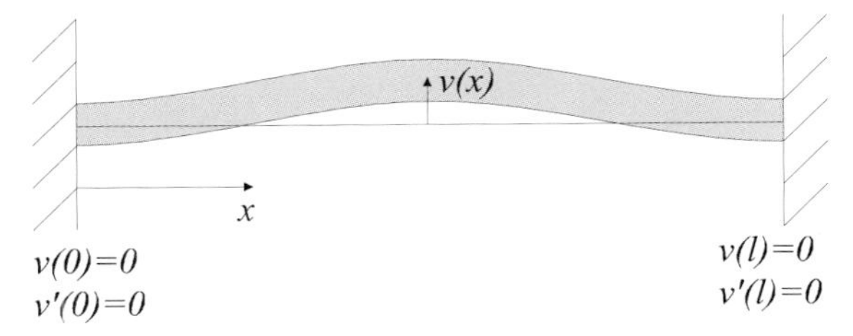

Fig. 9.2. Prismatic clamped-clamped beam

9.1.2 The Differential Equation of a Prismatic Microbridge

Small transversal movements of a viscously damped prismatic microbridge subject to an axial force N and an externally applied driving load $P_e(x,t)$ (see Fig. 9.3) are described by the following linear, inhomogeneous, partial differential equation, which is of the fourth order in space, and of the second order in time (Timoshenko 1974):

$$\hat{E}Iv''''(x,t) - Nv''(x,t) + \rho w_b h_b \ddot{v}(x,t) + c\dot{v}(x,t) = P_e(x,t) \tag{9.1}$$

The symbols have the following meaning:

- $v(x,t)$ — the deflection to be solved
- I — the moment of inertia ($I = w_b h_b^3 / 12$)
- c — the viscous drag parameter (force per unit length per unit velocity)
- w_b — the width of the beam
- h_b — the height of the beam
- $\hat{E}$ — the apparent Young's modulus of the beam material, which equals $\hat{E} = E/(1-\nu^2)$ for beams with $w_b >> h_b$, where ν is Poisson's ratio and E the Young's modulus.
- ρ — the density of the beam material

A prime and a dot denote differentiation to x and t, respectively. In (9.1) effects of rotary inertia and shear deformations, as well as nonlinear terms due to large amplitude vibrations have been neglected (Bernoulli assumptions).

In fact (9.1) describes the force equilibrium for every infinitesimal part dx of the beam. The various terms can be identified as follows. The first two terms describe the spring effect of the beam: its natural tendency to be uncurved. EI is called the flexural rigidity. Axially loaded beams ($N \neq 0$) show an additional term. The third term is correlated to the inertia of the beam: $\rho w_b h_b$ is the mass per unit length and $\ddot{v}(x,t)$ is the acceleration. The fourth term is related to the velocity-proportional friction forces acting on the beam. Frictional forces which are proportional to other powers of the velocity are not considered here. Finally, the right hand side is related to the driving force acting on the beam: the force which sets the beam in motion. This force is generated by one of the excitation mechanisms described in section 9.2.

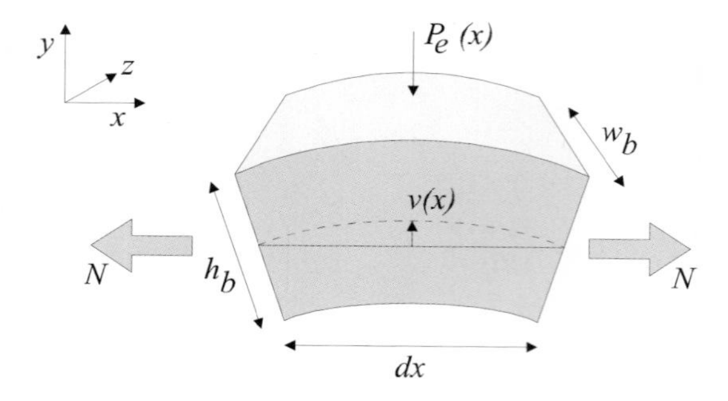

Fig. 9.3. An infinitesimal part *dx* of the beam

We assume that separation of space and time coordinates is possible and that the time dependent parts of v and P_e are harmonic functions (an eventual static deflection and load is not considered in this chapter):

$$v(x,t) = \overline{v}(x)e^{j\omega t} \tag{9.2.a}$$

$$P_e(x,t) = \overline{P}_e(x)e^{j\omega t} \tag{9.2.b}$$

The overstrike is meant to indicate the (complex) amplitude of an harmonic quantity. Both the modulus and phase of these quantities can depend on x. By factoring out the time dependent part, (9.1) reduces to:

$$\hat{E}I\overline{v}''''(x) - N\overline{v}''(x) + (\omega^2 \rho w_b h_b - j\omega c)\overline{v}(x) = \overline{P}_e(x) \tag{9.3}$$

Solving this equation is quite complicated. Nevertheless, it can be solved analytically if the right hand side is not too complicated. However, the resulting expressions for $v(x)$ are hopelessly complex mathematical functions and do not provide much insight in the physics of the beam. Therefore, in the next section, we will first solve the much simpler case of an undamped (c=0) and undriven (P_e=0) microbridge.

9.1.3 Solving the Homogeneous, Undamped Problem using Laplace Transforms

The differential equation for the homogeneous, undamped problem can be written as:

$$\hat{E}Iv''''(x) - Nv''(x) + (\omega^2 \rho w_b h_b)v(x) = 0 \tag{9.4}$$

This equation can be solved analytically, for example by the following procedure (Prak 1993):

- Laplace transformation of the differential equation
- Apply boundary conditions at x=0, which read $x(0)$=0 and $x'(0)$=0
- Inverse Laplace transform yields the solution with 2 integration constants

- Subtitution of the boundary conditions at $x=l$ ($x(l)=0$ and $x'(l)=0$) gives additional relations to eliminate the integration constants

The differential equation can be rewritten conveniently by introducing the parameters k and p:

$$v''''(x) - 2pk^2v''(x) - k^4v(x) = 0 \tag{9.5}$$

where:

$$k^4 = \omega^2 \rho w_b h_b / \hat{E}I \tag{9.6}$$

$$p = N / 2\hat{E}Ik^2 \tag{9.7}$$

Laplace transformation of (9.5) gives:

$$\begin{aligned} &s^4V(s) - s^3v(0) - s^2v'(0) - sv''(0) - v'''(0) - 2pk^2s^2V(s) + \\ &2pk^2sv(0) + 2pk^2v'(0) - k^4V(s) = 0 \end{aligned} \tag{9.8}$$

where $V(s)$ is the Laplace transform of $\overline{v}(x)$.

Applying the boundary conditions at $x = 0$, i.e. $v(x) = 0$ and $v'(x) = 0$, results in:

$$s\ V(s) - sv''(0) - v'''(0) - 2pk^2s^2V(s) - k\ V(s) = 0 \tag{9.9}$$

Eqn. (9.9) can be rewritten as follows:

$$V(s) = \frac{1}{k_1^2 + k_2^2}\left(\frac{1}{s^2 - k_2^2} - \frac{1}{s^2 + k_1^2}\right)\left(sv''(0) + v'''(0)\right) \tag{9.10}$$

where k_1 and k_2 are given by:

$$k_1{}^2 = k^2\left(\sqrt{p^2 + 1} - p\right) \tag{9.11}$$

$$k_2{}^2 = k^2\left(\sqrt{p^2 + 1} + p\right) \tag{9.12}$$

Inverse Laplace transformation of (9.10) yields:

$$v(s) = \frac{1}{k^2 + k_2^2}\left(C\ \left(\cosh(k_2x) - \cos(k\ x)\right) + C_2\left(k_2 \sin(k\ x) - k\ \sinh(k_2x)\right)\right) \tag{9.13}$$

Now, the boundary conditions at the other end of the beam, i.e. at $x = 0$, can be used to eliminate C_1 or C_2. For a clamped-clamped beam we have $v(l) = 0$ and $v'(l) = 0$, which results in the following set of equations:

$$\begin{aligned} &C_1\left(\cosh(k_2l) - \cos(k_1l)\right) = C_2\left(k_2 \sin(k_1l) - k_1 \sinh(k_2l)\right) \\ &C_1\left(k_2 \sinh(k_2l) - k_1 \sin(k_1l)\right) = C_2\left(k_1k_2 \cos(k_1l) - k_1k_2 \cosh(k_2l)\right) \end{aligned} \tag{9.14}$$

Solutions only exist if the following condition, i.e. the *characteristic equation* or *frequency condition*, is satisfied:

$$\cos(k_1 l)\cosh(k_2 l) - \frac{1}{2}\left(\frac{k_2}{k_1} - \frac{k_1}{k_2}\right)\sin(k_1 l)\sinh(k_2 l) = 1 \tag{9.15}$$

In this case, (9.14) results in the following relation between C_1 and C_2:

$$C_2 = C_1 \frac{\cosh(k_2 l) - \cos(k_1 l)}{k_1 \sinh(k_2 l) - k_2 \sin(k_1 l)} \tag{9.16}$$

Substitution of (9.16) into (9.13) yields the *mode shape* functions of the beam, which describe the shapes of the beam at resonance:

$$\begin{aligned} v(s) = \frac{C_1}{k_1^2 + k_2^2}\{&\cosh(k_2 x) - \cos(k_1 x) + \\ &\frac{\cosh(k_2 l) - \cos(k_1 l)}{k_1 \sinh(k_2 l) - k_2 \sin(k_1 l)}\left(k_2 \sin(k_1 x) - k_1 \sinh(k_2 x)\right)\} \end{aligned} \tag{9.17}$$

The characteristic equation (9.15) is satisfied for a discrete set of solutions yielding a discrete number of values for k, i.e. k_n for $n = 1,2,\ldots$. The corresponding resonance frequencies or eigenfrequencies ω_n are found from (9.6) and the associated mode shapes from (9.17). Because of the transcendental nature of the characteristic equation, a closed form expression for the eigenfrequencies as a function of the axial load N and the beam parameters cannot be obtained. If there is no axial force ($N = 0$), both k_1 and k_2 are equal to k and the following expression for the eigenfrequencies can be derived:

$$\omega_n = 2\pi n_i \frac{h}{l^2}\sqrt{\frac{\hat{E}}{\rho}} \tag{9.18}$$

For this case the values of n_i for the first five eigenfrequencies are listed in Table 9.1. These values as well as the values for structures with different boundary conditions can also be found in (Blevins 1979). The corresponding mode shape functions are shown in Fig. 9.4. The dependence of the eigenfrequencies on the axial load N, which is important if the beam is used as a resonant strain gauge, will be discussed in section 9.1.5.

Table 9.1. Coefficients for the first five eigenfrequencies of an axially unloaded microbridge.

i	n_i
1	1.028
2	2.834
3	5.555
4	9.182
5	13.72

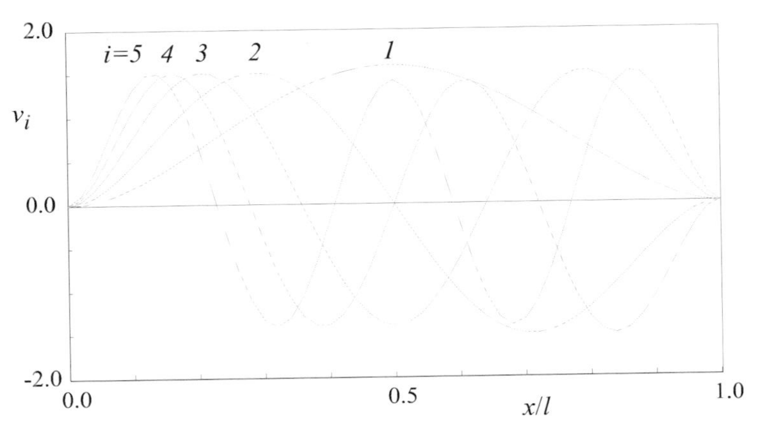

Fig. 9.4. First 5 normalized mode shapes of an unstrained microbridge. Note that the values of the extrema of the modes are nearly equal. For mode 1 to 5 they are 1.588, 1.510, 1.512, 1.512 and 1.512, respectively

9.1.4 Solving the Inhomogeneous Problem by Modal Analysis

In the previous section, we have seen how the homogeneous problem can be solved using Laplace transforms. In principle, the inhomogeneous case, i.e. when the beam is driven by a harmonic force or moment, can also be solved in this way, provided that the drive functions are not too complicated. However, as mentioned before, the results are hopelessly complex mathematical functions and do not provide much insight into the physics of the beam. Therefore, in this section, a much more elegant method is presented to describe the behaviour of a driven microbeam.

If a beam is driven (by some wild excitation function) and its vibration is damped, it can show any arbitrary (continuous) shape obeying the boudary conditions as illustrated in Fig. 9.5. It can be shown that the set of mode shape functions is an independent (orthogonal) set of functions: every arbitrary shape of the beam can be written as a weighed sum of the mode shape functions. This is a central statement in a technique called modal analysis (Meirovitch 1967). For example, the shape in Fig. 9.5 can be obtained by a combination of the first and second mode-shape. In general, the solution of the differential equation (9.3) can be written as the following expansion series:

$$\bar{v}(x) = \sum_{i=1}^{\infty} \bar{y}_i v_i(x) \tag{9.19}$$

In this equation, $v_i(x)$ are the *mode shapes*, and $\bar{y}_i$ are the *modal coordinates* (also called *generalized coordinates*).

The modal analysis strongly reminds us of the Fourier analysis. However, in modal analysis the mode shapes are used as a set of base functions instead of the sine and cosine functions in the case of a Fourier analysis. Just like a Fourier series, (9.19) has its inverse equation:

$$\overline{y}_i = \frac{1}{l}\int_0^l \overline{v}(x)v_i(x)dx \tag{9.20}$$

Very important properties of the mode shapes are that they are *orthogonal* and that they form a *complete* set, i.e.: *any* function satisfying the boundary conditions can be expanded in the mode shapes in one unambiguous way.

The differential equation (9.3) can be rewritten in a modal representation as follows (Meirovitch 1967):

$$M_i\ddot{\overline{y}}_i + R_i\dot{\overline{y}}_i + K_i\overline{y}_i = \overline{P}_{e,i}\, i = 1,2,\ldots \tag{9.21}$$

Thus, in this way we obtain an infinite number of simple second-order differential equations: one equation for every mode. This is illustrated by Fig. 9.6. Usually only a small number of modes play a dominant role and only these equations have to be taken into account. M_i, R_i and K_i denote the *modal mass, modal damping-constant* and *modal spring-constant*, respectively, and are defined as follows:

$$\begin{aligned} M_i &= \rho w_b h_b l \\ R_i &= cl \\ K_i &= \rho w_b h_b l \omega_i^2 \end{aligned} \tag{9.22}$$

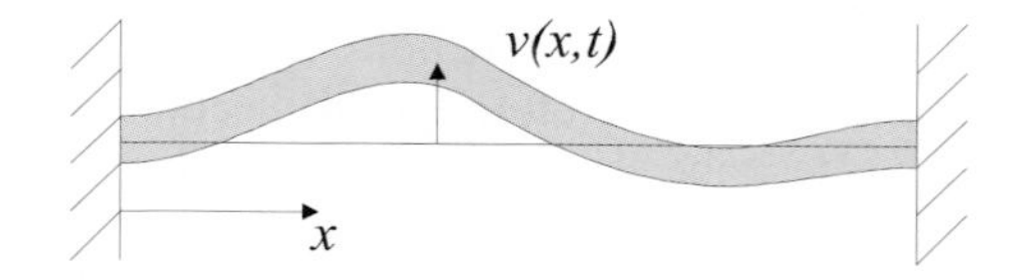

Fig. 9.5. Arbitrarily shaped microbridge

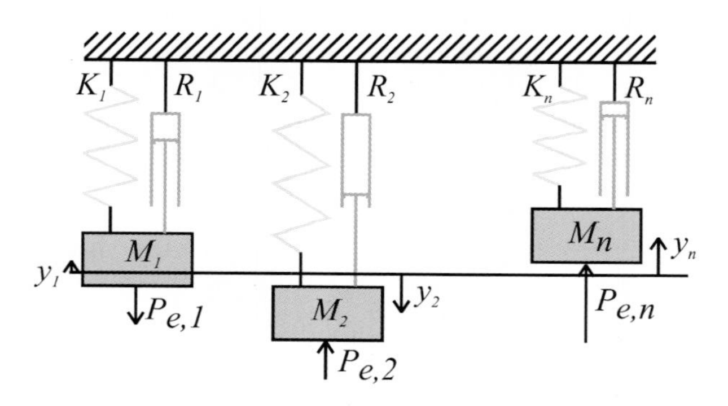

Fig. 9.6. Representation of the distributed beam by an infinite number of lumped elements

$\overline{P}_{e,i}$ are the *modal (generalized) loads*. They are a measure for the efficiency with which a specific mode is excited:

$$\overline{P}_{e,i} = \int_0^l \overline{P}(x) v_i(x) dx, \overline{P}(x) = \frac{1}{l} \sum_{i=1}^{\infty} \overline{P}_{e,i} v_i(x) \tag{9.23}$$

Summarizing, modal analysis allows us to describe the system as an infinite number of independent systems, each of which can be described by a standard equation of motion of second order in time. The solution of a particular mode can easily be found from (9.21) and is given by:

$$\overline{y}_i = \frac{\overline{P}_{e,i} / K_i}{1 + j \frac{1}{Q_i} \frac{\omega}{\omega_i} - \left(\frac{\omega}{\omega_i} \right)^2} \tag{9.24}$$

where Q_i is the *modal quality factor* defined by:

$$Q_i = \frac{\sqrt{M_i K_i}}{R_i} = \frac{M_i}{R_i} \omega_i \tag{9.25}$$

9.1.5 Response to Axial Loads

In Sect. 9.1.3 it was shown how the resonance frequency of an unloaded microbridge can be obtained by solving the characteristic equation (9.15). Unfortunately, a closed-form expression for the resonance frequency as a function of the axial load N could not be obtained. In order to obtain a relation between the resonance frequency and the axial force, we either have to solve the characteristic equation numerically or obtain an approximate solution, for example by employing Rayleigh's energy method (Shames 1985). The following expression can be obtained this way:

$$\omega \ (N) = \omega \ (0) \sqrt{1 + \gamma \ \frac{N}{E w_b h_b} \left(1 - \nu^2\right) \left(\frac{l}{h_b} \right)^2} \tag{9.26}$$

where

$$\omega_n(0) = \frac{\alpha_n^2}{\sqrt{12}} \frac{h_b}{l^2} \sqrt{\frac{E}{\rho \left(1 - \nu^2\right)}} \tag{9.27}$$

The coefficients α_n and γ_n are only dependent on the mode shape of mode n. They can be calculated using an approximate mode shape function (Shames 1985; Tilmans 1992) or fitted numerically. Numerically fitting of ω and $d\omega_n / dN$ at $N = 0$ results in the coefficients listed in Table 9.2. Fig. 9.7 shows a plot of the normalized angular resonance frequencies as a function of the normalized axial force. The error in the approximation given by (9.26) is smaller than 0.5% if

$\gamma_n Nl^2 / 12\hat{E}I < 1$ (Bouwstra 1991). Fig. 9.8 shows a plot of the error as a function of the normalized axial force. An empirical correction may be added to (9.26) in the form of an additional quadratic term in the axial force under the square root (Mullem 1991).

The normalized sensitivity of a sensor output signal with respect to the measurand is often referred to as the gauge factor. For resonant strain gauges, the gauge factor describing the sensitivity of the gauge to changes in the axial force N can be defined as:

$$G_{Nn} = \left[\frac{1}{\omega_n} \frac{\partial \omega_n}{\partial N} \right]_{N=N_0} \tag{9.28}$$

However, in most cases the axial force is a result of an axial elongation of the beam and it is more useful to define a gauge factor $G_{\varepsilon n}$:

$$G_{\varepsilon n} = \left[\frac{1}{\omega_n} \frac{\partial \omega_n}{\partial \varepsilon} \right]_{\varepsilon=\varepsilon_0} \tag{9.29}$$

Using (9.26) this can be written as:

$$G_{\varepsilon n} = \frac{1}{2} \left[\frac{\gamma_n \left(1-\nu^2\right) \left(\dfrac{l}{h_\mathrm{b}}\right)^2}{1+\gamma_n \varepsilon_0 \left(1-\nu^2\right) \left(\dfrac{l}{h_\mathrm{b}}\right)^2} \right] \tag{9.30}$$

The parameter l/h_b is very important with respect to the gauge factor. With today's fabrication technologies l/h_b ratios of 100 are easily obtained and ratios of 1000 are possible. This results in very high stress sensitivities as compared to the gauge factors of for example piezoresistive strain gauges, which are typically in the order of 100 for silicon strain gauges and much lower for metal strain gauges (see Chap. 5).

Table 9.2. Coefficients α_n and γ_n for a clamped-clamped microbeam (eqns. (9.26) and (9.27))

mode	α_n	γ_n
1	4.730	0.2949
2	7.854	0.1453
3	11.00	0.08119
4	14.14	0.05155
5	17.28	0.03553

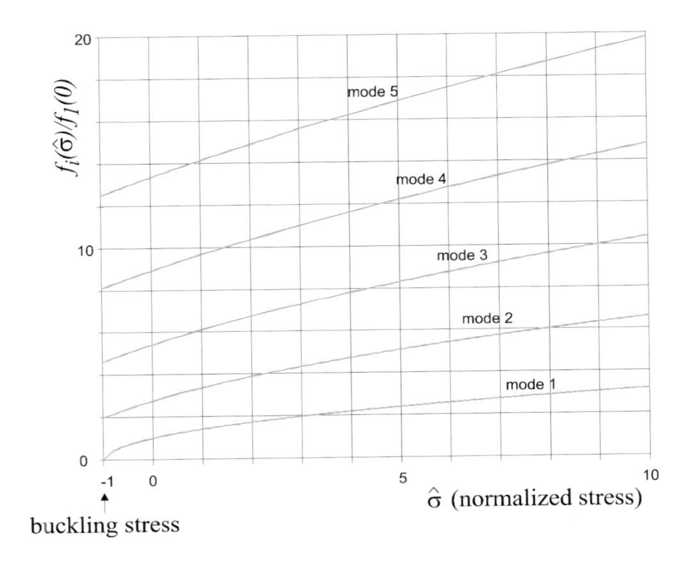

Fig. 9.7. Resonance frequency of a prismatic microbridge as a function of the normalized stress. The frequencies are normalized with respect to the stress free resonance frequency of the first mode

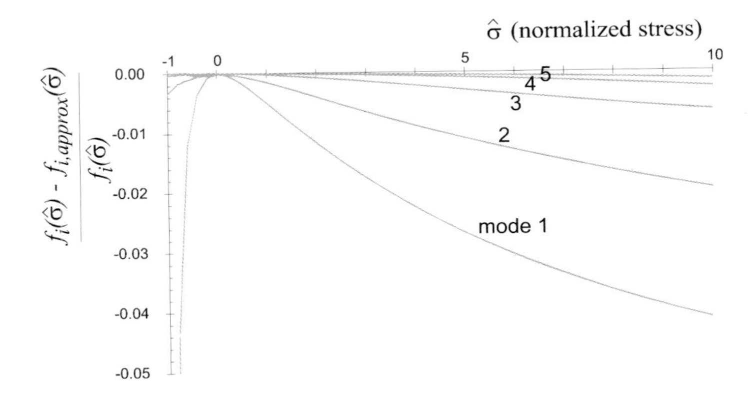

Fig. 9.8. Relative difference between the numerically calculated resonance frequency and the approximated resonance frequency for the first 5 modes of a microbridge according to Table 9.2

9.1.6 Quality Factor

The mechanical quality factor Q is a measure for the energy losses of the resonator or in other words, a measure for the mechanical damping. The Q-factor is defined as the ratio between the total energy stored in the vibration and the energy loss per cycle:

$$Q \equiv 2\pi \frac{\text{total energy stored in vibration}}{\text{dissipated energy per period}} \tag{9.31}$$

Low energy losses imply a high Q-factor. The Q-factor cannot be determined directly, but instead can be deduced from the response characteristics of the resonator. One common method of determining Q is from the steady-state frequency plot of a resonator excited by a harmonic force with constant amplitude:

$$Q \cong \frac{\omega_{\mathrm{res}}}{\Delta\omega_{-3\mathrm{dB}}} \tag{9.32}$$

where ω_{res} is the frequency with maximum frequency response and $\Delta\omega_{-3\mathrm{dB}}$ is the half-power bandwidth of the frequency response. Eqn. (9.32) indicates that Q is a measure for the sharpness of the frequency selectivity of the resonator. A high Q-factor means a sharp resonance peak and has several advantages:

- the energy required to maintain the vibration is low,
- minimal effect of the electronic circuitry on the oscillation frequency,
- low sensitivity to mechanical disturbances.

Usually, three mechanisms of energy loss are distinguished (Tilmans 1992): losses into the surrounding medium, losses into the mount used to support the resonator, and intrinsic damping due to energy losses inside the material of the resonator. Energy losses into the surrounding medium can be minimized by placing the resonator in an evacuated cavity (Ikeda 1988, Tilmans 1991, Guckel 1990). This is especially important if the resonator is close to another stationary surface and squeeze film damping occurs due to the pressure built-up in the intervening space (Howe 1986, Starr 1990). Energy losses into the mechanical support can be minimized by mechanically isolating the resonator from the mount or by using special balanced resonator structures (Stemme 1991) like the well-known double-ended tuning fork (Ueda 1985) and triple beam structure (Kirman 1983).

9.1.7 Nonlinear Large-Amplitude Effects

In the derivation of the differential equation (9.1) small vibration amplitudes were assumed. If the amplitude of vibration $v_{\max}$ becomes comparable to the radius of gyration, $r = \sqrt{I / w_{\mathrm{b}} h_{\mathrm{b}}}$, of the cross section of the beam, the dependence of the axial force on the amplitude of the vibration has to be taken into account, resulting in a non-linear term in (9.1) (Eisley 1964). If the distance between the ends of the beam is rigidly fixed, a tensile axial force will be developed by the transverse deflection. This will result in an increase of the resonance frequencies of the beam and is known in the literature as the *hard-spring effect*. To get some idea of the impact of this effect on the natural frequencies, an additional potential energy term is included in Rayleigh's quotient. The additional stretching of the mid-plane of the beam results in a potential energy change U_{NL} given by:

$$U_{NL} = \frac{1}{2}\int_0^l \frac{Ew_b h_b}{4}\left(\frac{dw}{dx}\right)^4 dx \tag{9.33}$$

Including this term in Rayleigh's quotient (Shames 1985) results in a modified expression of eqn. (9.26):

$$\omega_n(N) = \omega_n(0)\sqrt{1+\gamma_n \frac{N}{Ew_b h_b}\left(1-\nu^2\right)\left(\frac{l}{h_b}\right)^2 + \beta_n\left(1-\nu^2\right)\left(\frac{v_{max}}{h_b}\right)^2} \tag{9.34}$$

where β_n is a constant depending on the mode shape of mode n.

The dependence of the resonance frequency on the vibrational amplitude of the fundamental mode of a rigidly clamped-clamped beam, together with a typical large-amplitude forced response plot, is shown in Fig. 9.9. This Figure also indicates the effect of the magnitude of the driving force on the response plot (Landau 1964). If the magnitude of the driving force exceeds a critical value F_{crit}, the amplitude becomes three-valued within a range of frequencies. This frequency interval defines the region of hysteresis, with two stable points, one at large amplitude and the other at small amplitude. To avoid hysteresis, the magnitude of the driving force should be smaller than the critical load. It can be shown that the critical load is inversely proportional to $Q^{3/2}$, where Q is the quality factor of the resonator. At the critical load, the amplitude of the vibration is defined as the critical amplitude $v_{max,crit}$, being inversely proportional to $Q^{1/2}$. Considering a clamped-clamped beam to behave as a simple spring-mass system with a restoring force having a cubic dependence on the amplitude, it can be shown (Landau 1964) that the critical amplitude can be approximated by:

$$\frac{v_{max,crit}}{h_b} \approx \sqrt{\frac{2}{Q\beta_n(1-\nu^2)}} \tag{9.35}$$

Hence, even though a high Q has several advantages as indicated in the previous section, it enhances the chance of hysteresis occurring.

Noise or other instabilities of the vibrational amplitude will limit the ultimate frequency resolution or stability of the gauge. Any frequency change caused by an amplitude change must be small compared to the minimum frequency change one wishes to resolve. It can be derived from Eqns. (9.34) and (9.29) that the changes in amplitude Δv_{max} should be smaller than:

$$\frac{\Delta v_{max}}{v_{max}} << \left(\frac{\gamma_n}{2\beta_n G_{\varepsilon n}}\right)\left(\frac{l}{h_b}\right)^2\left(\frac{h_b}{v_{max}}\right)^2\left(\frac{\Delta\omega}{\omega}\right)_{min} \quad (v_{max} < v_{max,crit}) \tag{9.36}$$

where $(\Delta\omega/\omega)_{min}$ is the minimum resolution required. For a clamped-clamped beam with an aspect ratio $l/h = 200$, $h/v_m = 500$, zero residual strain ($\varepsilon_0 = 0$) and a minimum frequency resolution of 10^{-6}, (9.36) demands a relative variation of the amplitude much smaller than 0.5. If the beam thickness is $h_b = 1$ µm, this means that v_{max} is 2 nm and Δv_{max} much smaller than 0.5 nm. The critical amplitude in this case is approximately equal to 20 nm for $Q = 10000$.

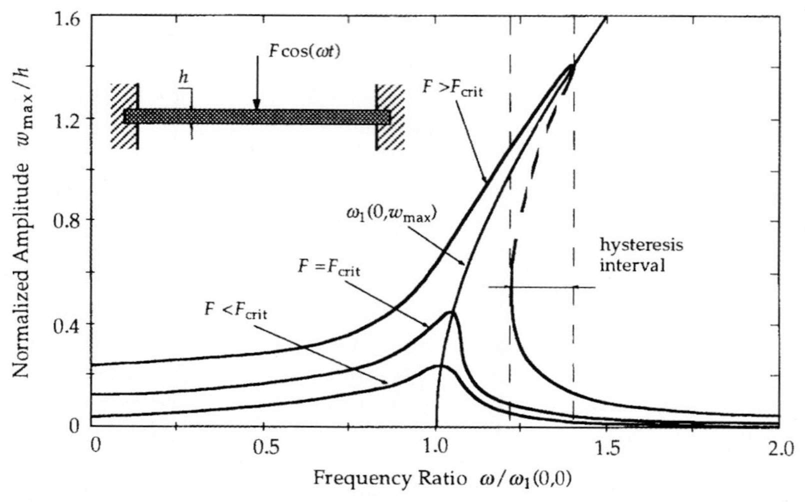

Fig. 9.9. Forced response plots, showing the normalized amplitude vs. The normalized frequency for a clamped-clamped beam with a rectangular cross section. If the magnitude F of the excitation force exceeds a critical value F_{crit}, the amplitude becomes three-valued for a particular frequency interval. The dashed part of the curve in this interval indicates unstable points (Landau 1964)

9.2 Excitation and Detection Mechanisms

In order to determine the resonance frequency, the resonator has to be brought into vibration and the vibration has to be detected. Usually an oscillator circuit is realized where the resonator is the frequency determining element (see Chapter 10). Six pairs of excitation and detection mechanisms have been proposed:

- electrostatic excitation and capacitive detection
- magnetic excitation and detection
- piezoelectric excitation and detection
- electrothermal excitation and piezoresistive detection
- optothermal excitation and optical detection
- dielectric excitation and detection

The first two excitation principles produce a transversal force on the resonator while the latter four produce bending moments.

9.2.1 Electrostatic Excitation and Capacitive Detection

Electrostatic excitation is simply accomplished by the attracting force between two capacitor plates, one being the resonator and the other being the substrate. Capacitive detection is based on the fact that a charged capacitor causes a current if the capacitance fluctuates. Electrostatic excitation and capacitive detection of a microbridge can be realized relatively easy: the electrodes for excitation and detection can be diffused into the substrate and the microbridge crossing the electrodes is connected to ground. A problem is the squeeze film damping caused by the narrow gap between the substrate and the microbridge. For a high Q-factor operation in vacuum is required. Another problem is the small capacitance value which results in a very small detection signal. The detection signal may further be obscured by electrical feedthrough due to parasitic capacitances.

9.2.2 Magnetic Excitation and Detection

Magnetic excitation and detection is easily realized with the help of an external magnetic field. An electrical current flowing through the microbridge results in a Lorentz force. Usually an H-shaped resonator is used where one beam of the H is used for excitation and the other for detection (magnetic induction). An example of such a sensor was already discussed in Chap. 6 (Ikeda 1990).

If low doped silicon is used for the material from which the microbridges are made, this transduction has the important advantage that one can make use of silicon alone, thus without spoiling the superior material properties of single crystalline silicon by the deposition of thin films. A problem is that the signals are very small and that it is difficult to integrate the magnet in the sensor. There is also a problem with the heat dissipated by conducting a current through the microbridge, which causes compressive stress in the microbridge due to thermal expansion.

9.2.3 Piezoelectric Excitation and Detection

Piezoelectric excitation and detection is widely used for quartz resonators since quartz itself is piezoelectric. For the piezoelectric transduction of silicon based devices, other materials have to be used, preferably in the form of thin films. A commonly used material is ZnO, which can be sputtered (reactive magnetron sputtering). If a ZnO film is sandwiched between two electrodes, a voltage gives rise to both a change in thickness and a change in lateral dimensions of the film. The latter effect is exploited to force bridges and membranes into bending mode vibrations. The problem with piezoelectric transduction lies in the piezoelectric material. Sputtered ZnO is a semiconductor with an overshoot of Zn atoms and the conductivity poses a lower limit to the frequency below which no electric field can exist in the material. This problem can be overcome by depleting the thin film, e.g. in a MOS structure (Blom 1990). Other problems with ZnO are that it is sensitive to humidity, light and the history of thermal treatment and that it is not IC-compatible. Furthermore, the deposition of a ZnO layer on the resonator reduces the Q-factor and increases the temperature sensitivity due to differences in thermal expansion.

9.2.4 Electrothermal Excitation and Piezoresistive Detection

Electrothermal excitation and piezoresistive detection are easily realized. The two transduction mechanisms form a pair because both can be produced using the same material (doped polysilicon thin films) in the same processing steps. The excitation mechanism is based on thermal expansion. A heat source e.g. on top of a beam gives rise to a temperature gradient along and normal to the beam, the latter causing a bending because of a gradient in the thermal expansion. This transduction mechanism is optimal if there is a temperature gradient across the whole beam - in other words if the penetration depth of the thermal wave into the beam is of the order of the thickness of the beam or larger. The induced mechanical bending moment is then independent of the frequency. If the penetration depth is smaller than the thickness of the beam, the induced moment becomes inversely proportional to the frequency. This applies for high frequencies.

Resistors of the same material as for the excitation can be used for the detection by exploiting the relatively large piezoresistive effect of polysilicon. The sensitivity is low, but the electrical feedthrough can be made very small. The heat dissipated in the microbridge introduces thermal compressive stresses so that the microbridges can even buckle (Lammerink 1990). This causes problems in the performance of the sensor. Even if buckling can be avoided the thermal compressive stress must be well controlled.

Electrothermal excitation is very useful for sensors which explicitly make use of the thermal domain such as Bouwstra's mass flow sensor (Bouwstra 1990, 1989) which is based on an anemometer principle. In other applications the thermal stress is too disturbing.

9.2.5 Optothermal Excitation and Optical Detection

Optothermal excitation and optical detection provides the important advantage of the possibility to avoid any electrical voltages at the sensor. It is of importance for the use of sensors in explosive regions and in regions with high electric fields. The excitation mechanism is similar to electrothermal excitation with the difference that the heat source is provided by the absorption of light. For optical detection several methods have been described: variation of transmission through a gap in a wave guide by means of a shutter (Lammerink 1987), use of an optical proximity sensor (Horsthuis 1982) or by integrated interferometry using the surface of the vibrating beam and the end of a glass fiber as mirrors (Warkentin 1987).

9.2.6 Dielectric Excitation and Detection

Dielectric excitation is based on the lateral deformation of a dielectric thin film, sandwiched between a top and bottom electrode, due to an electrostatic force arising if a voltage is applied across the electrodes. The lateral deformation will cause bending moments. The detection is capacitive, based on the change of the capacitance of a dielectric capacitor if the dielectric is deformed. So far, the method seems promising, but the signals appear to be extremely small. Only if materials with a very high dielectric constant, such as PZT, having a reported dielectric constant as large as 2500 for a thin film (Böffgen 1990), are used is this method interesting for future applications.

9.3 Examples and Applications

In this section some examples of resonant sensors will be discussed. A global overview will be given, for processing details reference is made to the literature.

Several technologies have been developed over the past three decades to fabricate micromechanical resonators. In general, two main technologies can be distinguished: bulk micromachining and surface micromachining. Combinations of both have also been used. Examples of bulk micromachined resonating structures are resonating pressure sensors (Greenwood 1984, Lammerink 1987, Thornton 1990, Buser 1991, Petersen 1991), resonating force sensors (Blom 1989, Van Mullem 1991, Funk 1995), resonating accelerometers (Tudor 1988, Satchell 1989, Chang 1990), a resonating mass-flow sensor (Bouwstra 1990, Legtenberg 1991) and a vibration sensor (Benecke 1985). The first surface micromachined resonator was fabricated by Nathanson *et al.* in 1965 (Nathanson 1965, 1967). They used metal films for both the construction material and the sacrificial layer. More recently 'construction material/sacrificial layer' combinations of polysilicon/oxide (Howe 1986, Putty 1989, Guckel 1990), polysilicon/nitride (Guckel 1989), nitride/(poly)silicon (Bouwstra 1990, Legtenberg 1991, Guckel 1986, Sugiyama 1986), oxide/silicon (Parameswaran 1989) and polyimide/aluminium (Schmidt 1987) have been used. Prototypes of resonating structures include a resonant vapour sensor (Howe 1986), a resonating mass-flow sensor (Bouwstra 1990), a resonant force sensor (Sniegowski 1990, Zook 1991), a resonating accelerometer (Chang 1990) and resonating polysilicon microbridges (Howe 1986, Putty 1989, Guckel 1990, Linder 1990).

A large number of *resonant pressure sensors* have been proposed. The first designs were based on resonating membranes as indicated in Fig. 9.10(a) (Smits 1983, Lammerink 1985, Bouwstra 1987). These devices seemed to be ideal for pressure sensing (a pressure difference across a membrane stiffens it). However, They suffered from the problem that the frequency also shifted due to the mass of the gas in the vicinity of the membrane. Hence, the resonance frequency was not only dependent on the pressure, but also on the sort of gas and its temperature. This concept was soon abandoned in favor of a new generation in which the membrane itself does not vibrate, but instead a resonator is fabricated on the membrane, which senses the static deflection of the membrane (Fig. 9.10(b)). By the end of the 1980's resonators became sealed in local vacuums, fabricated using reactive sealing techniques compatible with micromachining. This concept results in high quality factors and will probably be used in most future resonant pressure sensors.

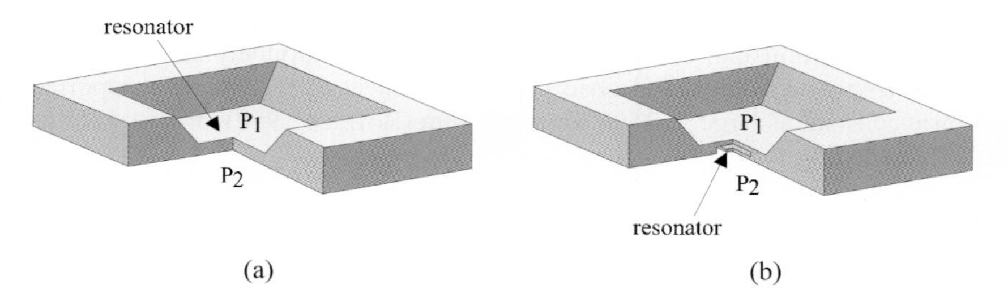

Fig. 9.10. Two basic types of resonant pressure sensors: (a) resonant membrane and (b) resonator integrated in membrane

One of the first silicon pressure sensors with a resonator integrated on a membrane was developed by Greenwood (Greenwood 1984, 1988) and is commercially available from Druck Ltd.. This sensor does not use a local vacuum around the resonator. Instead, a vacuum is realized in the package, which also serves as a reference pressure for the membrane (see Chap. 6, Figs. 6.34 and 6.35). A sensor with a local vacuum cavity around the resonator was developed by Ikeda et al. at Yokogawa Electric Corporation in 1988 (Ikeda 1988, 1990). This sensor and it's fabrication process was extensively discussed in Chap. 6 (Fig. 6.37). A monocrystalline silicon resonator was realized by using dopant selective etching.

An alternative way of fabricating sealed resonators has been described by Guckel et al. (Guckel 1989, 1990). They realized polysilicon resonators with a process based on sacrificial layer etching and reactive sealing techniques (Guckel 1984). Hermetically sealed cavities can be formed with a low residual pressure, as indicated by the obtained quality factors of around 35 000. Fine-grained tensile polysilicon, grown using low-pressure chemical vapour deposition (LPCVD), is used as a construction material and silicon oxide and/or nitride can be used as a sacrificial layer. The sacrificial layer etchant used is hydrofluoric acid. 'Sticking' of the resonator and the sealing cap to the substrate is avoided by freeze-sublimation procedures (Guckel 1989).

A *resonant mass-flow sensor* was proposed by Bouwstra et al. (Bouwstra 1990). The sensor consists of a bridge across a V-shaped groove, as schematically illustrated in Fig. 9.11. Polysilicon resistors on the bridge are used for thermal excitation and piezoresistive detection. The gas flow controls the temperature of the heated bridge by convective cooling and since the mechanical stresses of the vibrating bridge are dependent upon the temperature, a sensor with a flow dependent resonance frequency is obtained.

A very interesting and completely different mass-flow sensor was presented by Enoksson *et al.* (Enoksson 1997). They realized a channel inside the resonator. A liquid mass-flow passing through this channel induces a Coriolis force, resulting in a transfer of energy to a second vibration mode. The amplitude of the second vibration is a measure for the mass-flow and is detected optically.

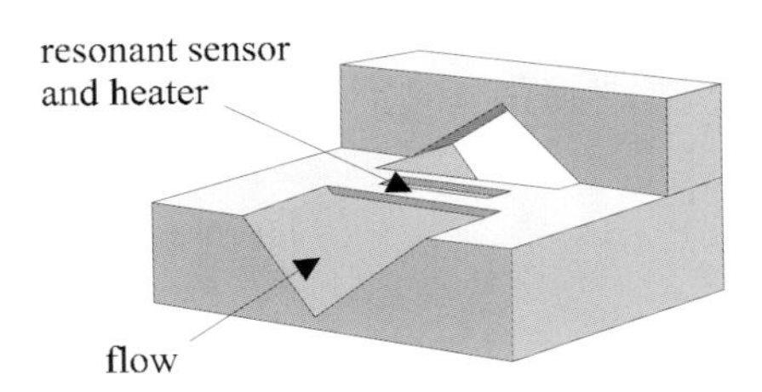

Fig. 9.11. Structure of a resonant mass flow sensor (Bouwstra 1990)

Fig. 9.12 shows the structure of a resonant vapor sensor, which was presented by Howe and Muller (Howe 1986). The sensor consists of a polysilicon beam with a polymer layer on top of it. The polysilicon beam is made by surface micromachining using a sacrificial silicon dioxide layer. Absorption of chemicals from the surrounding medium results in a change of the mass of the resonator, which results in a change in resonance frequency.

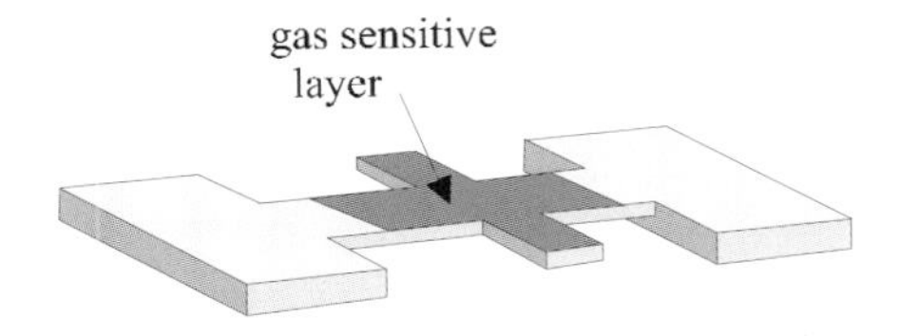

Fig. 9.12. Vapor sensor consisting of resonant microbridge with gas sensitive layer (Howe 1986)

Fig. 9.13 shows the basic structure of a *two-axis resonant accelerometer* (Chang 1990). A proof mass is suspended by four polysilicon microbridges. Acceleration in either direction will result in differential stresses in the corresponding pair of resonators. The difference between the resonance frequencies is a measure for the acceleration.

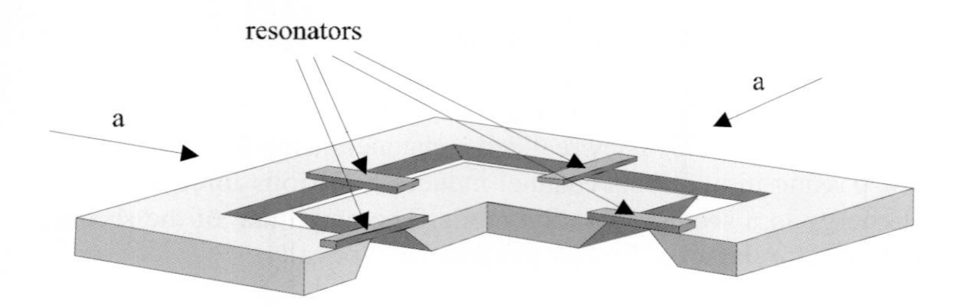

Fig. 9.13. Resonant acceleration sensor (Chang 1990)

10. Electronic Interfacing

In this chapter we will discuss some basic topics with respect to the electronic interface circuits needed for microsensors. The chapter is divided in three parts, corresponding to the three most frequently used readout mechanisms, namely piezoresistive, capacitive and resonant.

First, in Sect. 10.1 we will discuss piezoresistive sensors. The section starts with a short explanation of the piezoresistive effect and the use of bridge circuits to measure the relatively small changes in resistance. Next, some amplifier circuits are presented which are commonly used to amplify the bridge output voltage. The section concludes with some more advanced topics like feedback control loops, switching techniques to reduce low frequency noise and offset and some basic analog-to-digital conversion techniques.

In Sect. 10.2 capacitive sensors are discussed. Several ways to measure capacitance are presented, including bridge and oscillator circuits. A problem with measuring small capacitances is the influence of parasitic capacitance. Therefore, a large part of the section is devoted to techniques to eliminate this influence.

Resonant sensor interface circuits are discussed in Sect. 10.3. Resonant sensors are usually connected in an oscillator circuit. The actual implementation of the circuit depends very much on the actuation and detection method of the resonator. For simple one-port capacitive resonators a standard crystal oscillator circuit may be suitable. Other resonators require much more complicated circuits. An important issue is the so-called hard-spring effect (see Chap. 9), which occurs at large driving amplitudes and is an important source of hysteresis and nonlinearity. Furthermore, the resonance frequency is usually dependent on the driving amplitude thus accurate amplitude control is extremely important.

An interesting point of discussion is whether the electronic circuits have to be integrated on the sensor chip. There may be several reasons for doing this. Most important is the fact that it may improve the sensor's performance. Especially in the case of capacitive and resonant sensors parasitic capacitance or electromagnetic interference can easily adversely affect the performance. Obviously, these effects are minimized if the circuits and the sensor structure are on the same chip. Another motivation for monolithic integration can be the increased reliability. Integrating the electronics with the sensor reduces the number of interconnections and, hence, increases the reliability. On the other hand, monolithic integration imposes severe restrictions on the sensor design, as the sensor fabrication process has to be compatible with the fabrication process of the electronics. For example, the high electric fields associated with anodic bonding may not be compatible with a CMOS process.

In some cases, where the sensor electronics require a much larger die area than the mechanical part of the sensor and when the sensor can be fabricated with relatively few additional processing steps, integrating the sensor on the electronics chip may reduce the fabrication costs.

For optimal performance, a hybrid solution - where the sensor and electronics are realized on separate chips but mounted in the same package - will often be the best choice. In this way the sensor and electronics can be optimized independently from each other.

10.1 Piezoresistive Sensors

The resistance of a piece of material is proportional to the ratio of the length l and the cross section A:

$$R = \rho \cdot \frac{l}{A} \tag{10.1}$$

The constant ρ is called the specific resistance and is dependent on the type of material. In Chap. 5 we found that a strain results in a relative change in resistance (5.4):

$$\frac{dR}{R} = (1+2\gamma)\varepsilon_l + \pi_l E \varepsilon_l + \pi_t E \varepsilon_t \tag{10.2}$$

where π_l and π_t are the piezoelectric coefficients in the longitudinal and transversal directions, respectively.

Both the specific resistance ρ and the piezoresistive coefficients π_l and π_t are temperature dependent. Therefore, piezoresistive sensors always require some sort of temperature compensation. Especially in monocrystalline silicon the dependence of ρ on the temperature is very large. The resistivity increases with temperature due to the decreasing electron/hole mobility. Polycrystalline strain gauges are composed of small grains of silicon crystals and the increasing resistivity of the grains can be compensated by a decreasing resistivity at the grain boundaries by suitably choosing the doping level. On the other hand monocrystalline strain gauges generally show better performance with respect to noise and drift.

10.1.1 Wheatstone Bridge Configurations

A well-known technique to compensate for temperature variations is to use a Wheatstone bridge configuration as indicated in Fig. 10.1. The output voltage of the bridge is given by:

$$V_{\text{bridge}} = \frac{R_4}{R_2 + R_4} \cdot V_{\text{supply}} - \frac{R_3}{R_1 + R_3} \cdot V_{\text{supply}} \,, \tag{10.3}$$

which can easily be rewritten as:

$$V_{\text{bridge}} = \left(\frac{R_1 R_4 - R_2 R_3}{(R_2 + R_4)(R_1 + R_3)} \right) \cdot V_{\text{supply}} \tag{10.4}$$

Normally all resistors are chosen equal to each other, resulting in zero output voltage. For sensor applications at least one of the resistors has to be dependent on the measurand. In this case, e.g. when $R_1 = R_2 = R_3 = R$ and $R_4 = R + \Delta R$, the bridge output voltage can be expressed as:

$$V_{\text{bridge}} = \left(\frac{R \cdot (R + \Delta R) - R \cdot R}{(2R + \Delta R) \cdot 2R} \right) \cdot V_{\text{supply}} = \left(\frac{\Delta R / R}{4 + 2 \cdot \Delta R / R} \right) \cdot V_{\text{supply}} \tag{10.5}$$

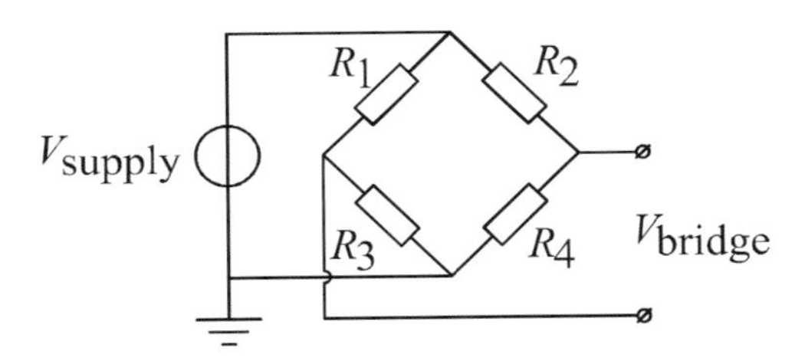

Fig. 10.1. Wheatstone bridge configuration

It follows that for small changes in resistance, i.e. $\Delta R << R$, the output voltage changes approximately linear with ΔR:

$$V_{\text{bridge}} \approx \frac{1}{4} \cdot \frac{\Delta R}{R} \cdot V_{\text{supply}} \tag{10.6}$$

For large resistance variations the bridge output is non-linear as is apparent from (10.5) because of the presence of ΔR in the denominator. This nonlinearity is due to the fact that the change in resistance causes a change in the current flowing through the bridge. An improvement by a factor of 2 can be obtained by biasing the bridge with a current source instead of a voltage source. In this case the bridge output voltage becomes:

$$V_{\text{bridge}} = \left(\frac{\Delta R}{4 + \Delta R / R} \right) \cdot I_{\text{supply}} \tag{10.7}$$

The nonlinearity can be eliminated completely by using two resistors that are dependent on the measurand, e.g. $R_2 = R_3 = R$ and $R_1 = R_4 = R + \Delta R$, in combination with current biasing. Fig. 10.2 shows a possible implementation. In this circuit the operational amplifier controls the bridge voltage such that the voltage drop across R_s equals V_{bias}. Thus, the bridge current is given by $I_{\text{supply}}=V_{\text{bias}}/R_s$. The bridge output voltage is now given by:

$$V \quad = \left(\frac{\Delta R}{2}\right) \cdot I \quad = \left(\frac{\Delta R}{2 \cdot R_s}\right) \cdot V_{\text{a}} \tag{10.8}$$

We see that besides the fact that the response is now linear the sensitivity has increased by a factor of 2.

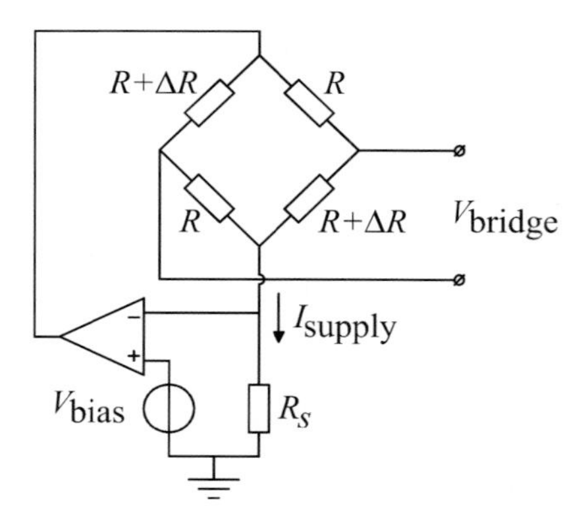

Fig. 10.2. Wheatstone bridge with two active elements and current biasing

Often it is possible to dimension and position the strain gauges in such a way that 2 gauges have an increasing resistance with the measurand and two other gauges have a decreasing resistance. This is indicated in Fig. 10.3. Compared to a bridge with two active elements the sensitivity has again doubled:

$$V_{\text{bridge}} = \Delta R \cdot I_{\text{supply}} = \left(\frac{\Delta R}{R}\right) \cdot V_{\text{supply}} \tag{10.9}$$

From this equation we see that the response is linear for both current and voltage biasing. This is because the total bridge resistance is now independent of the measurand and the supply voltage and bridge current are directly related to each other. However, current and voltage biasing do give quite different results when the temperature dependence is considered. From both (10.8) and (10.9) we see that when using current biasing the bridge output voltage is proportional to the change in resistance ΔR and independent of the total strain gauge resistance R. Using voltage biasing the bridge output is proportional to the relative change in resistance $\Delta R/R$. Often the temperature dependence of the specific resistance ρ and the piezoresistive coefficients π_l and π_t in (10.2) is such that current biasing results in first order temperature compensation. Another simple and frequently used temperature compensation is to use voltage biasing with a temperature dependent resistor in series with the bridge.

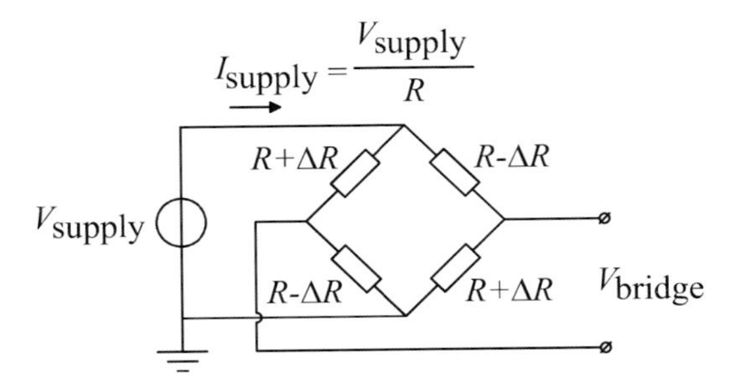

Fig. 10.3. Wheatstone bridge with 4 active elements

10.1.2 Amplification of the Bridge Output Voltage

In the previous section we have seen that the bridge output voltage is proportional to the relative change in resistance. As the change in resistance is generally in the order of 0.01 to 0.1%, the bridge output voltage is quite small and needs to be amplified. Fig. 10.4 shows a simple differential amplifier circuit that can be used for this purpose. An important property of this circuit is that the differential input voltage

$$V_{diff} = V_{diff+} - V_{diff-} \tag{10.10}$$

is amplified and the common mode input voltage

$$V_{common} = \frac{(V_{diff+} + V_{diff-})}{2} \tag{10.11}$$

is rejected. V_{diff+} and V_{diff-} denote the input voltages of the circuit with respect to ground.

In Fig. 10.4, the voltages at the non-inverting (+) and inverting (–) inputs of the operational amplifier are given by:

$$\begin{aligned} V_+ &= \left(\frac{R_4}{R_3 + R_4}\right) \cdot V_{diff-} \\ V_- &= V_{out} + \left(\frac{R_2}{R_1 + R_2}\right) \cdot (V_{diff+} - V_{out}) \end{aligned} \tag{10.12}$$

The operational amplifier amplifies it's differential input voltage ($V_+ - V_-$) with a gain which is normally in the order of 10^5. Because of this large gain, the input voltages are effectively forced equal to each other:

$$\left(\frac{R_4}{R_3 + R_4}\right) \cdot V_{diff-} \approx V_{out} + \left(\frac{R_2}{R_1 + R_2}\right) \cdot (V_{diff+} - V_{out}) \tag{10.13}$$

This can be rewritten as follows:

$$V_{\text{out}} \approx -\frac{1}{1-\left(\dfrac{R_2}{R_1+R_2}\right)} \cdot \left[\left(\frac{R_2}{R_1+R_2}\right) \cdot V_{\text{diff}+} - \left(\frac{R_4}{R_3+R_4}\right) \cdot V_{\text{diff}-}\right] \tag{10.14}$$

Using (10.10) and (10.11) we can rewrite this in terms of V_{diff} and V_{common}:

$$V_{\text{out}} \approx -\frac{1}{1-\left(\dfrac{R_2}{R_1+R_2}\right)} \cdot \left[\begin{array}{c}\left(\dfrac{R_2}{R_1+R_2} - \dfrac{R_4}{R_3+R_4}\right) \cdot V_{\text{common}} + \\ \dfrac{1}{2}\left(\dfrac{R_2}{R_1+R_2} + \dfrac{R_4}{R_3+R_4}\right) \cdot V_{\text{diff}}\end{array}\right] \tag{10.15}$$

We see that the output voltage is dependent only on the differential input voltage and not on the common mode input voltage when the resistor ratio's R_2/R_1 and R_4/R_3 are chosen equal to each other. In this case (10.15) reduces to:

$$V_{\text{out}} \approx -\frac{\left(\dfrac{R_2}{R_1+R_2}\right)}{1-\left(\dfrac{R_2}{R_1+R_2}\right)} \cdot V_{\text{diff}} = -\frac{R_2}{R_1} \cdot V_{\text{diff}} \tag{10.16}$$

Thus, the voltage gain is given by R_2/R_1. The quality of the common-mode rejection is dependent on the matching of the resistors.

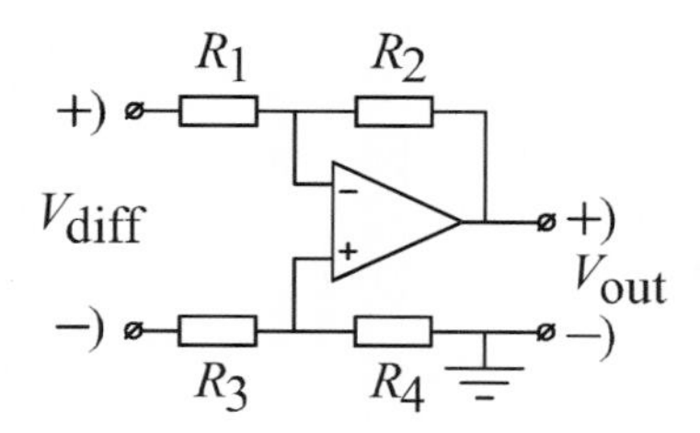

Fig. 10.4. Simple differential amplifier using a single operational amplifier

A disadvantage of the circuit in Fig. 10.4 is the rather low input impedance, which results in a reduction of the bridge output voltage. Therefore, a differential preamplifier as shown in Fig. 10.5 often precedes the circuit. This circuit is usually called an instrumentation amplifier. In this circuit the two operational amplifiers at the input will change their output in order that their differential input voltages become virtually zero. In Fig. 10.5 this means that the input voltage v_{diff} is copied to resistor R_6. As the current through R_5 and R_7 must be equal to the current through R_6 (assuming that no current flows into the inputs of the amplifiers) it can easily be shown that:

$$V_x = I_{R_{5,6,7}} \cdot (R_5 + R_6 + R_7) = \frac{V_{\text{diff}}}{R_6} \cdot (R_5 + R_6 + R_7) = \left(\frac{R_5 + R_7}{R_6} + 1 \right) \cdot V_{\text{diff}} \quad (10.17)$$

Usually R_5 and R_7 are chosen equal and when also $R_2/R_1 = R_4/R_3$ the response of the complete circuit becomes:

$$V_{\text{out}} = -\left(\frac{2R_5}{R_6} + 1 \right) \cdot \left(\frac{R_2}{R_1} \right) \cdot V_{\text{diff}} \quad (10.18)$$

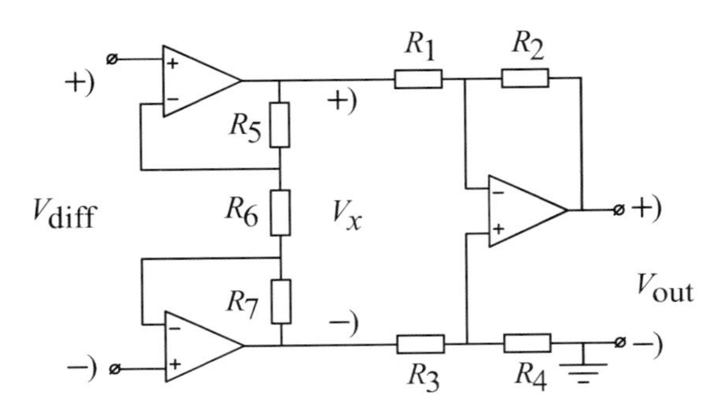

Fig. 10.5. Instrumentation amplifier with high input impedance

10.1.3 Noise and Offset

As mentioned before, the change in resistance of piezoresistors is very small and the offset and noise introduced by the amplifier can easily obscure the sensor signal. Especially drift, i.e. a slowly changing offset voltage, is difficult to distinguish from the sensor signal. Fig. 10.6 shows the principle of a so-called chopper amplifier. In a chopper amplifier the input voltage is modulated around a carrier frequency f_c. The amplifier amplifies the resulting ac-signal and after amplification the signal is demodulated. In Fig. 10.6, the modulation of the input voltage is illustrated by the switch, which results in a square-wave signal with an amplitude equal to the input voltage. The square-wave signal is amplified and demodulation is done by a multiplier. The offset and low frequency noise of the amplifier are effectively filtered from the output signal.

Of course, the principle of the chopper amplifier is only suitable for slowly changing input signals (with a frequency well below the chopper frequency f_c). Therefore, a chopper amplifier is often combined with a normal operational amplifier to form a so-called chopper-stabilized amplifier. Furthermore, several similar techniques have been developed using switches to periodically sample and store the amplifier offset and subtract it from the output voltage. One of these techniques, Correlated Double Sampling (CDS) will be discussed in section 10.2. Various types of auto-zero amplifiers and chopper-stabilized amplifiers are available as standard electronic building blocks, just like operational amplifiers.

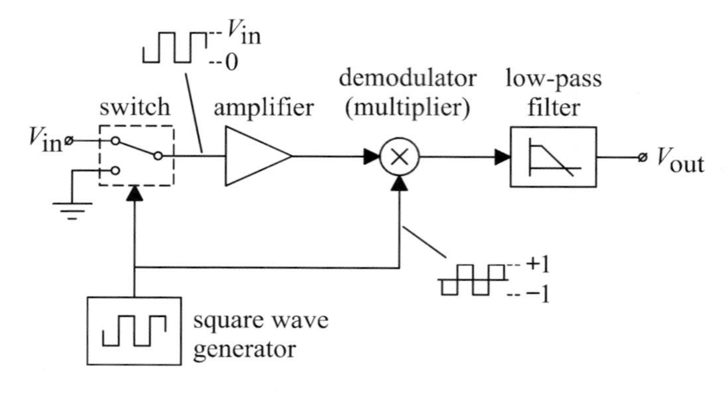

Fig. 10.6. Simple chopper type amplifier

10.1.4 Feedback Control Loops

In the previous section we have seen the use of feedback to obtain an accurate amplifier gain. Sometimes it is useful to include the sensor element in the feedback loop. We have already seen some examples of this in Chaps. 6 and 7, where force compensation is used to increase the bandwidth and accuracy of pressure sensors and accelerometers. Another example are thermal anemometer type flow sensors which were discussed in Chap. 8. Fig. 10.7 shows a typical control circuit used for thermal anemometers (Fingerson 1996). In the circuit the anemometer is connected in a Wheatstone bridge. When a flow passes the anemometer, its temperature will drop and its resistance will decrease. Thus, the voltage at the inverting input of the amplifier will decrease. As a result, the output voltage of the amplifier will rise and the current through the sensor will increase. The increased current heats the sensor until the system is again in equilibrium. The temperature sensor indicated in Fig. 10.7 can be added to compensate for fluctuations of the temperature of the flow medium. When correctly dimensioned, the circuit will keep the anemometer at fixed temperature difference with the surrounding medium. For high frequency operation the dashed capacitor can be added. Usually the capacitor value is adjusted to obtain a "maximally" flat frequency response.

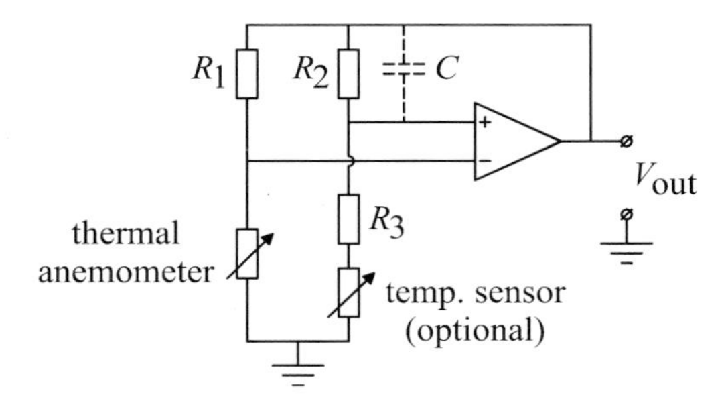

Fig. 10.7. Constant temperature anemometer control circuit

10.1.5 Interfacing with Digital Systems

The circuits discussed so far all have an analog output signal in the form of a voltage. However, most modern sensor systems require some sort of digital signal processing. Thus, the analog voltage has to be converted into a shape that is compatible with digital electronics. One way to do this is to use an analog-to-digital converter (ADC). Another way is to convert the voltage into a frequency or time delay, which can be easily measured by digital circuits using a reference clock frequency. An important advantage of a digital or frequency signal is that it can be transported over long distances with virtually no loss (see Fig. 10.8).

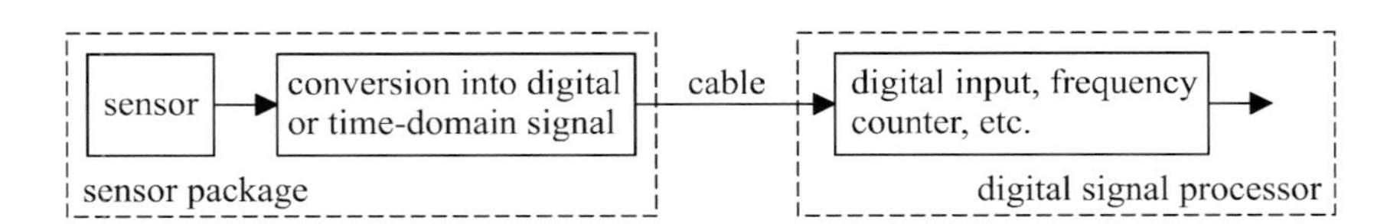

Fig. 10.8. Digital and time-domain signals can be transported over long distances with no loss, therefore a signal converter is often included in the sensor package

10.1.5.1 Analog-to-Digital Conversion

A lot of research has been done on fast and accurate analog-to-digital conversion and many ADC's are now commercially available as a standard building block. In this book we will restrict ourselves to two types of converters, namely the dual-slope converter and the ΣΔ-modulator, because these techniques combine high accuracy with relatively simple analog circuits. Both techniques are rather slow, but this normally no problem, as most sensor signals tend to change slowly.

Dual-slope AD converters

Fig. 10.9 illustrates the operation of a dual-slope AD converter. The circuit basically consists of a switch, an integrator, a comparator and some control circuitry. At the start of a conversion cycle, the input of the integrator is connected to the unknown input voltage V_{in} and the voltage is integrated during a fixed period of time T_1. The output voltage of the integrator at the end of the integration interval is proportional to V_{in}:

$$V_{int,max} = \frac{1}{RC} \cdot T_1 \cdot V_{in} \tag{10.16}$$

Next, the integrator input is switched to a reference voltage V_{ref} and the integrator output voltage decreases with a fixed slope until it has returned to zero (see Fig. 10.9(b)). During this interval, the integrator output voltage has to change by the same (but opposite) amount, thus:

$$V_{int,max} = \frac{1}{RC} \cdot T_1 \cdot V_{in} = \frac{1}{RC} \cdot T_2 \cdot V_{ref} \tag{10.17}$$

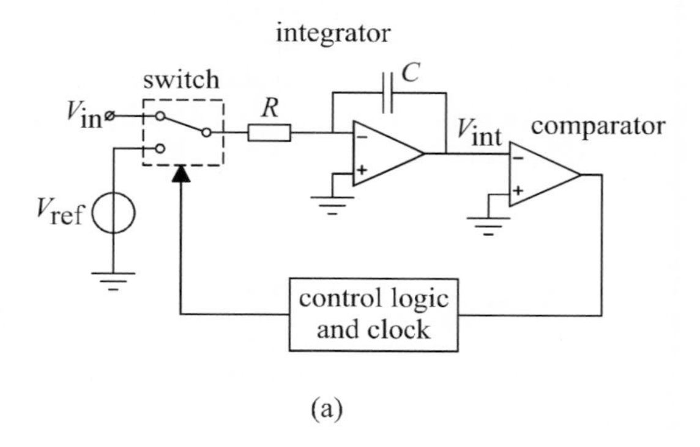

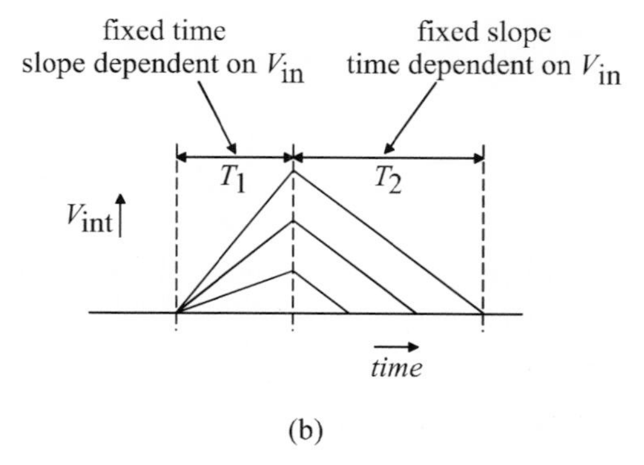

Fig. 10.9. Principle of dual-slope AD conversion: (a) circuit, (b) shape of integrator output voltage

Rearranging gives:

$$T_2 = \frac{V_{in}}{V_{ref}} \cdot T_1 , \tag{10.18}$$

showing that T_2 is a measure for the input voltage V_{in}.

Sigma-delta modulation

Fig. 10.10(a) illustrates the principle of sigma-delta modulation (see e.g. (Gray 1987)). The digital output signal *y(t)* is fed back and subtracted from the input signal *x(t)*. The difference between the signals is low-pass filtered and fed to a rough quantizer. Usually, the quantizer is simply a comparator to avoid nonlinearity errors in the subsequent D/A conversion. In this case the output signal has the shape of a bit-stream and the D/A converter reduces to a switch between two levels. The low-pass filter is necessary to ensure that the quantizer evaluates the average of the difference between the input signal and the output bit stream. The mean value of the bit stream will be a highly accurate representation of the instantaneous value of the (relatively slowly changing) input signal. Usually, the bit stream is processed by a digital decimation filter to obtain a multi-bit digital output.

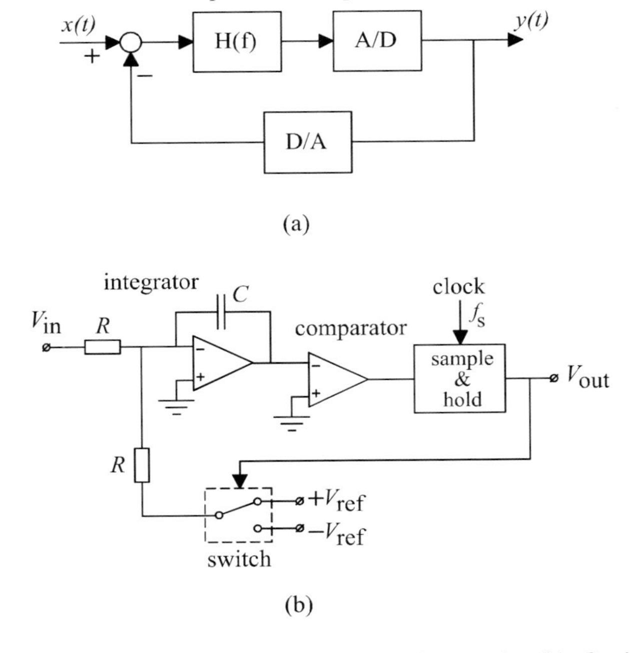

Fig. 10.10. Principle (a) and simple implementation (b) of a sigma-delta modulatior

Fig. 10.10(b) shows a simple implementation of a sigma-delta modulator. The low-pass filter is implemented as an integrator and the quantizer is a comparator followed by a sample and hold circuit.

10.1.5.2 Voltage to Frequency Converters

Instead of performing a complete analog to digital conversion it is often sufficient to convert the analog quantity into a signal in the time domain like a frequency or a pulse-width. Several voltage-to-frequency converters have been proposed, e.g. (Gilbert 1976). However, a frequency signal by itself is not enough as it can change due to process variations, temperature variations, etc.. A solution is to generate two frequencies, whose ratio represents the signal information. Usually, a single voltage-to-frequency converter is used and the voltage input is switched periodically between the sensor signal and a reference voltage.

Fig. 10.11 shows an oscillator circuit which is particularly suited for sensor bridges. (Huijsing 1986). The center frequency of oscillation is defined by R_0 and C_0. A bridge imbalance causes a deviation from this frequency:

$$f = \frac{1}{4R_0C_0}\frac{R_1}{R_2}\left[1+\frac{R_0}{R_3}\left(1-\frac{R_2}{R_1}\frac{R_3}{R_4}\right)\right] \tag{10.19}$$

The bridge supply voltage is inverted twice per period, which has the advantage that the offset of the integrator amplifier is automatically canceled. Furthermore, the oscillation frequency is independent of the bridge supply voltage.

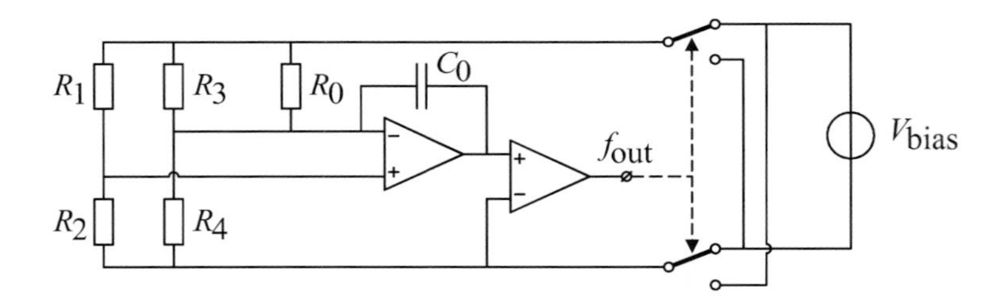

Fig. 10.11. Operational amplifier based bridge ratio to frequency converter (Huijsing 1986)

An alternative approach is to modulate the duty-cycle instead of the frequency (Huijsing 1986). Fig. 10.12 shows a simple circuit which was proposed by Spencer et al. (Spencer 1985). The circuit produces a square-wave output signal whose duty-cycle depends only on the input voltage and the temperature and is independent of component values and matching.

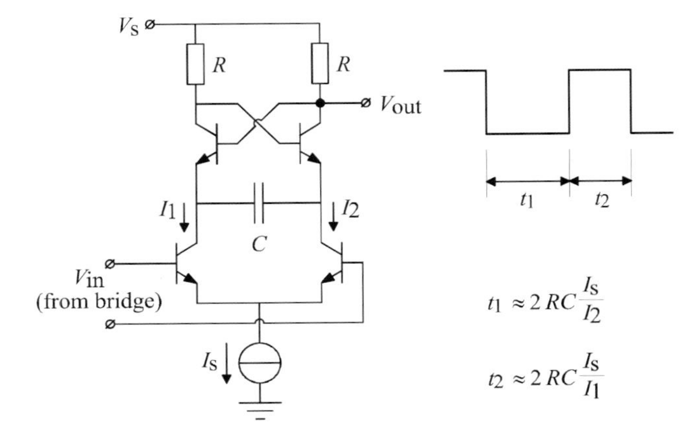

Fig. 10.12. Voltage controlled duty-cycle oscillator (Spencer 1985)

10.2 Capacitive Sensors

Two techniques are commonly used to measure capacitance changes:

- Just like resistance changes, capacitance changes can be measured by connecting the capacitance in a bridge configuration.
- The capacitance change can be used to control the oscillation frequency of a sinusoidal or relaxation oscillator circuit.

A bridge configuration usually results in a very simple circuit and, therefore, this technique may be preferred when the electronic circuit has to be integrated on the sensor die. On the other hand, a frequency signal can be easily interfaced with digital electronics and transported over large distances with virtually no loss.

10.2.1 Impedance Bridges

Just like resistors, capacitors can be compared by connecting them in a bridge configuration. The difference is that we now have to apply an alternating supply voltage. Fig. 10.13 shows a simple capacitive voltage divider, which is driven symmetrically. The output voltage is given by (neglecting parasitic capacitance C_p and resistance R):

$$V_{out} = \left(\frac{2C_2}{C_1 + C_2} - 1 \right) \cdot V_{in} \tag{10.20}$$

Or in the common situation of a differential change in capacitance (C_1=C-ΔC and C_2=C+ΔC):

$$V_{out} = \frac{\Delta C}{C} \cdot V_{in} \tag{10.21}$$

Note that the exact shape of the voltages is not important. A square wave signal is often used instead of the sine wave indicated in the figure.

In micromachined devices the sensor capacitances are usually very small and the influence of parasitic capacitance C_p and resistance R cannot be neglected. For example, the resistance R can only be neglected if it is much larger than the impedance of the sensor capacitance at the signal frequency f, thus:

$$R >> \frac{1}{2\pi \cdot f \cdot C} \tag{10.22}$$

For a typical capacitance value of 1 pF and a signal frequency of 100 kHz this means that R should be significantly larger than 1.6 MΩ. Still, R is usually required to define the dc output voltage. However, instead of being a fixed resistor it can also be a switch, which is periodically closed as will be shown later.

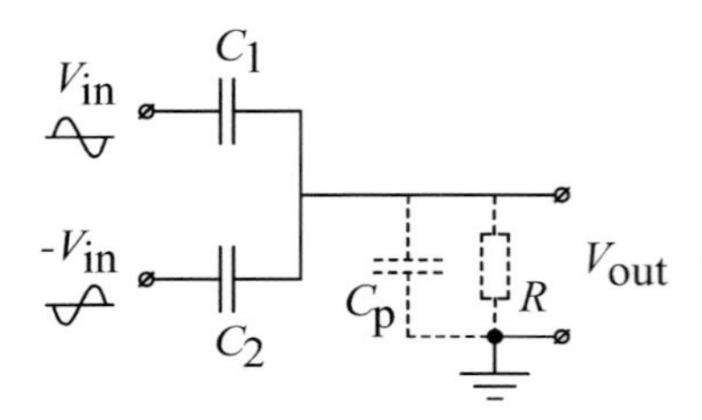

Fig. 10.13. Capacitive voltage divider

The influence of parasitic capacitance can be largely eliminated by a technique called active guarding or bootstrapping. This is indicated in Fig. 10.14. The output node of the circuit is shielded and a voltage buffer is added with its output connected to the shield. Due to the shielding the parasitic capacitance to ground (C_{p1} in the figure) is significantly reduced. Instead, we have a parasitic capacitance C_{p2} to the shield electrode, however the influence of this capacitor is eliminated because the voltage across it is kept zero by the buffer amplifier and, consequently, there is no current flowing through C_{p2}. An important drawback of bootstrapping is that the circuit may become instable if the buffer gain becomes larger than 1.

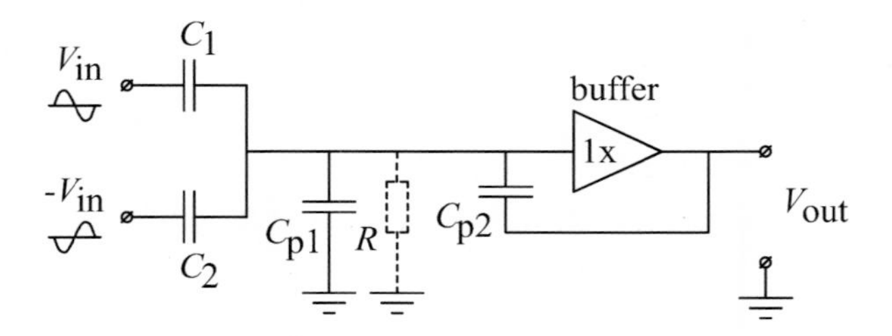

Fig. 10.14. Elimination of parasitic capacitor C_{p2} by bootstrapping

Another technique to eliminate the influence of parasitic capacitance is shown in Fig. 10.15. Here the capacitive voltage divider is followed by a so-called charge amplifier consisting of an operational amplifier with a feedback capacitor C_f. The operational amplifier will adjust it's output voltage such that the differential input voltage remains zero. As a result, the voltage across the parasitic capacitor C_p remains virtually zero and the capacitor is effectively eliminated. For simplicity, the circuit is analyzed using square wave input signals. In that case, the rising edge of V_{in} causes a charge equal to $(C_2-C_1)V_{in}$ to be injected in the circuit. As no current can flow into the amplifier input, the amplifier will adjust it's output voltage such that the charge is transferred to the feedback capacitor C_f. Thus, square wave input signals with amplitude V_{in} will result in an output amplitude given by:

$$V_{out} = \frac{C_2 - C_1}{C_f} V_{in} \tag{10.23}$$

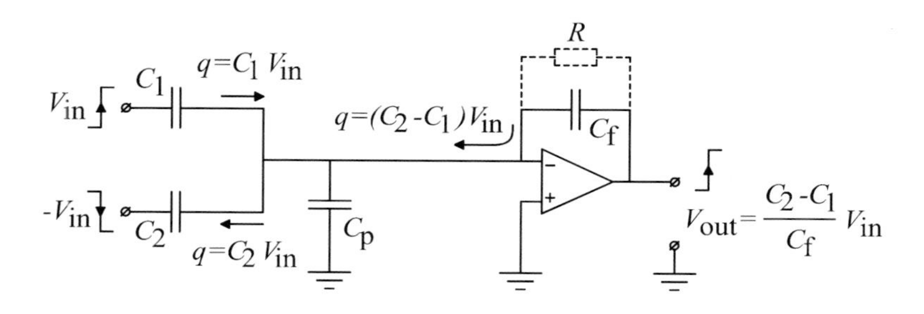

Fig. 10.15. Principle of charge sensing

A commonly used variation of the circuit in Fig. 10.15 is shown in Fig. 10.16 (Puers 1990, Peeters 1992). Now the unknown sensor capacitor is placed in the feedback path of the amplifier. As a result, the amount of charge on the capacitor remains constant, which can be very useful when electrostatic forces cannot be neglected.

Fig. 10.17 shows a typical switched capacitor implementation of the circuit from Fig. 10.15 (Park 1983). The anti-phase input signals are easily realized by switches implemented by MOS transistors. Thus the amplitude of the input signals is equal to the supply voltage V_s and the output amplitude is equal to $\left(\left(C_{sensor} - C_{reference}\right)/C_f\right) \cdot V_s$ (see eqn. (10.23)). Note that the dashed resistor in Fig. 10.15 has been replaced by a transistor. This transistor is used to periodically discharge the feedback capacitor in order to obtain a well-defined dc voltage level at the input of the amplifier.

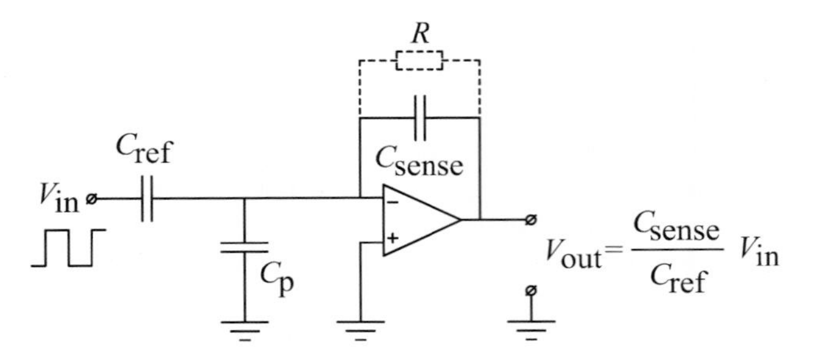

Fig. 10.16. Placing the sensor capacitor in the feedback path of the amplifier results in a constant charge on the capacitor, which can be useful to avoid nonlinearities due to electrostatic forces (Puers 1990, Peeters 1992)

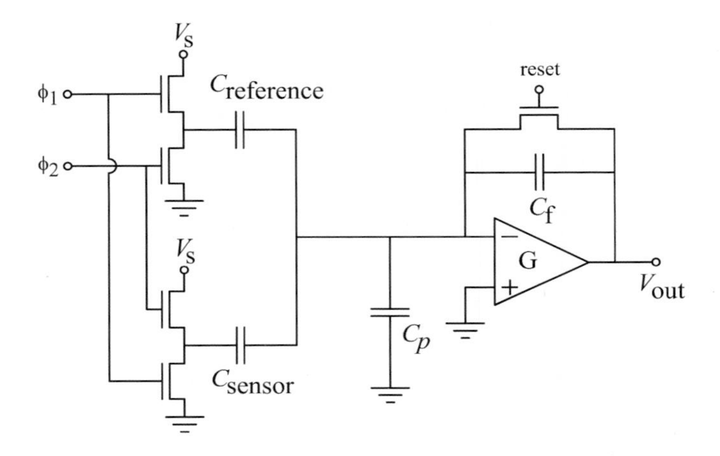

Fig. 10.17. Typical switched-capacitor type readout circuit for capacitive sensors (Park 1983)

To eliminate noise and drift from the amplifier, the circuits discussed so far are often combined with a synchronous detection circuit as illustrated in Fig. 10.18. The resulting circuit is similar to the chopper amplifier discussed in section 10.1.3. The output of the amplifier is multiplied by a square wave with the same frequency and in phase with the input signal and subsequently low-pass filtered.

Another well-known technique to eliminate noise and drift from the amplifier is Correlated Double Sampling (CDS) (Degrauwe 1985, Lu 1995). Fig. 10.19 shows the basic circuit structure. The circuit contains a number of switches, which are controlled by two non-overlapping clock signals. When the first clock signal is active, the switches marked ϕ_1 are closed and all capacitors and the input of the charge amplifier are connected to ground. The voltage at the output of the amplifier is measured and compensated by the second amplifier. The voltage required for compensation is stored on capacitor C_{az}. Next, ϕ_2 becomes active and the circuit responds exactly like the circuit in Fig. 10.15, with the difference that the offset and low frequency noise of the amplifier is still stored on C_{az} and subtracted from the output voltage.

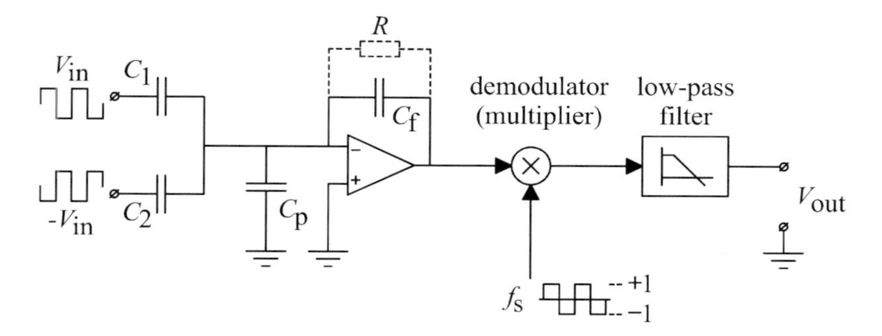

Fig. 10.18. Elimination of noise and drift from the amplifier by synchronous detection

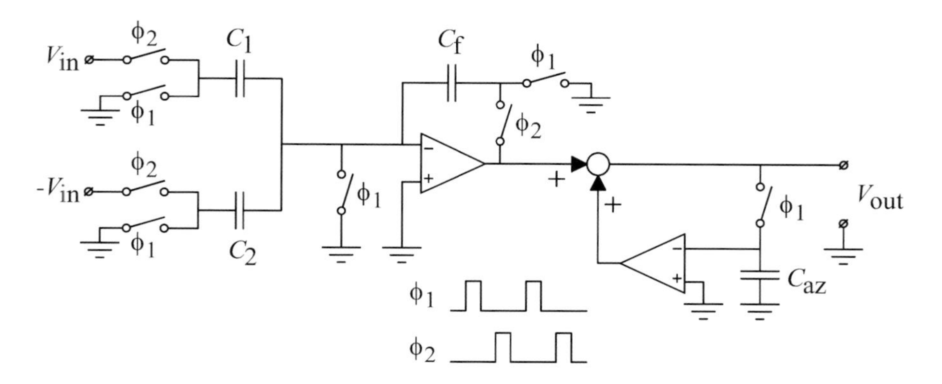

Fig. 10.19. Principle of Correlated Double Sampling (CDS)

10.2.2 Capacitance Controlled Oscillators

As mentioned before, it is often useful to have an output signal in the form of a frequency. Therefore, in this Section we will discuss some simple oscillator circuits with an output frequency which is dependent on a capacitor value. One of the simplest circuits which quite frequently used is shown in Fig. 10.20 (Huijsing 1986, Seo 1993, Chatzandroulis 1997, Matsumoto 1999). The circuit consists of a Schmitt trigger, a capacitor and two current sources. The capacitor is loaded back and forth between the two voltage levels of the Schmitt trigger. The oscillation period is simply calculated as the sum of the rise and fall times:

$$T_{\mathrm{tot}} = T_{\mathrm{rise}} + T_{\mathrm{fall}} = C\frac{V_{\mathrm{high}} - V_{\mathrm{low}}}{I_1} + C\frac{V_{\mathrm{high}} - V_{\mathrm{low}}}{I_2}, \tag{10.24}$$

where V_{high} and V_{low} denote the switching levels of the Schmitt trigger.

Usually the current sources are chosen equal to each other, however this is no necessity and the current sources may used to put additional information in the duty cycle of the output wave form. The current sources may also be replaced by a single feedback resistor between the input and output of the Smitt trigger, however this will introduce some nonlinearity because the current flowing into the capacitor will then be dependent on the capacitor voltage.

A problem in the Schmitt trigger oscillator is that any parasitic capacitance from the input of the Schmitt trigger to ground will influence the oscillation frequency. In sensor applications this parasitic capacitance may be significantly larger than the sensor capacitance. Therefore, when C_{sense} is not necessarily connected to ground, the circuit in Fig. 10.21 is preferred (Matsumoto 1999). Now the capacitor is driven at one side by the voltage sources V_{high} and V_{low}. By driving with a voltage the influence of parasitic capacitance C_{p1} is effectively

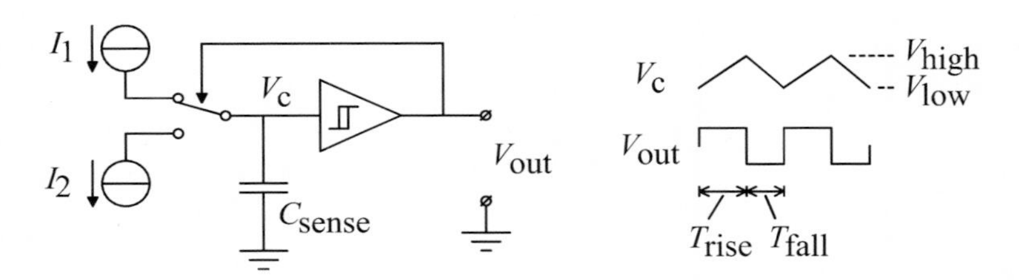

Fig. 10.20. Simple Schmitt trigger type capacitance-to-frequency converter

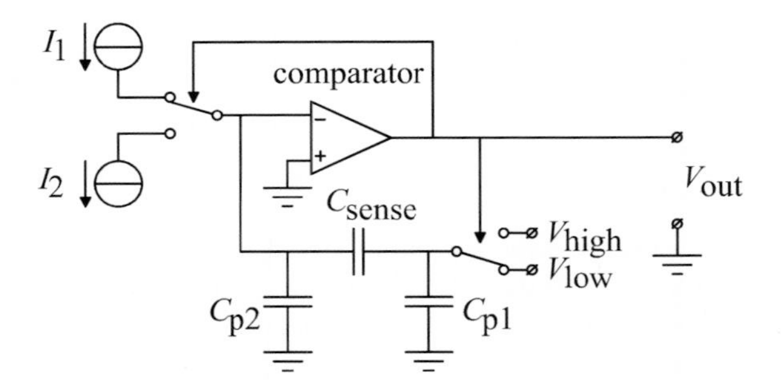

Fig. 10.21. Capacitance-to-frequency converter with insensitivity to parasitic capacitance's C_{p1} and C_{p2}

eliminated. The other side of C_{sense} is connected to a comparator. This comparator will always switch at the same voltage level at the input, thus, at the moment of switching, the voltage across the second parasitic capacitance C_{p2} will always be the same. Therefore, C_{p2} has no direct influence on the oscillation frequency, although it does influence the voltage amplitude at the comparator input since C_{sense} and C_{p2} form a capacitive voltage divider. The oscillation period is the same as for the simple Schmitt trigger circuit and is given by (10.24).

A further modification to the circuit results in a simple implementation of the current sources as indicated in Fig. 10.22. A charge amplifier has been inserted at the input of the comparator. As a result the voltage at the left side of C_{sense} is kept at ground level and any charge injected at this node is transported to C_f. The fact that the voltage at the left side of C_{sense} is now constant can be exploited by replacing the current sources by a single resistor. The circuit obtained this way is very similar to a switched capacitor oscillator proposed by Martin (Martin 1981). A modification of this circuit, called the *modified Martin oscillator*, was used by Toth and Meijer (Toth 1992) to realize the capacitance-controlled oscillator shown in Fig. 10.23. Operation of this circuit is similar to the circuit in Fig. 10.22, except that the switches are replaced by digital NAND gates switching between 0 and 5V. Furthermore, C_{sense} is replaced by a fixed capacitor C_{offset} and two additional NAND gates are used to connect a reference capacitor C_{ref} and the sensor capacitor C_{sense} virtually in parallel with C_{offset}. This has the important advantage that 3 different period measurements can be performed, namely for the capacitance values C_{offset}, $C_{offset}+C_{sense}$ and $C_{offset}+C_{ref}$. It can be shown that the capacitance ratio C_{sense}/C_{ref} is given by:

$$\frac{C_{sense}}{C_{ref}} = \frac{T(C_{offset} + C_{sense}) - T(C_{offset})}{T(C_{offset} + C_{ref}) - T(C_{offset})}, \tag{10.25}$$

where all additive and mutiplicative errors due to nonidealities and temperature dependencies in the circuit have been eliminated. Only nonlinearity and random errors remain (Toth 1997).

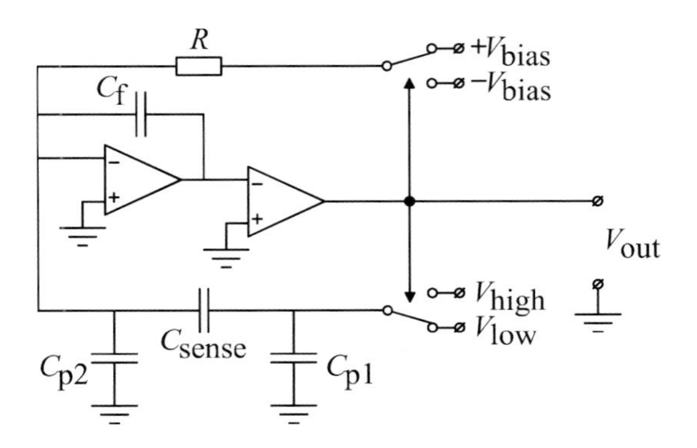

Fig. 10.22. The current sources in Fig. 10.21 can be replaced by a charge amplifier and a resistor

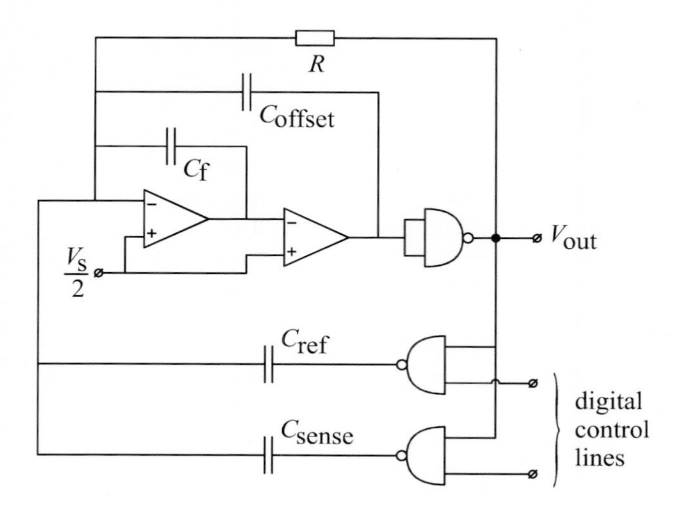

Fig. 10.23. Modified Martin oscillator (Toth 1992, 1997)

10.3 Resonant Sensors

10.3.1 Frequency Dependent Behavior of Resonant Sensors

Close to the resonance frequency the mechanical resonance in most resonant sensors can be modeled electrically by either a series resonant circuit or a parallel resonant circuit, as indicated in Fig. 10.24. Fig. 10.25 shows a plot of the impedance of these circuits as a function of frequency. At the resonance frequency ω_0, the series resonant circuit has a minimum impedance equal to R_s and zero phase shift (at ω_0 the impedance of the combination of C_s and L_s is zero). The parallel resonant circuit has a maximum impedance equal to R_p (at the resonance frequency the parallel combination of C_p and L_p results in an infinite impedance).

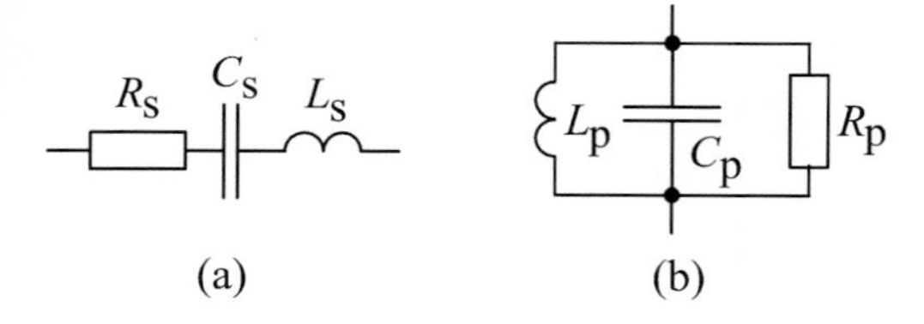

Fig. 10.24. A series resonant circuit (a) and a parallel resonant circuit (b)

10.3.2 Realizing an Oscillator

In practical systems, the easiest and fastest way to measure the resonance frequency of a resonant sensor is by building an electronic oscillator circuit in which the sensor determines the oscillation frequency. The oscillation frequency is close to the resonance frequency but not necessarily equal to it, due to the frequency dependent behavior of the electronic circuitry.

A generally used model of an oscillator circuit is shown in Fig. 10.26. It consists of an amplifier with a feedback network, which contains the frequency-determining element, i.e. the resonant sensor. For a stable oscillation, the following conditions have to be fulfilled:

$$\left|A(j\omega)\cdot H(j\omega)\right|_{\omega=\omega_{\mathrm{osc}}} = 1 \tag{10.26}$$

$$\arg(A(j\omega)\cdot H(j\omega))_{\omega=\omega_{\mathrm{osc}}} = 0 + k\cdot 360^{o} \qquad (k = 0,\ 1,\ \ldots) \tag{10.27}$$

These conditions are known as the *Barkhausen criterion for oscillation.* In these equations, $A(j\omega)H(j\omega)$ is the open loop frequency response of the circuit. Eq. (10.26) states that the gain in the loop has to be exactly equal to 1. Eq. (10.27) states that the phase shift in the loop must be zero or a multiple of 360 degrees. The oscillation frequency ω_{osc} is not exactly equal to the resonance frequency ω_0, because a phase shift caused by the amplifier $\arg(A(j\omega))$ has to be compensated by a phase shift in the resonator.

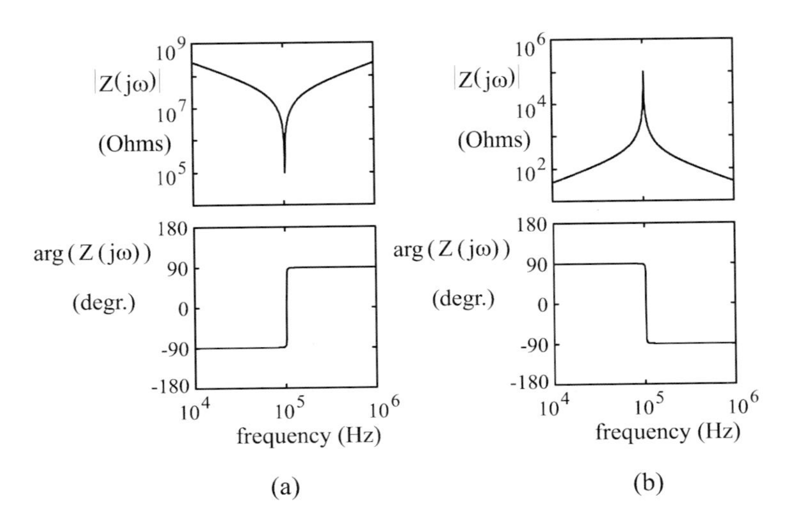

Fig. 10.25. Frequency dependent impedance of (a) a series resonant circuit with R_s=100 kΩ, C_s=0.006 pF and L_s=400 H and (b) a parallel resonant circuit with R_p=100 kΩ, C_s=4 nF and L_s=0.6 mH

For illustration, Fig. 10.27 shows two possible circuit implementations. In Fig. 10.27(a), the resonator forms a voltage divider with resistor R. When R is chosen approximately equal to R_s this results in an attenuation by a factor of 2 at the resonance frequency. Thus, the amplifier should have a voltage gain of at least 2 to allow oscillation. In Fig. 10.27(b) the resonator is driven by a voltage and the resulting current is sensed. The current is converted into a voltage and after amplification the voltage is fed back to the resonator. Again, the loop gain, given by $\frac{A \cdot R}{R_s}$, should be larger than 1.

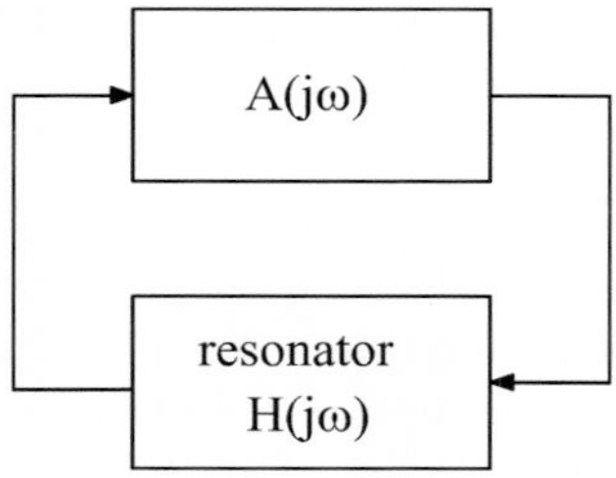

A(jω)

resonator
H(jω)

Fig. 10.26. Oscillator consisting of an amplifier with frequency dependent feedback

10.3.3 One-Port Versus Two-Port Resonators

To excite and detect the vibrational motion of a resonant sensor two approaches can be used. One uses a single element, which combines the excitation and detection of the structure and is designated as a one-port resonator (Fig. 10.28(a)). The other uses separate elements for excitation and detection, resulting in a two-port resonator (Fig. 10.28(b)). For the excitation and detection several transduction mechanisms are available, as discussed in Chap. 9.

The performance of (resonant) sensors is degraded by several unwanted effects. In case of a one-port resonator these are due to the parasitic parallel loads, which can obscure the mechanical resonance. For a two port resonator, a similar effect is caused by the electrical cross-talk between the driving port and the detection port. The electrical cross-talk is a result of capacitive and/or resistive coupling.

In the remainder of this chapter we will discuss two examples. The first example, which will be discussed in Sect. 10.3.4, is based on a one-port resonant beam resonator with capacitive excitation and detection. The second example is discussed in Sect. 10.3.5 and uses an H-shaped two-port resonator with electro-dynamic excitation and detection.

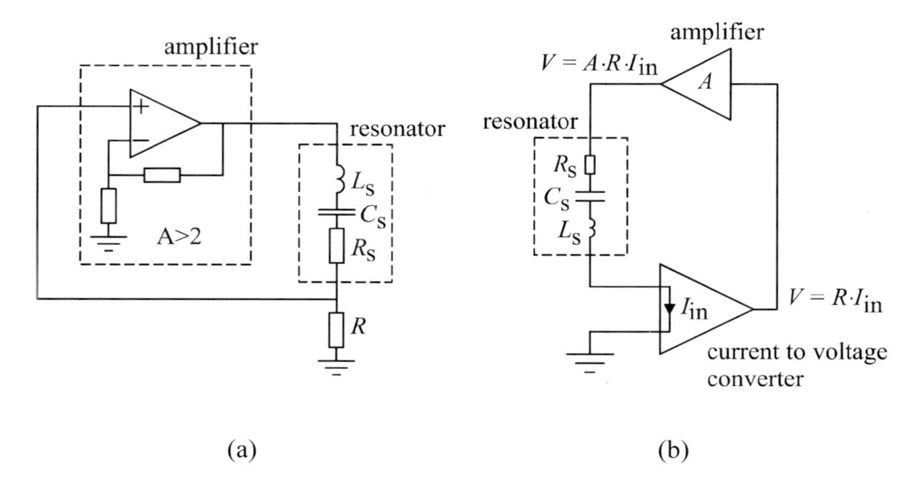

Fig. 10.27. Possible circuit implementations based on the principle indicated in Fig. 10.26. The frequency-defining element is a series resonant circuit.

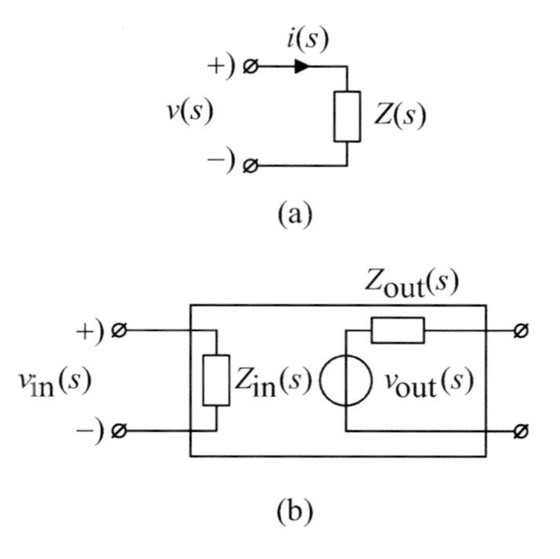

Fig. 10.28. One-port resonator (a) and two-port resonator (b)

10.3.4 Oscillator Based on One-Port Electrostatically Driven Beam Resonator

Fig. 10.29 shows the structure of a simple clamped-clamped beam resonator that could easily be fabricated in a silicon surface micromachining process. An electrode is diffused in the substrate beneath the beam and the beam itself consists of highly conductive doped polysilicon. In this way a capacitor is realized between the beam and the electrode. The beam is brought in vibration by applying an alternating voltage superimposed on a dc bias voltage to the electrodes. A bias voltage is necessary because only attractive electrostatic forces can be realized, thus an alternating voltage without a dc bias would result in the beam being excited at twice the signal frequency. Furthermore, the bias voltage defines the equilibrium deflection of the beam around which the smaller dynamic vibration takes place. For an increasing bias voltage, the influence of the ac voltage on the beam will also increase, resulting in a larger vibration. However, the bias voltage cannot be made arbitrarily large since above a certain value V_{PI}, the pull-in voltage, the beam will become unstable and the capacitor will collapse as described in Chapter 5.

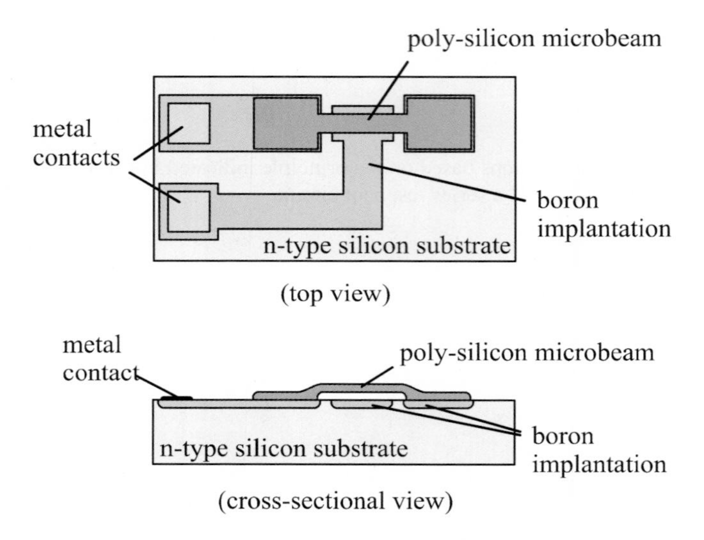

Fig. 10.29. Simple one-port electrostatically driven beam resonator

Before we can design an oscillator circuit we need to obtain an adequate electrical model of the resonator. For the structure in Fig. 10.29 this is quite simple. Fig. 10.30(a) shows the structure with the corresponding electrical model. The mechanical resonance is modeled by a series resonant circuit as described in Sect. 10.3.1. The transformer models the coupling between the electrical and the mechanical domain. The electrical capacitance between the beam and the electrode in the substrate is modeled by C_0. The parasitic capacitance's from the resonator electrodes to the substrate are modeled by C_{p1} and C_{p2}. The model can be simplified as shown in Fig. 10.30(b) by omitting the transformer and adjusting the values of L_s, C_s and R_s.

We now see that the electrical equivalent is not at all the simple series resonance circuit that we discussed in Sect. 10.3.1. We may be able to eliminate C_{p1} and C_{p2} by techniques similar to the ones we used in Sect. 10.2 for the measurement of small capacitances. However, we will never be able to eliminate C_0. Thus, without modification the simple circuits shown before in Fig. 10.27 are not suitable for driving this resonator.

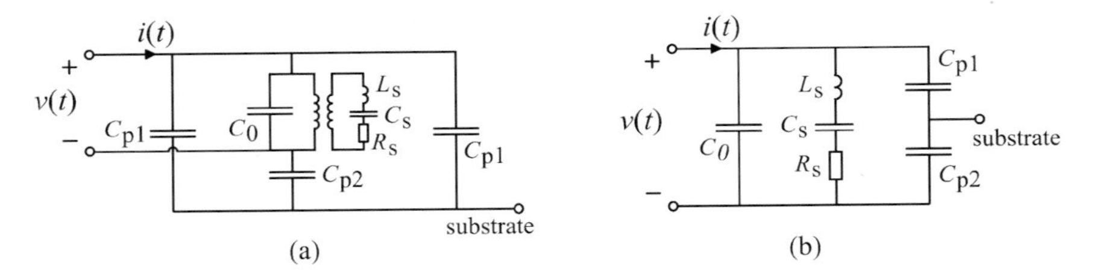

Fig. 10.30. Electrical model of the resonator in Fig. 10.29

Fig. 10.31 illustrates the impact of C_0 on the impedance of the resonator. Besides the expected minimum in impedance we now also have a maximum at a slightly higher frequency. For higher frequencies the total impedance is dominated by C_0 and the impedance decreases while the phase returns to -90 degrees. In an oscillator circuit this can easily result in unwanted high-frequency oscillations due to the combination of a small resonator impedance and additional phase shift in the electronic circuit.

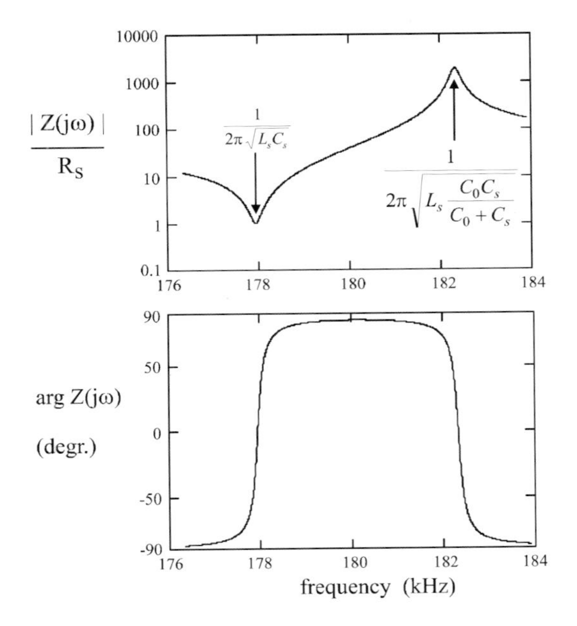

Fig. 10.31. Frequency dependent impedance of the equivalent circuit in Fig. 10.30(b)

The resonator model depicted in Fig. 10.30(b) is essentially the same as the model of a quartz crystal. Therefore, it should be possible to use a (modified) crystal oscillator circuit for the resonator. This was proposed by Bienstman et al., who adapted a standard three-point crystal oscillator circuit for use with an electrostatically driven one-port resonator (Bienstman 1996). Fig. 10.32 shows a basic three-point crystal oscillator (see for example (Vittoz, 1988)) and the principle of the circuit used by Bienstman et al.. The main difference is the addition of resistor R_b and capacitor C_k, which are needed to bias the resonator at a voltage V_{bias} independent of the biasing of the circuit. We see that in this way a very compact oscillator circuit can be realized. For details about the operation of the oscillator we refer to (Vittoz 1988) and (Bienstman 1996).

A problem occurs when C_0 becomes very large, which can be the case for very small resonators. Fig. 10.33 illustrates the effect on the resonator impedance. We see that with increasing C_0 the two resonance peaks approach each other and, more importantly, the maximum phase shift decreases. In this case the design of an oscillator circuit becomes rather complicated. A possible circuit is indicated in Fig. 10.33. The resonator is connected in the feedback loop of an amplifier where it forms a voltage divider together with the parallel connection of R_{in} and C_{in}, similar to the basic circuit of Fig. 10.27(a). The parasitic capacitances to ground, C_{p1} and C_{p2}, are effectively eliminated as one is driven by the output of the amplifier and the other is in parallel with capacitor C_{in}, which in fact provides compensation for C_0.

Fig. 10.35 shows a plot of the open-loop frequency response of the circuit. We see that 4 frequency regions can be distinguished. For very low frequencies, region (1), the gain is completely defined by the resistors in the circuit. Resistor R_{par} (not shown in Fig. 10.34) is the leakage resistance parallel to the resonator. For somewhat higher frequencies, region (2), the gain increases due to capacitor C_0 until we reach the cut off frequency defined by R_{in} and C_{in}. Above that

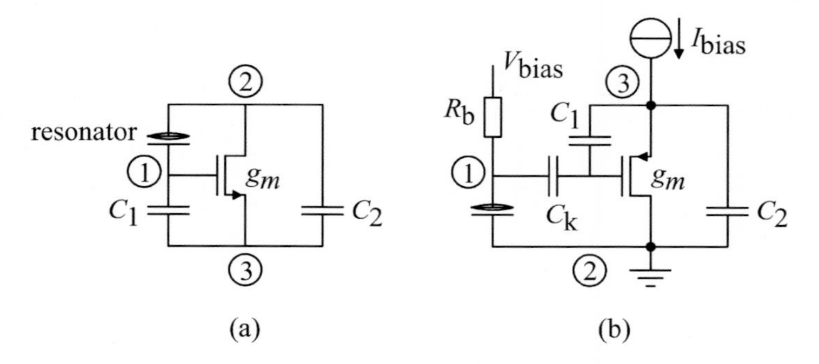

Fig. 10.32. One-port resonator in a basic three-point crystal oscillator: (a) basic circuit and (b) circuit with provisions to apply a dc bias voltage across the resonator (Bienstman 1996)

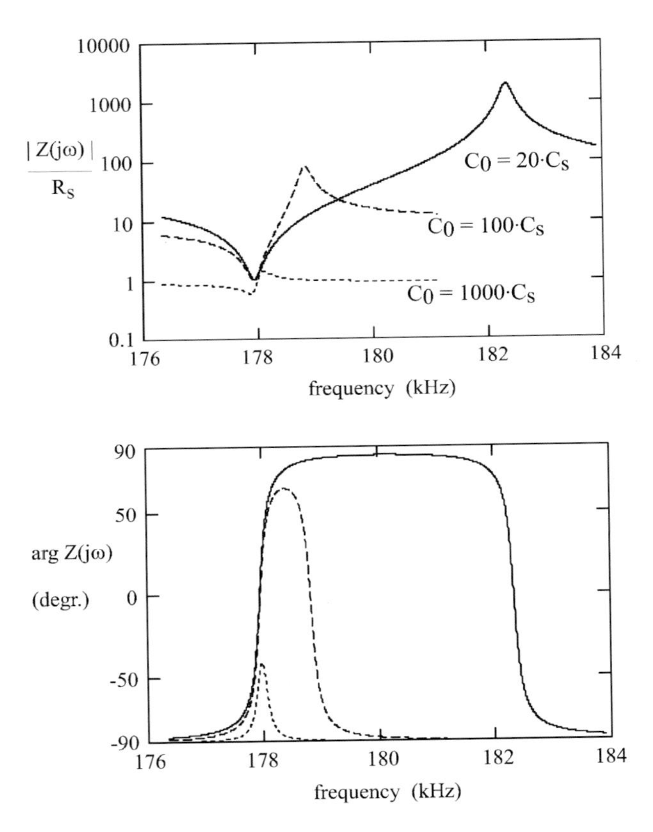

Fig. 10.33. Frequency dependent impedance of the equivalent circuit in Fig. 10.30(b) for various values of the parallel capacitance C_0

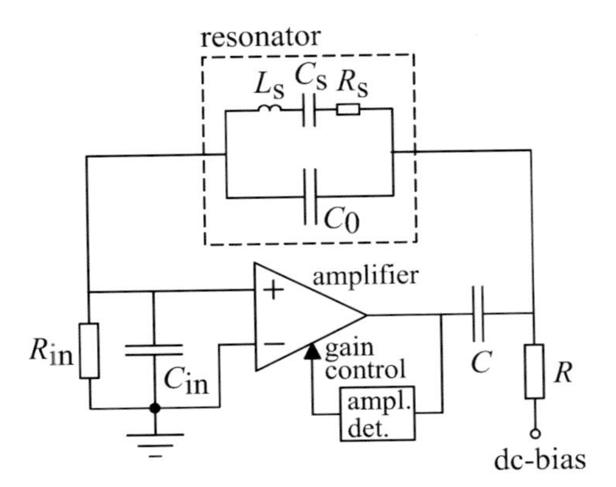

Fig. 10.34. Principle of an oscillator suitable for a one-port electrostatically driven beam resonator with large parallel capacitance C_p

frequency we have a flat part (region (3)) in the amplitude response as the feedback path is now dominated by the capacitive voltage divider consisting of C_0 and C_{in}. By adjusting R_{in} and C_{in}, the circuit should be dimensioned such that the resonance peak of the resonator lies within this region and that the phase shift reaches zero degrees. In this way the highest gain will occur at the resonance peak. In combination with the zero phase shift this will result in an oscillation at a frequency very close to the resonant frequency of the resonator. The gain control loop indicated in Fig. 10.34 will adjust the gain of the amplifier such that the closed loop gain (at the oscillation frequency) is exactly equal to 1. As the highest gain occurs at the resonance peak, no unwanted high frequency oscillations can occur due to additional phase shift caused by the amplifier (region (4)).

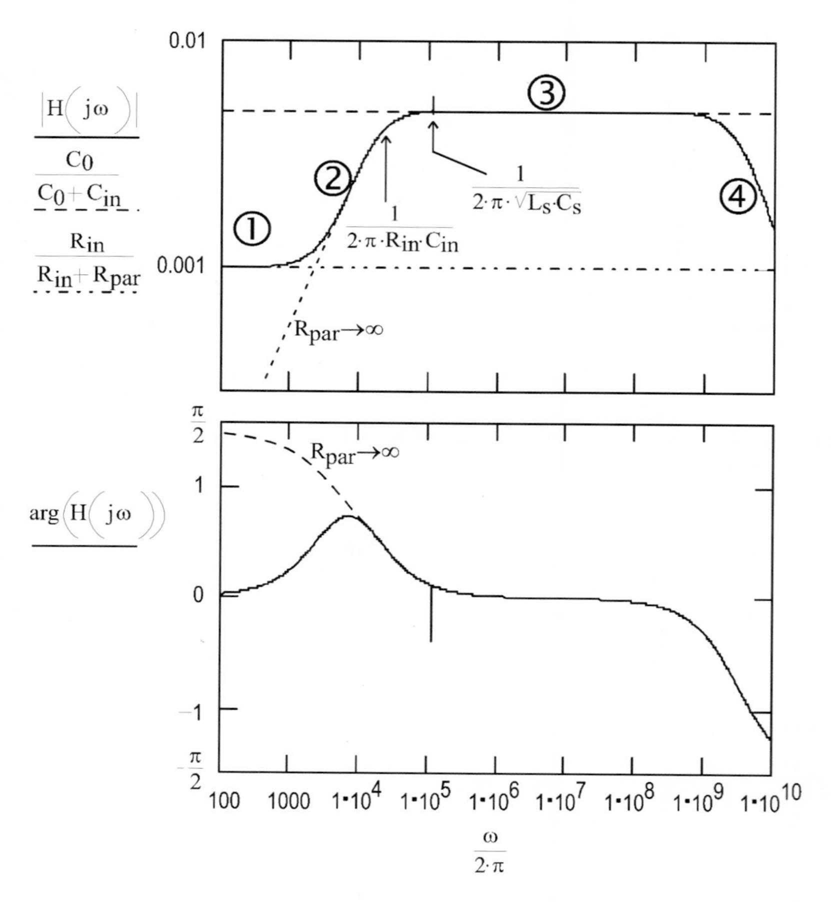

Fig. 10.35. Open-loop frequency characteristic of the circuit in Fig. 10.34. Four frequency ranges can be distinguished (see text)

10.3.5. Oscillator Based on Two-Port Electrodynamically Driven H-shaped Resonator

Fig. 10.36 illustrates the basic structure and operation of an H-shaped resonator with electrodynamical excitation and detection (Shirley 1993). A simplified electrical model is shown in Fig. 10.37. We see that the mechanical resonance is now modeled by a parallel resonance circuit consisting of L, C and R_T. Series resistance in the beam and the connecting wires is represented by R_{o1} for the drive beam and R_{o2} for the sense beam. The output voltage due to electromagnetic inductance is modeled by the transformer TX.

Fig. 10.38 shows the block schematic of a complete oscillator (Shirley 1993). The voltage from the sense beam is first amplified by a differential amplifier with gain G. The amplitude of resulting voltage is a measure for the amplitude of vibration. Therefore, the signal is further amplified and the amplitude is detected and compared to a reference value. A deviation in the detected amplitude results in an adjustment of the gain in the amplifier K which provides the current through the drive beam of the resonator. As the resonance frequency is dependent on the vibration amplitude due to the hard spring effect (see Chap. 9), the amplitude should be controlled very precisely. Thus the circuitry detecting the amplitude, including the preamplifier G, should have an extremely low drift and temperature dependence. Any deviation in the detected amplitude will immediately be reflected by a change in the resonance frequency.

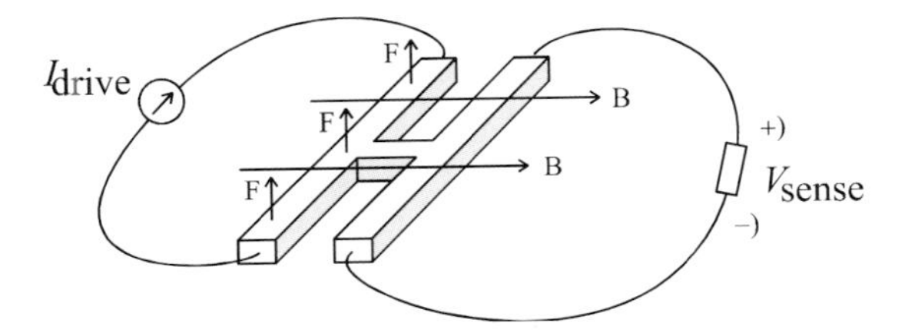

Fig. 10.36. Electrodynamically driven H-shaped two-port resonator (Shirley 1993)

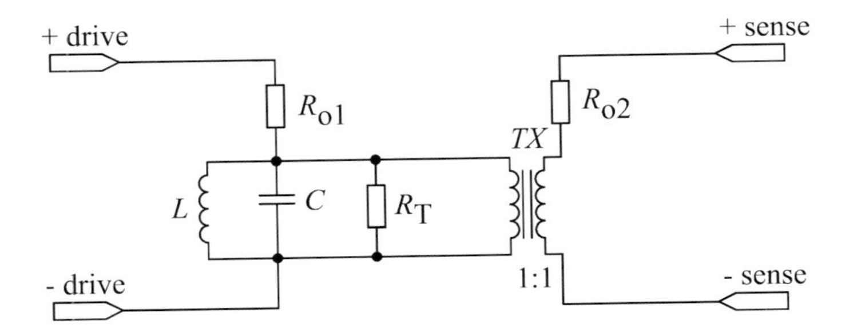

Fig. 10.37. Simplified model of the resonator in Fig. 10.36 (Shirley 1993)

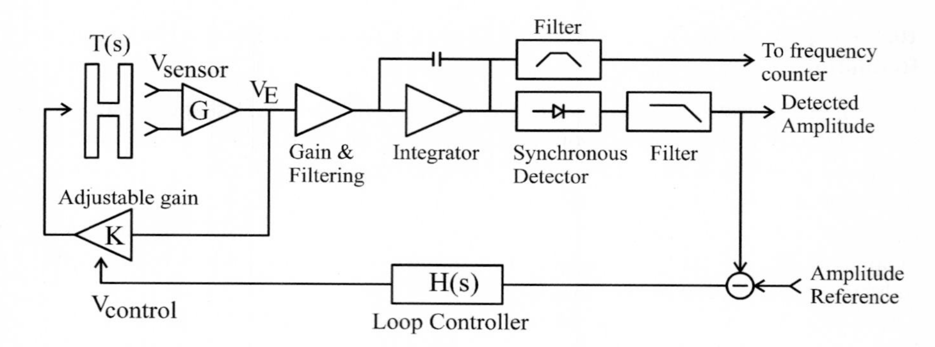

Fig. 10.38. Block schematic of a complete oscillator circuit for an electrodynamically driven H-shaped two-port resonator (Shirley 1993)

11. Packaging

Packaging of microsensors presents special problems: part of the sensor requires environmental access while the rest may require protection from environmental influences. This is schematically illustrated in Fig. 11.1 (Senturia 1988). Due to this special role of the package, it is important that the package is designed simultaneously with the sensor. At the very least some time should be spent on examining the feasibility of a package. Otherwise, one may end up with a sensor that requires an extremely expensive package or cannot be packaged at all.
Usually, 4 levels of packaging are distinguished (Reinert 1995):

- Level 1: the attachment of a bare semiconductor/sensor chip to a package with connections to the pin-outs.
- Level 2: the attachment of the packaged chip to the substrate or board.
- Level 3: the attachment of the substrate or board to a backplane or interconnecting bus.
- Level 4: system integration.

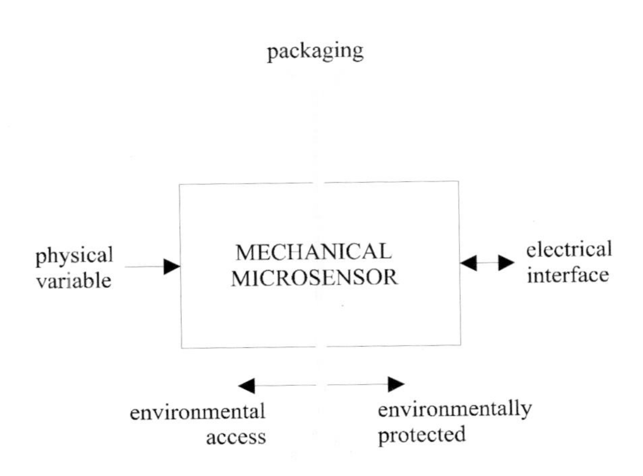

Fig. 11.1. Schematic illustration of the special role of packaging for mechanical microsensors

For obvious reasons, we will restrict ourselves to the first level of packaging. Within this level, we can distinguish between wafer-level packaging and component-level packaging. Wafer level packaging means that an entire wafer with sensors is bonded to another – usually silicon or glass – wafer which acts as a support and protection for the sensor chips during the remainder of the packaging process.

In this chapter we will first discuss some general packaging techniques in Sect. 11.1. Next, in Sects. 11.2, 11.3 and 11.4 we will concentrate on the packaging of three categories of mechanical sensors, namely pressure sensors, inertial sensors and flow sensors, respectively.

11.1 Packaging Techniques

11.1.1 Standard Packages

The most commonly used packages for sensors are usually based on derivatives of conventional semiconductor packaging. In this section we will discuss a few basic packages, namely plastic, ceramic and metal can packages. Each of these has been adapted in one form or another for the packaging of silicon sensors.

Molded plastic packages

Fig. 11.2 shows cross-sectional views of two basic plastic packages. The chip is attached to a metal lead frame. In a postmolded package (Fig. 11.2(a)), the chip is first mechanically bonded to the die attach pad in the center of the lead frame. Next, the electrical connections are made from the chip bond pads to the

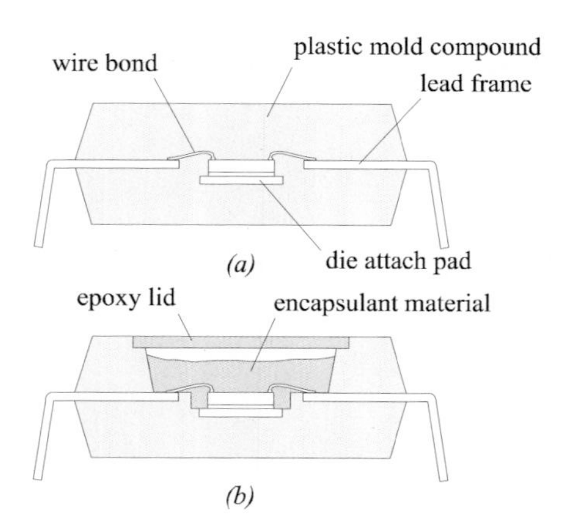

Fig. 11.2. Cross sectional view of (a) a postmolded and (b) a premolded plastic dual in line package

corresponding lead frame fingers. Finally, the package is formed by plastic molding around the lead frame and chip assembly. Postmolding technology is generally not suitable for sensor packaging because of the high associated (thermal) stress and possible damage due to the flow of the molding compound. Still, Cotofana et al. demonstrated a low-cost packaging solution by protecting the sensitive sensor surface during the molding process with a gel-coated nipple (Cotofana 1998).

The alternative to postmolding is the use of a premolded package as shown in Fig. 11.2(b). In this case the package is first molded around the lead frame. Next, the chip is placed and the electrical connections are made. The chip is protected by a suitable encapsulant.

Ceramic packages

Ceramic packages are often used when hermetic sealing is required. Fig. 11.3 shows the structure of a standard CERDIP, or ceramic DIP, and a window frame package (Cohn 1997). In a CERDIP package, the lead frame is attached to the ceramic base, the chip is mounted on the lead frame and the package is closed by a ceramic cap. Glass sealing results in a hermetic seal, however it requires a relatively high sealing temperature, typically more than 400 °C. An alternative is the window-frame package shown in Fig. 11.3(b). Here the glass sealing between the ceramic base and window frame is done before mounting the chip. Final sealing can be done by a low-temperature solder seal. This type of package can easily be adapted for sensor applications. Furthermore, this package can be useful in the development phase of a sensor because the chip is easily accessible for inspection and measurements.

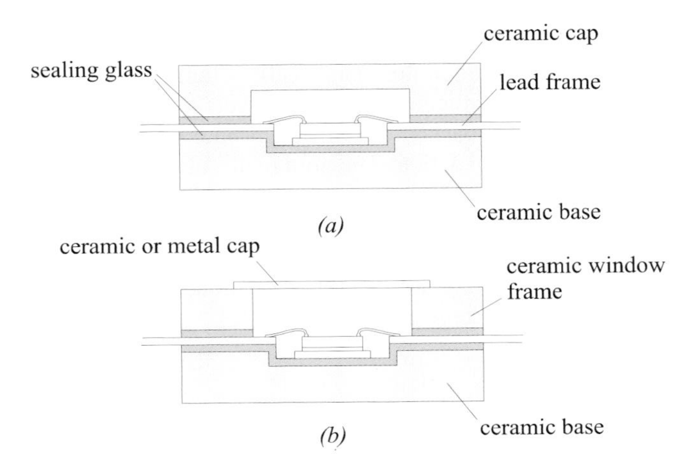

Fig. 11.3. Structure of (a) CERDIP and (b) ceramic window frame package.

Metal can packaging

Metal cans are cylindrically shaped packages that have an array of leads extending through the base of the package as shown in Fig. 11.4. Glass seals are used between the leads and the base. The metal cap is brazed or welded to the base. Metal can packages provide an excellent hermetic sealing and are especially useful for sensors that have to operate in hostile environments.

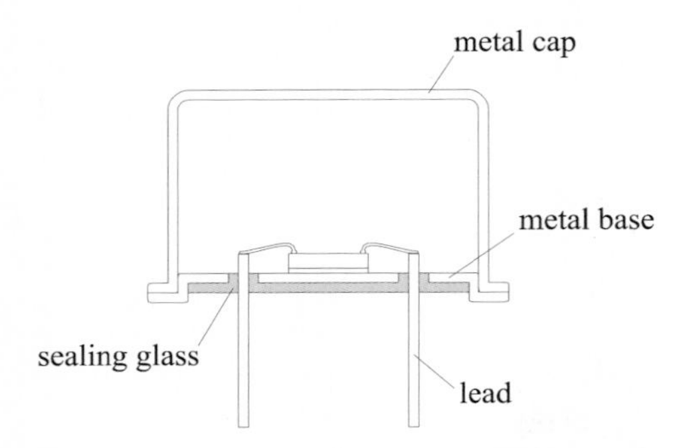

Fig. 11.4. Structure of metal can package

11.1.2 Chip Mounting Methods

One of the first steps in packaging is the attachment of the sensor die to the package. The following methods are commonly used (Reichl 1991, Reinert 1995):

- AuSi eutectic bonding

  Gold and silicon form a eutectic bond at 363 °C. Usually a thin layer of gold is deposited on the back side of the silicon wafer and alloyed at a temperature between 370 and 400 °C. The gold-backed die is brought into contact with a gold plated package with a scrubbing action. The gold melts, attaching the die to the package. The main problem is that it places high thermal stresses on the chip because of the high process temperature. Using ceramic substrate materials, such as AlN and SiC, whose coefficients of thermal expansion closely matches that of silicon (see Table 11.1), will reduce the thermal stresses which can be detrimental to large chips (Cohn 1997).

- Epoxy bonding

  Depending on the type of filler material, epoxy adhesives may produce an electrically conductive or an insulating bond. The die is attached by pressing it onto a small amount of the selected epoxy resin. Epoxies need curing at about 175 °C before the package can be moulded or sealed. Because of the lower process temperature and the low Young's modulus of epoxy the resulting thermal stress is much lower and epoxy adhesives are commonly used for large chips. The use of silver-fillers makes the epoxy both electrically conductive, to provide low resistance between the chip and the substrate, and thermally conductive to allow a good thermal path between the chip and the rest of the package. In general, epoxy chip bonds are as good or better than their metal counterparts, except in the most demanding applications, that is, applications that require high temperatures, high current through the chip bond, and critical thermal performance requirements (Cohn 1997).

- Glass bonding

  In this technique the die attach material consists of a silver-loaded glass system in an organic medium and is applied as a paste. After drying and organic burn-out the system should be completely inorganic, consisting of 80% silver and 20% glass. Glass bonding is usually used in ceramic packages where the final curing is combined with the high temperature glass sealing operation.

Table 11.1. Parameters of different substrate materials (from: Reichl 1991).

Parameters	Material				
	Si	Al_2O_3	AlN	SiC	Pyrex
Thermal expansion coefficient (10^{-6} K^{-1})	2.6	5.6	2.9	3.7	2.9
Heat conductivity (W/m K)	150	10–35	130–170	70	1.1
Young's modulus (Gpa)	170	300–380	300	480	63

11.1.2 Wafer Level Packaging

As mentioned in the introduction, for sensor systems the first step in the packaging process is often a wafer bonding step. The bonded wafer can have several functions:

- Protect delicate sensor parts during the remainder of the packaging process. A cover wafer is bonded to one or both sides of the sensor wafer and after bonding the wafer package can be handled in a manner similar to a standard integrated circuit wafer.
- Serve as an intermediate wafer between the sensor wafer and the package to isolate the sensor from package induces stress.
- Cavities can be formed at the bond surface between the two wafers. An example is the pressure sensor shown in Fig. 11.5. In this case a low-melting temperature glass is used as an intermediate layer. The realized cavity can

contain a reference pressure or vacuum for absolute pressure sensors or it can be connected to the outside world by a pressure port for differential pressure measurements. Another important application of cavities is to realize the vacuum required by resonant sensors. With two electrodes the cavities can be used to realize the capacitors needed in capacitive sensors. To obtain a high vacuum inside cavities it may be necessary to evacuate the cavity after bonding through openings which are sealed later, e.g. by evaporation of Al or SiO_2. Another option is to use nonevaporable getter (NEG) materials inside the cavity (Henmi 1994).

A variety of bonding techniques can be used (Ko 1985, Schmidt 1998, Brand 1999):

- Anodic bonding

  A silicon wafer can be anodically bonded to a pyrex wafer or to another silicon wafer with a thin sputtered layer of pyrex, as described in Section 3.4.1.

- Silicon fusion bonding

  Silicon fusion bonding involves the bonding of two bare silicon wafers as discussed in Section 3.4.2. An advantage is the absence of an intermediate layer between the silicon wafers, resulting in minimal stress due to thermal expansion. A problem is the required processing temperature of 700 ^{o}C, which inhibits the use of standard metal interconnect layers.

- Thermocompression bonding

  Thermocompression bonding involves the bonding of two wafers at elevated temperatures under an applied pressure. Adhesion is achieved by using an intermediate layer. Usually this adhesion layer is applied to one of the wafers prior to the bonding process. Several materials can be used for the adhesion layer.

  In so-called low-temperature glass bonding (Frank 1994) the intermediate layer consists of a low-melting-point inorganic oxide glass, such as lead oxide or boric oxide. Usually a glass paste consisting of glass frit and an organic binder is applied to one of the wafers. Next, this wafer is heated to burn off the organic binder and sinter the powdered glass. Finally, the two wafers are aligned and contacted and the bond is made by thermocompression at a temperature above the softening point of the glass (typically 450 ^{o}C).

  Instead of glass also gold, resulting in a eutectic bond (Ko 1985), and solder are used as the intermediate layer. The main advantage of solder bonding is that it can be performed at a low process temperature, depending on the solder material (Brand 1999). A disadvantage is that it is difficult to obtain a uniform spacing between the wafers.

- Adhesive bonding

  Adhesives are an easy way to bond wafers together, however the resulting bond is not hermetic and a uniform spacing between the wafers is hard to obtain.

Table 11.3 summarizes the differences between the various bonding methods.

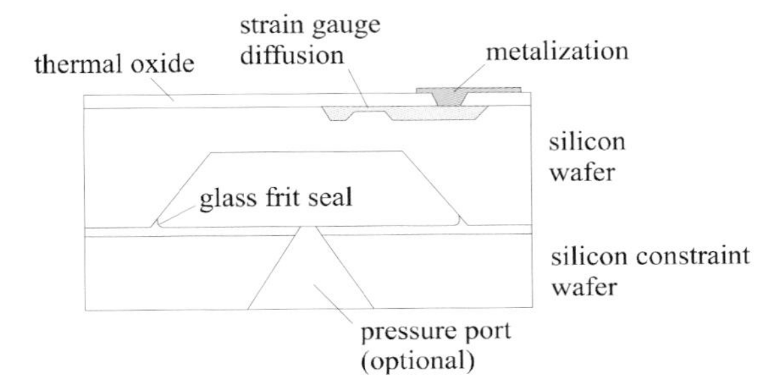

Fig. 11.5. Motorola pressure sensor using glass frit wafer bonding for packaging (Gragg 1984)

11.1.3 Interconnection Techniques

Wire bonding is still the most common technique for electrically connecting the die. There are two main technologies used to form a wire bond, which differ in the mechanical stress applied to the die, the minimum spacing, the used wire alloys and the possibility of adjusting a wire (Reinert 1995). Ultrasonic wire bonding brings an aluminium wire into contact with the pad on the die and a pulse of ultrasonic energy scrubs the two surfaces together, breaking the oxides and forming a good bond. This technique is normally used for hermetic packages. Thermosonic wire bonding uses a combination of heat and ultrasonic energy. The ultrasonic energy decreases the amount of external required heat and is therefore less stressing to the die. Most often this technique is used for gold wire bonding in plastic packages. Due to the higher process speed of this technique and the good corrosion resistance this is used in high volume, high pincount applications where the die is joint to a leadframe. Wire bonding can be fully automated and is suitable for various metalizations and substrates.

In Tape Automated Bonding (TAB) the dies are attached to a flexible tape containing the interconnection leads. The TAB process starts with the so-called Inner Lead Bonding (ILB), i.e. each pad on the chip is connected to a corresponding lead on the tape. This can be done simultaneously for all leads or sequentially. Either the chip or the tape should contain bumps. Next, the chip is

Table 11.2. Wafer-bonding techniques (from: Brand 1999)

Parameters	Bonding methods					
	Anodic	Silicon fusion	Low melting-T glass	Eutectic (e.g. Au-Si)	Solder (e.g. Sn-Pb)	Adhesive
Temperature (°C)	300–450	200–1100	≤ 450	380	180–350	< 150
Pressure	no	little	yes	yes	yes	yes
Electric field	yes	no	no	no	no	no
Wafer material	$Si\text{-}SiO_2$	Si-Si	various	Si-	various	various
Intermediate layer	no	no	glass frit	Au film	solder	polymer
Surface roughness	$< 1\ \mu m$	< 1 nm	non critical		non critical	non critical
Hermeticity	yes	yes	yes	yes	yes	no

encapsulated and tested while still on tape. The tape containing the tested devices can be stored on reel. Finally, the chips are connected to a substrate by the Outer Lead Bonding (OLB) process, which includes excising the chips from the tape, lead forming, attaching the chip to the substrate and bonding the leads to the substrate. TAB technology offers several advantages over conventional wire bonding, like higher pin counts, smaller bonding pads on the chips, superior electrical and high frequency performance, more predictable electrical performance due to the controlled lead geometry and the test and burn-in possibility prior to final assembly.

In Flip Chip (FC) technology the chips are bonded face down to a substrate via bumps. Several bump materials are used, including solder, gold, copper and nickel. Commonly used substrates are ceramics, silicon, laminates and glass. IBM has been using FC packaging since the mid 1960's (Totta 1997). This technology is known as Controlled-Collapse Chip Connection (C4) and was developed as a method for achieving extremely high interconnection densities. For C4 attachment, solder bumps are deposited on metal pads on the surface of the chip. The chip bumps are aligned with corresponding metal pads on the ceramic substrate and the solder is reflowed to simultaneously form all electrical and mechanical connections between the chip and the substrate.

In interesting interconnection technique is the use of a conductive seal. This technique is similar to that used for the connection of LCD displays. This is illustrated in Fig. 11.6 which shows the construction of a Honeywell pressure sensor (Dyrbye 1996). Thanks to the use of the conducting seal packaging consists of simply clicking the five parts together.

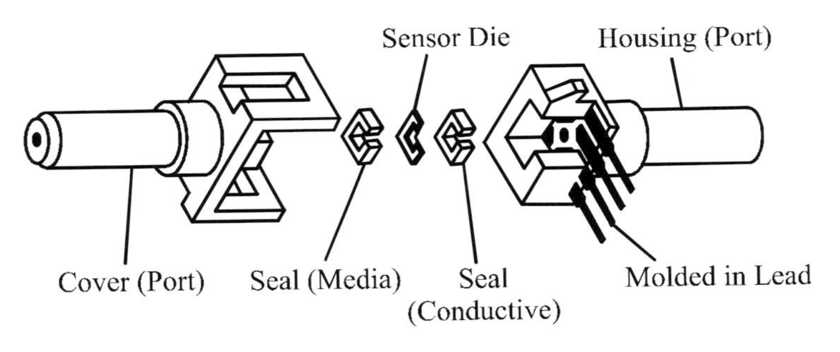

Fig. 11.6. The assembly of a pressure sensor from Honeywell (Dyrbye 1996)

11.1.4 Multichip Modules

Multichip module (MCM) technology offers an attractive approach to integrating and packaging MEMS devices because of the ability to support a variety of die types in a common substrate without requiring changes or compromises to either the MEMS or electronics fabrication processes (Butler 1998). MCMs can assume many different appearances depending on factors such as the type of substrate and method of interconnection between the dies. Commonly used substrate materials include ceramics, silicon and laminates used in printed wiring boards. The dies can be attached to the surface of the substrate or embedded in the substrate. Interconnection between the dies can be made through a variety of methods such as wirebonding, flip-chip solder bumps, or direct metallization. Besides the obvious size and weight benefits of MCM packaging, the close proximity of the dies allows for improved system performance by providing low-noise wiring and, in some cases, eliminating unnecessary interconnections.

As an example, Fig. 11.7 shows the fabrication sequence for a high density interconnect process (Butler 1998, Daum 1993). In this process the dies are embedded into cavities in the substrate and a thin-film interconnect structure is fabricated on top of the components. The process was originally developed for microelectronics, however Butler et al. demonstrated that it can be used for packaging MEMS devices with a few modifications.

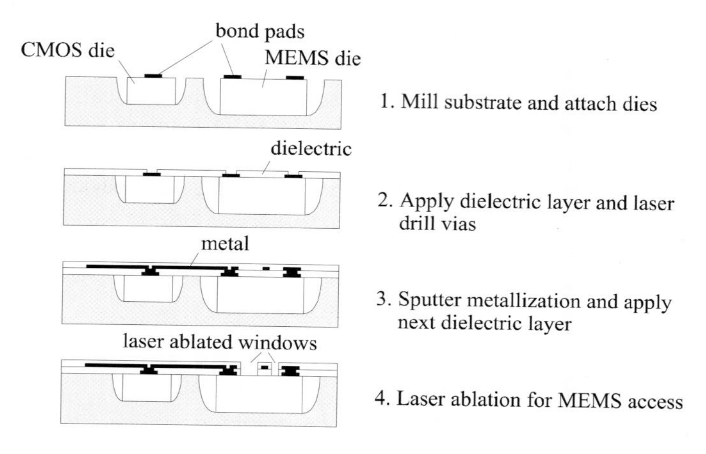

Fig. 11.7. High density interconnect (HDI) process adapted for packaging MEMS devices (Butler 1998)

Another example of multichip packaging of microsystems was demonstrated Morrissey et al. (Morrissey 1998). They realized 3D packaging of a microsystem by building a stack of plastic leadless chip carriers (PLCCs) as shown in Fig. 11.8. The PLCC is a chip carrier containing a cavity with walls. The chip sits on a die paddle in the center of the cavity and wirebonds are used for interconnection. The cavity is filled with an epoxy encapsulant to protect the die and wire bonds. The interconnections between the different chip levels are realized by plating the outer surfaces with a copper/nickel/gold layer and patterning this layer with an excimer laser.

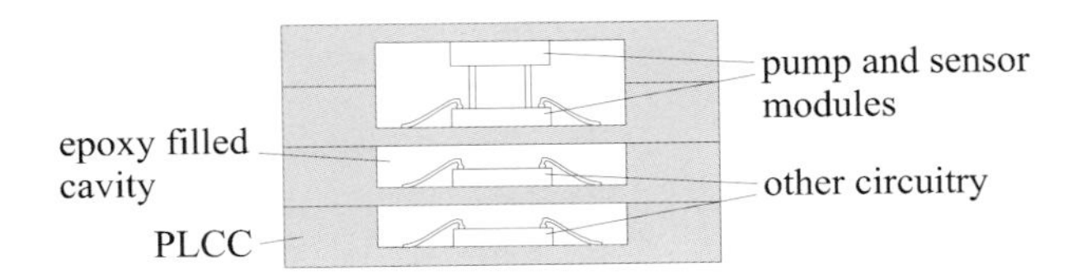

Fig. 11.8. 3D packaging using a stack of standard plastic leadless chip carriers (PLCCs) (Morrissey 1998)

11.1.5 Encapsulation Processes

Encapsulation is used to protect the sensor die against adverse influences from the environment like moisture, contaminants, mechanical vibration and shock. To provide adequate protection, the encapsulant should fulfill the following requirements (Brand 1999):

- Environmental protection of the device (no cracks, very dense, good adhesion to substrate).
- Suitable mechanical, thermal, and electrical properties (e.g. minimal mechanical stress, mathing thermal expansion coefficients).
- Predictable properties from batch to batch, tolerant to small process changes.

Commonly used encapsulants are epoxies, silicones and polyurethanes. Prior to encapsulation the surfaces should be thoroughly cleaned to remove traces of organic and ionic contaminants.

11.2 Stress Reduction

One of the main sources of errors in mechanical sensors is stress. Mechanical stress can have an obvious cause like mechanical loading of the package (for example during mounting) or less obvious causes like swelling or shrinkage of the encapsulant (due to humidity or during curing). Thermomechanical stresses can arise when parts of the sensor have different thermal expansion coefficients.

Stress reduction can already be accomplished at the chip level through the use of special decoupling zones (Buser 1990, Spiering 1993, Schneider 1997). Fig. 11.9 shows a corrugated membrane structure proposed by Spiering et al. (Spiering 1993). The structure was meant for application in strain gauge based pressure sensors. Spiering et al. demonstrated that the corrugation effectively reduced package-induced stresses and significantly lowered the residual stress in the membrane. As a result a larger pressure sensitivity and a reduced temperature dependence were obtained.

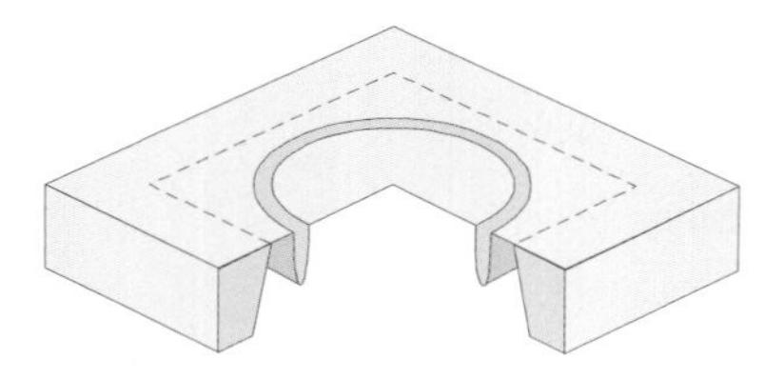

Fig. 11.9. Membrane with deep circular corrugation for package stress reduction (Spiering 1993)

Other effective techniques for on-chip stress reduction include the use of structures with inherent compensation (e.g. folded support beams), using differential sensor configurations (e.g. Wheatstone bridge) and connecting sensitive parts to the substrate in only one point. Examples of the latter technique can be found in inertial sensors in which the springs supporting the proof mass are connected to a single point on the substrate, e.g. (Choi 1996, Ohlckers 1997, Geiger 1999).

To minimize stress induced by the package, the sensor chip can be attached using special adhesives with a low temperature expansion coefficient, a low Young's modulus and a good heat conductance to minimize temperature differences. Attaching the chip to the package at only one end, such that the other end can freely move, will further reduce the induced stress. However, these techniques are only applicable when a hermetic connection between the chip and the package is not necessary.

In the case that a hermetic connection is required a common technique is to bond the sensor wafer to a thick silicon support wafer, either by silicon fusion bonding or by anodic bonding with a thin intermediate Pyrex layer (Hanneborg 1990).

11.3 Pressure Sensors

Pressure sensors are available in a large variety of packages, ranging from standard TO-8 or Ceramic DIP packages with an additional pressure port to custom made plastic packages. A commonly used first step in packaging is bonding the sensor wafer to a support wafer as indicated in Fig. 11.5. To minimize stress due to thermal expansion the support wafer should also be a silicon wafer. Fig. 11.10 shows a typical sensor structure where the sensor wafer is isolated from the metal package by a 2 mm thick silicon support wafer (Yamada 1983).

An even better isolation from the package can be obtained by bonding the silicon support wafer to a Pyrex tube as proposed by Hanneborg and Øhlckers and indicated Fig. 11.11 (Hanneborg 1990). The entire sensor consists of silicon (except for the thin Pyrex layers) and the only connection with the outside world is the thin Pyrex tube.

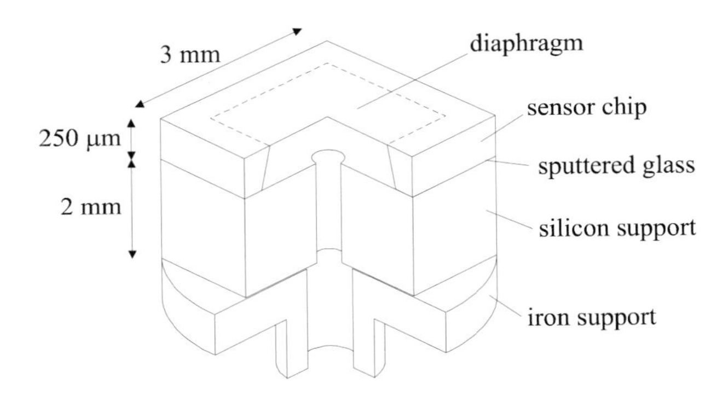

Fig. 11.10. Pressure sensor structure with low package induced stress due to a thick silicon support wafer (Yamada 1983)

In pressure-sensing applications it is often desirable to expose the sensor to agressive/corrosive media. In order to protect the silicon sensing element an intermediate material is added. The material must be non-compressive and soft and have low thermal expansion. Silicone oil is commonly used and is separated from the external media by a flexible membrane. This membrane has to be designed very carefully so as not to induce time or temperature dependent stress to the sensing element (Dyrbye 1996). Fig. 11.12 shows a typical sensor structure. A different approach was used at Motorola (Czarnocki 1998). They exploited the fact that the backside of the sensor die is already quite resistant to harsh media. By using two sensor dies in one package and subtracting the sensor output signals a differential pressure sensor can be realized with only the backside of the sensors being exposed to the environment.

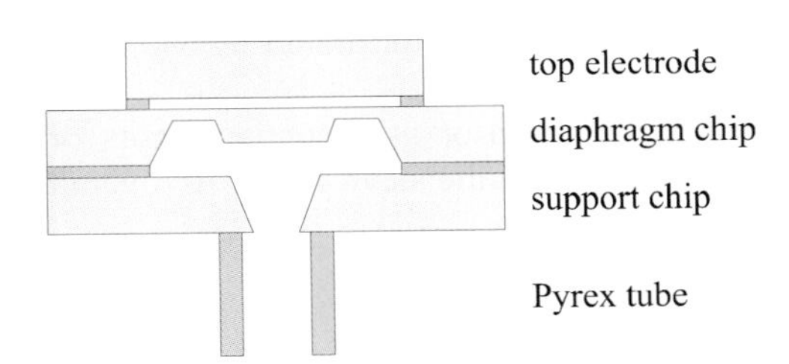

Fig. 11.11. Structure of a capacitive pressure sensor with low temperature dependence (Hanneborg 1990)

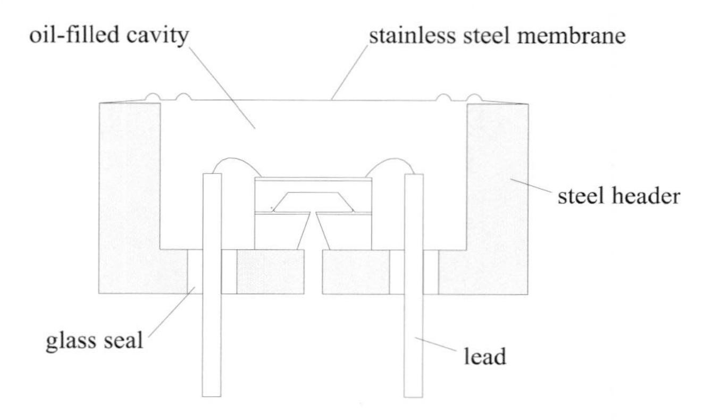

Fig. 11.12. Pressure sensor enclosed in an oil-filled compartment with a stainless steel diaphragm

11.4 Inertial Sensors

Acceleration and angular rate sensors do not require a physical contact with the outside world, thus the package can be hermetically sealed. The main function of the package is to protect the sensor without introducing significant stress or drift. Also, the package should not degrade the sensors frequency response and temperature dependence. As mentioned in section 11.2, package induced stress can be minimized by using a soft adhesive for the attachment of the sensor die to the package and by only using adhesive at one side of the sensor die.

Both metal can (e.g. (Chau 1996, Rudolf 1990)) and multilayer ceramic hermetic packages (Koen 1995) have been used to house the sensor and its interface circuit. The overall packaging cost can be reduced and the performance can be improved by packaging the sensor/circuit at the wafer level using capping glass or silicon wafers that are bonded to the device wafer, then using plastic injection moulding for the final external package, as demonstrated by Motorola (Kniffin 1996) and Ford (Stalnaker 1997).

A point of attention is the alignment of the sensor as mounting errors or misalignment directly affect the sense direction of the device and its overall performance (Yazdi 1998).

11.5 Thermal Flow Sensors

Thermal flow sensors require a package that fulfills demands of a short thermal path to the media, good decoupling of thermal shunting effects, quick response time and relatively low power consumption (Dyrbye 1996). A problem is that the sensitive side of the sensor chip usually also contains the interconnection pads. Thus, exposing the sensor to the flow will also expose the interconnection wires. A solution to this problem is to use flip-chip bonding on a thin substrate with the substrate having an opening at the position of the sensor elements. This approach was used by Mayer et al. (1997). Fig. 11.13 shows a photograph of their sensor. The sensor chip is flip-chip bonded on a thin flexible substrate which is then folded inside the package such that the chip is aligned with the flow channel.

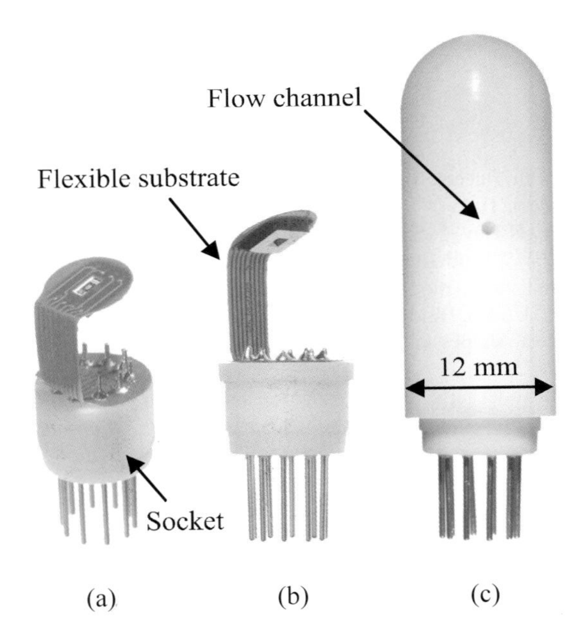

Fig. 11.13. Packaged micromachined flow sensor: chip/flex/socket unit (a), (b) before and (c) after insertion into the plastic housing (Mayer 1997)

Another option is to realize the sensor elements on a thin membrane and expose the backside of the membrane to the flow. Several techniques can be used to minimize the distance between the membrane and the flow channel. An elegant solution is to realize the sensor elements in an n-type epitaxial layer on a p-type substrate, then bonding the epitaxial layer to a Pyrex support wafer and etching away the p-type substrate (Trautweiler 1997).

References

Analog Devices, "ADXL05-monolithic accelerometer with signal conditioning," Analog Devices, Norwood, MA, data sheet, 1995

Ashauer, M., H.Glosch, F.Hedrich,. N.Hey, H.Sandmaier, W.Lang, "Thermal flow sensor for liquids and gases", *Proc. MEMS*, Heidelberg 1998, pp. 351-355

Ayazi F., K. Najafi, "Design and fabrication of of a high-performance polysilicon vibrating ring gyroscope," *Proc. MEMS*, Heidelberg 1998, pp. 621-626

Bäcklund Y. and L. Rosengren, "New shapes in (100) Si using KOH and EDP etches", *J. Micromech. Microeng.* **2** (1992) 75-79

Bäcklund Y., K. Ljungberg and A. Söderbärg, "A suggested mechanism for silicon direct bonding from studying hydrophilic and hydrophobic surfaces", *J. Micromech. Microeng.* **2** (1992) 158

Bäcklund Y., L. Rosengren, B. Hök, "Passive silicon transensor intended for biomedical, remote pressure monitoring," *Sensors and Actuators* **A21-A23** (1990) 58-61

Bellhouse, B.J., Schultz, L, "Determination of mean and dynamic skin friction separation in low-speed flow with a thin film heated element", *J. Fluid Mechanics* **24** (1966) 379-400

Benecke W., L. Csepregi, A. Heuberger, K. Kühl, H. Seidel, "A frequency-selective piezoresistive silicon vibration sensor," *Proc. 3rd Int. Conf. Solid-State Sensors and Actuators (Transducers '85), Philadelphia, PA, USA*, 1985, pp. 103-108

Berberig, O., K. Nottmayer, J. Mizuno, Y. Kanai, T. Kobayashi, "The Prandtl micro flow sensor (PMFS): A novel silicon diaphragm capacitive sensor for flow velocity measurement", *Proc. Transducers* '97, (Chicago, June 16-19, 1997), p. 155-158

Berenschot J.W., J.G.E. Gardeniers, T.S.J Lammerink and M. Elwenspoek, "New applications of r.f.-sputtered glass films as protection and bonding layers in silicon micromachining", *Sensors & Actuators* A **41-42** (1994) 338

Bergqvist J., F. Rudolf, "A new condenser microphone in silicon," *Sensors and Actuators* **A21-A23** (1990) 123-125

Bethe, K, D. Baumgarten, J. Frank, "Creep of sensor's elastic elements: Metals versus non-metals," *Sensors and Actuators* **A21-A23** (1990) 844-849

Bienstman J., H.A.C. Tilmans, E.J.E.A. Peeters, M. Steyaert, R. Puers, "An oscillator circuit for electrostatically driven silicon-based one-port resonators," *Sensors and Actuators A* **52** (1996) 179-186

Blevins R.D., *Formulas for natural frequencies and mode shapes*, Van Nostrand Reinhold, New York, 1979

Blom F.R., D.J. Yntema, F.C.M. van de Pol, M. Elwenspoek, J.H.J. Fluitman, Th.J.A. Popma, "Thin-film ZnO as a micromechanical actuator at low frequencies," *Sensors and Actuators* **A21-A23** (1990) 226-228

Blom F.R., S. Bouwstra, J.H.J. Fluitman, and M. Elwenspoek, "Resonating silicon beam force sensor," *Sensors and Actuators* **17** (1989) 513-519

Boer, H.J., "A new approach for measuring (very) small liquid flows", *Proc. Sensor* 95, p. 97-102

Böffgen M., H. Smitt, "Sputtering of thin films of Pb(ZrTi)O_3," *Ferroelectrics* **108** (1990) 15-20

Boillat, M.A., A.J. van der Wiel, A.C. Hoogerwerf, N.F. de Rooij, "A differential pressure liquid flow sensor for flow regulation and dosing systems", *Proc. Transducers*, Stockholm, Sweden, June 1995, p. 350-352

Boser B.E., R.T. Howe, "Surface micromachined accelerometers," *IEEE J. Solid-State Circuits* **31** (1996) 366-375

Bouwstra S., and B. Geijselaers, "On the resonance frequencies of microbridges," *Proc. 6th Int. Conf. Solid-State Sensors and Actuators (Transducers '91), San Fransisco, June 24-27*, 1991, 538-542

Bouwstra S., E. de Weerd and M. Elwenspoek, "In situ phosphorus-doped polysilicon for excitation and detection of vibrations of micromechanical devices", *Sensors and Actuators* **A24** (1990) 227-235

Bouwstra S., P. Kemna, R. Legtenberg, "Thermally excited resonating membrane mass flow sensor," *Sensors and Actuators* **A20** (1989) 213-223

Bouwstra S., R. Legtenberg, H.A.C. Tilmans, and M. Elwenspoek, "Resonating microbridge mass flow sensor," *Sensors and Actuators* **A21-A23** (1990) 332-335

Bouwstra S., *Resonating Microbridge Mass Flow Sensor*, PhD Thesis, Universiteit Twente (1990)

Brand O., M. Wälti, M. Hornung, F. Mayer, *Microsensor Packaging*, FSRM, Switzerland, 1999.

Branebjerg, J. O.S. Jensen, N.G. Laursen, O. Leistico, "A micromachined flow sensor for measuring small liquid flows", *Proc. Transducers*, San Francisco, USA, June1991, p. 41-44

Bruun, H.H., *Hot Wire Anemometry, principles and signal analysis*, Oxford, University press, Oxford, 1995.

Burger J.F., G.-J. Burger, T.S.J. Lammerink, S. Imai, J.H.J. Fluitman, "Miniaturised friction force measurement system for tribological research on magnetic storage devices", *Proc. MEMS '96*, San Diego, CA, USA, February 11-15, 1996, pp. 99-104

Burger, J., E. Berenschot, G.-J. Burger, H. Visscher, M. Elwenspoek, "Piezoresistive friction force sensor for tribological research", *J. Micromechanics Microeng.* **8** (1998) 138-140

Buser R.A., N.F. de Rooij, "Very high Q-factor resonators in monocrystalline silicon," *Sensors and Actuators* **A21-A23** (1990) 323-327

Buser R.A., N.F. de Rooij, and L. Schultheis, "Silicon pressure sensor based on resonating element," *Sensors and Actuators A* **25-27** (1991) 717-722

Butler J.T., V.M. Bright, J.H. Comtois, "Multichip module packaging of microelectromechanical systems," *Sensors and Actuators A* **70** (1998) 15-22

Büttgenbach, S., *Mikromechanik*, Teubner Studienbücher, Stuttgart, 1991

Chang S.C., M.W. Putty, D.B. Hicks, C.H. Li, and R.T. Howe, "Resonant-bridge two-axis microaccelerometer," *Sensors and Actuators* **A21-A23** (1990) 342-345

Chatzandroulis S., D. Goustouridis, P. Normand, D. Tsoukalas, "A solid-state pressure-sensing microsystem for biomedical applications," *Sensors and Actuators A* **62** (1997) 551-555

Chau H.-L., K.D. Wise, "An ultraminiature solid state pressure sensor for a cardiovascular catheter", *IEEE Trans. Electron Devices* **35** (1988) 2355

Chau K.H.-L., S.R. Lewis, Y. Zhao, R.T. Howe, S.F. Bart, R.G. Marcheselli, "An integrated force-balanced capacitive accelerometer for low-g applications," *Sensors and Actuators A* **54** (1996) 472-476

Cho S.T., K. Najafi, E. Lowman, K.D. Wise, "An ultrasensitive silicon pressure-based microflow sensor," *IEEE Trans. Electron Devices* **39** (1992) 825-835

Choi J., K. Minami, M. Esashi, "Application of deep reactive ion etching for silicon angular rate sensor," *Microsystem Technologies* **2** (1996) 186-190

Christel L., K. Petersen, P. Barth, F. Pourahmadi, J. Mallon, J. Bryzek, "Single-crystal silicon pressure sensors with 500x overpressure protection," *Sensors and Actuators* **A21-A23** (1990) 84-88

Chun K.J., K.D. Wise, "A capacitive tactile imaging array," *Rec. of the 3rd Int. Conf. on Solid-State Sensors and Actuators*, 1985, 22-25

Clark W.A., R.T. Howe, R. Horowitz, "Surface micromachined, z-axis, vibratory rate gyroscope," in *Tech. Dig. Solid-State Sensor and Actuator Workshop (Hilton Head `96)*, Hilton Head Island, SC, 1996, 283-287

Cohn C., M.T. Shih, "Packaging and interconnection of integrated circuits," in *Electronic Packaging and Interconnection Handbook*, C.A. Harper (ed.), McGraw-Hill, 1997, Chapter 6

Cotofana C., A. Bossche, P. Kaldenberg, J. Mollinger, "Low-cost plastic sensor packaging using the open-window package concept," *Sensors and Actuators A* **67** (1998) 185-190

Curle, N., The Laminar Boundary Equations, Oxford University Press 1962.

Czarnocki W.S., "Media isolated sensor," *Sensors and Actuators A* **67** (1998) 142-145

Daum W., W. Burdick Jr., R. Fillion, "Overlay high-density interconnect: A chips-first multichip module technology," *IEEE Comput.* **26** (1993) 23-29

De Bree H.-E., Leussink P., Korthorst T., Jansen H., Lammerink T., Elwenspoek M., "The μ-flown, a novel device measuring acoustical flows", *Proc. Transducers* 1995, Vol. 1, 536-539

De Bree, H.-E., "The microflown", PhD Thesis, University of Twente (1997)

De Bree, H.-E., H.V. Jansen, T.J.S Lammerink, G.J.M. Krijnen, M. Elwenspoek, "Bi directional fast flow sensor with large dynamic range", *Proc. MME* '98, Ulvic (Norway) June 3-5, 1998, 194-197

De Bree, H.-E., W.F. Druyvestyn, E. Berenschot, M. Elwenspoek, "Three-dimensional sound intensity measurements using microflown particle velocity sensors, *Proc. MEMS*, Orlando, Fl., USA, Jan. 1999, 124-129

Degrauwe M. *et al.*, "A micropower CMOS instrumentation amplifier," *IEEE J. Solid-State Circuits*, 1985, 805-807

Deheij, B. *Micromachining in Stainless Steel*, Master thesis, University of Twente, The Netherlands 1996

Di Giovanni M., Flat and Corrugated Diaphragm Design Handbook, Marcel Decker, Inc, New York, 1982, p. 130ff

Donk A.G.H. van der, P.R. Scheeper, W. Olthuis, P. Bergveld, "Amplitude-modulated mechanical feedback system for silicon condenser microphones," *J. Micromech. Microeng.* **2** (1992) 211-214

Dudaicevs H., M. Kandler, Y. Manoli, W. Mokwa, E. Spiegel, "Surface micromachined pressure sensors with integrated CMOS readout electronics," *Sensors and Actuators A* **43** (1994) 157-163

Dyrbye K., T.R. Brown, G.F. Eriksen, "Packaging of physical sensors for aggressive media applications," *J. Micromech. Microeng.* **6** (1996) 187-192

Ebefors, T., E. Kälvesten, G. Stemme, "Three dimensional silicon triple-hot-wire anemometer based on polyimide joints", *Proc. MEMS* '98, Heidelberg, Germany, January 25-29, 1998, pp. 93-98

Eckert, E.R.G., Physical laws of fluid mechanics, in: *Fluid mechanics measurements*, ed. R.J. Goldstein, Taylor & Francis, Bristol 1996), pp. 65ff.

Eisley J.G., "Nonlinear vibration of beams and rectangular plates," *J. Appl. Math. Phys.* **15** (1964) 167-175

Elwenspoek M.C., T.S.J. Lammerink, R. Miyake, J.H.J. Fluitman, "Towards integrated micro liquid handling systems", *J. Micromechanics Microeng.* **4** (1994) 227-245

Elwenspoek, M. and H.V. Jansen, *Silicon Micromachining*, Cambridge University Press, Cambridge, New York, 1999

Elwenspoek, M. and T.S.J. Lammerink, "Theory of thermal flow sensors", Eurosensors 13, 13-15 September, 1999, The Hague, The Netherlands

Enoksson P., G. Stemme, E. Stemme, "A silicon resonant sensor structure for Coriolis mass-flow measurements," *J. Micromechanical Systems* **6** (1997) 119-125

Eyre, B., J. Blosiu, D. Wiberg, "Taguchi optimisation for the processing of Eron SU-8 resist," *Proc. MEMS '98*, Heidelberg, Germany, January 25-29, 1998, p218-222

Feynman, R.P, R.B. Leighton, M.L. Sands, *The Feynman lectures on physics*, Vol 2 (Addison-Wesley Pub, Redwood City, CA, USA, 1965, ch. 40, 41

Fingerson L.M., F. Freymuth, Thermal Anemometers, in: *Fluid mechanics measurements*, ed. R.J. Goldstein, Taylor & Francis, Bristol 1996), pp. 115ff.

Fingerson L.M., P. Freymuth, "Thermal anemometers," in *Fluid Mechanics Measurement*, R.J. Goldstein (ed.), Taylor & Francis, 1996, Chapter 3.

Fluitman, J.H.J., "Microsystmes technology: objectives," *Sensors and Actuators A* **56** (1996) 151-166

Frank R., M.L. Kniffin, L. Ristic, "Packaging for sensors," in *Sensor Technology and Devices*, L. Ristic (ed.), Artech House, 1994, pp. 203-238.

Franke, H. *dtv-Lexikon der Physik*, Deutsche Taschenbuch Verlag (München 1970) Vol. 9, p.33 f.

French P.J., A.G.R. Evans, "Piezoresistance in polysilicon and its applications to strain gauges", *Solid State Electronics* Vol 32 (1989) 1-10

Funk K., T. Fabula, G. Flik, F. Lärmer, "Piezoelectrical driven resonant force sensor: fabrication and crosstalk," *J. Micromech. Microeng.* **5** (1995) 143-146

Funk, J., G. Wachutka, H. Baltes, "Scaling laws for optimised design of integrated thermopiles", *Proc. Transducers*, Yokohama, Japan, June 1993, p. 742-745

Gass, V., H. van der Schoot, N.F. de Rooij, "Nanofluid handling by micro-flow-sensor based on drag force measurement", *Proc, MEMS*, Ft. Lauderdale, FL, USA, Feb. 1993a, 167-172

Gass, V., H. van der Schoot, S, Jeanneret, N.F. de Rooij, "Integrated flow-regulated silicon micropump", *Proc. Transducers*, Yokohama, Japan, June 1993b, 1048-1051

Geiger W., B. Folkmer, J. Merz, H. Sandmaier, W. Lang, "A new silicon rate gyroscope," *Sensors and Actuators A* **73** (1999) 45-51

Gilbert B., "A versatile monolithic voltage-to-frequency convertor," *IEEE J. Solid-State Circuits* **SC-11** (1976) 852-864

Gogoi B.P., C.H. Mastrangelo, "A low-voltage force-balanced pressure sensor with hermetically sealed servomechanism," *Proc. MEMS'99*, Orlando, USA, 1999, 493-498

Goldberg, H.D., Breuer. K.S., Schmidt, M.A., A silicon wafer-bonding technology for microfabricated shear stress sensors with backside contacts", *Solid state sensor and actuator workshop*, Hilton Head Island, 1996, 111-115

Goldstein, H. Classical Mechanics, Addison-Wesley Pub. Reading, MA, 1959

Gragg J.E., W.E. McCulley, W.B. Newton, C.E. Derrington, "Compensation and calibration of a monolithic four terminal silicon pressure transducer," in *Tech. Dig. IEEE Solid-State Sensor Workshop*, Hilton Head, SC, June 1984, 21-27.

Gravesen, P., J. Branebjerg and O.S. Jensen, "Microfluidics - A review", *J. Micromechanics Microeng.* **3** (1993) 168-182

Gray R.M., "Oversampled sigma-delta modulation," *IEEE Trans. Commun.* **COM-35** (1987) 481-489

Greenwood J.C, D.W. Satchel, "Miniature silicon resonant pressure sensor," *IEE Proceedings D* **135** (1988) 369-372

Greenwood J.C., "Etched silicon vibrating sensor," *J. Phys. E: Sci. Instrum.* **17** (1984) 650-652

Greenwood J.C., D.W. Satchell, "Miniature silicon resonant pressure sensor," *IEE Proceedings* **135** (1988) 369-372

Greenwood J.C., T. Wray, "High accuracy pressure measurement with a silicon resonant sensor," *Sensors and Actuators A* **37-38** (1993) 82-85

Guckel H, D.K. Showers, D.W. Burns, C.K. Nestler, C.R. Rutigliano, Proc. IEEE Solid-State Sensors and Actuators Workshop, Hilton Head Island, SC June 1986 (no page numbers)

Guckel H., "Surface micromachined physical sensors," *Sensors and Materials* **4** (1993) 251-264

Guckel H., C. Rypstat, M. Nesnidal, J.D. Zook, D.W. Burns, D.K. Arch, "Polysilicon resonant microbeam technology for high performance sensor applications," *Proc. IEEE Solid-State Sensors and Actuators Workshop*, Hilton Head Island, 1992, 153-156

Guckel H., D.K. Showers, D.W. Burns, C.K. Nesler, C.R. Rutigliano, "Deposition techniques and properties of strain compensated LPCVD silicon nitride films," *IEEE Solid-State Sensors Workshop, Hilton Head Island, SC, USA, Dec. 7-10*, 1986.

Guckel H., D.W. Burns, "Planar processed polysilicon sealed cavities for pressure transducer arrays," *Proc. IEEE Int. Electron Devices Meeting, San Fransisco, USA, Dec. 9-12*, 1984, 223-225

Guckel H., D.W. Burns, C.G. Visser, H.A.C. Tilmans, D.W. DeRoo, "Fine grained polysilicon films with built-in tensile strain," *IEEE Trans. Electron Devices* **ED-35** (1988) 18-19

Guckel H., D.W. Burns, C.R. Rutigliano, Proc. IEEE Solid-State Sensors and Actuators Workshop, Hilton Head Island, SC June 1986 (no page numbers)

Guckel H., D.W. Burns, H.A.C. Tilmans, D.W. DeRoo and C.R. Rutigliano, Techn. Dig. IEEE Solid State Sensor and Actuator Workshop, Hilton Head SC, June 1988, p. 96

Guckel H., D.W. Burns, H.A.C. Tilmans, D.W. DeRoo, C.R. Rutigliano, "Mechanical properties of fine grained polysilicon - the repeatability issue," *IEEE Solid-State Sensor and Actuator Workshop*, 1988, p. 398

Guckel H., J. J. Sniegowski and T. R. Christenson, Proc. IEEE Micro Electro Mechanical Systems Workshop, 1989, p. 71

Guckel H., J.J. Sniegowski, T.R. Christenson, F. Raissi, "The application of fine-grained, tensile polysilicon to mechanically resonant transducers," *Sensors and Actuators* **A21-A23** (1990) 346-351

Guckel H., J.J. Sniegowski, T.R. Christenson, S. Mohney, T.F. Kelly, "Fabrication of micromechanical devices from polysilicon films with smooth surfaces," *Sensors and Actuators* **20** (1989) 117-122

Guckel, H., D. Burns, C. Rutigliano, E. Lovell, B. Choi, "Diagnostic microstructures for the measurement of intrinsic stress in thin films," *J. Micromechanics Micreng.* **2** (1992) 86-95

Gui C. *Direct Waferbonding with chemical mechanical polishing*, Ph.D. thesis, University of Twente, The Netherlands 1998

Haisma J. B.A.C.M. Spierings, U.K.P. Biermann, A.A. van Gorkum, "Diversity and feasibility of direct bonding: a survey of a dedicated optical technology", *Appl. Opt.* **33** (1994) 1154-1169

Hanneborg A, P. Øhlckers, "A capacitive silicon pressure sensor with low T_{co} and high long-term stability," *Sensors and Actuators* **A21-A23** (1990) 151-154

Hanneborg A., "Silicon waferbonding techniques for assembly of micromechanical elements", Proc. IEEE Workshop on MEMS, Nara, Japan, Jan. 30 - Feb.2, 1991, p.92

Hanratti, T.J., Campbell, J.A. "Measurement of wall stress", in: *Fluid mechanic measurement*, ed. Goldstein, 2nd edition, (Francis & Taylor, Washington D.C. 1996), 575-648

Harendt C., B. Höfflinger, H.-G. Graf and E. Penteker, "Silicon direct bonding for sensor applications: Characterisation of the bond quality", *Sensors and Actuators* A **25-27** (1991) 87

Hashimoto M., C. Cabuz, K. Minami, M. Esashi, "Silicon angular rate sensor using electromagnetic excitation and capacitive detection," *J. Micromech. and Microeng.* **5** (1995) 219-225

Hayashi, T., Proc. MEMS'94, Oiso, Japan, January 25 - 28, 1994, 39

Henmi H., S. Shoji, K. Yoshimi, M. Esashi, "Vacuum packaging for microsensors by glass-silicon anodic bonding," *Sensors and Actuators* **43** (1994) 243

Heuberger A., "*Mikromechanik*", Springer Verlag, Heidelberg, 1989

Hohenstatt, M. "Thermal mass flow meters", in: J. Scholz and T. Ricolfi (eds.) *Sensors, a Comprehensive Survey*, Vol. 4. VZH, Weinheim 1990, Ch.9, pp. 325-330

Horsthuis W.G.H., J.H.J. Fluitman, "The development of fiber optic microbend sensors," *Sensors and Actuators* **43** (1982/1983) 99-110

Howe R.T., "Resonant microsensors," *Rec. of the 4th Int. Conf. Solid-State Sensors and Actuators (Transducers '87)*, 1987, 843-848

Howe R.T., and R.S. Muller, "Resonant-microbridge vapor sensor," *IEEE Trans. Electron Devices* **ED-33** (1986) 499-506.

Howe R.T., University of California at Berkeley, 1988, private communication

Huijsing J.H., "Signal conditioning on the sensor chip," *Sensors and Actuators* **10** (1986) 219-237

Huysing, J.H., "monolithic flow sensors, a survey", *Proc. Vaste Stof Sensoren Conference*, Delft, The Netherlands (1980) 39-48

Huzák, M, ."Fluidic oscillator for flowmeter", *Proc. MME* '98, Ulvic (Norway) June 3-5, 1998, 147-150

Ikeda K., H. Kuwayama, T. Kobayashi, T. Watanabe, T. Nishikawa, and T. Yoshida, "Silicon pressure sensor with resonant strain gauge built into diaphragm," *Tech. Digest, 7th sensor Symp., Tokyo, Japan*, 1988, 55-58

Ikeda K., H. Kuwayama, T. Kobayashi, T. Watanabe, T. Nishikawa, T. Yoshida, and K. Harada, "Silicon pressure sensor integrates resonant strain gauge on diaphragm," *Sensors and Actuators* **A21-A23** (1990) 146-150

Ikeda K., H. Kuwayama, T. Kobayashi, T. Watanabe, T. Nishikawa, T. Yoshida, and K. Harada, "Three-dimensional micromachining of silicon pressure sensor integrating resonant strain gauge on diaphragm," *Sensors and Actuators* **A21-A23** (1990) 1007-1010

Ikeda, K., H. Kuwayama, T. Kobayashi, T. Watanabe, T. Nishikawa, T. Yoshida, *Sensors and Actuators* **A21-A23** (1990) 193

Jansen H.V., M. de Boer, R. Legtenberg and M. Elwenspoek, Proc. Micro Mechanics Europe, Pisa, Italy, Sept. 1994, p.60.

Jansen H.V., M.J. de Boer, B. Otter and M. Elwenspoek, Proc. IEEE workshop on Micro Electro Mechanical Systems, Amsterdam, Jan. 29 - Feb. 2, 1995, p. 88.

Jansen H.V., *Plasma etching in microtechnology*, Ph.D. Thesis, University of Twente, The Netherlands (1996)

Jiang, F. Y.C. Tai, C.H.Ho, R. Karan, M. Garstenauer, "Theoretical and experimental studies of the micromachined hot-wire anemometer", *IEDM*, San Francisco, 1994, 139-142

Jiang, F. Y.C. Tai, J.B. Huang, C.H. Ho, "Polysilicon structures for shear stress sensors", *Digest IEEE TENCON'95*, Hong Kong, Nov. 1995

Jiang, F.-K., Tai, Y.-C., Gupta, B., Goodman, R., Tung, S., Huang, J., Ho, C.-M., "A surface micromachined Shear stress Imager", *Proc. MEMS* 1996, 110-115

Johnson, R.G., R.E. Higashi: "A highly sensitive silicon chip microtransducer for air flow and differential pressure sensing applications", *Sensors and Actuators* **11** (1987) 63-72

Jouwsma, W., Proc. Sensor 88, Nürnberg, Germany, May 1988, Wunstorf 2, *ACS Organisations*, 1988, p. 219-232

Jouwsma, W. "Marketing and design in flow sensing", *Sensors and Actuators* A **37-38** (1993) 274-279

Juneau T., A.P. Pisano, J.H. Smith, "Dual axis operation of a micromachined rate gyroscope," in *Tech. Dig. 9th Int. Conf. Solid-State Sensors and Actuators (Transducers'97)*, Chicago, IL, 1997, pp. 839-842

Kälvesten, E. *Pressure and wall shear stress sensors for turbulence measurements*, PhD thesis, KTH, Stockholm, Sweden 1996

Kälvesten, E., C. Vieider, L. Löfdahl, G. Stemme, "An integrated pressure-flow sensor for correlation measurements in turbulent gas flows", *Sensors and Actuators* A **52** (1996) 51-58

Kälvesten, E., L. Löfdahl, G. Stemme, "A small-size silicon microphone for measurements in turbulent gas flows", *Sensors and Actuators* A **46** (1994a) 103-108

Kälvesten, E., L. Löfdahl, G. Stemme, "Small piezoresistive silicon microphones specially designed for the characterisation of turbulent gas flows", *Sensors and Actuators* A **46** (1994b) 151-155

Kampen R.P. van, R.F. Wolffenbuttel, "Modeling the mechanical behavior of bulk-micromachined silicon accelerometers," *Sensors and Actuators A* **64** (1998) 137-150

Kämper, K.-P., J. Döpper, W. Ehrfeld, S. Oberbeck, "Self-filling low-cost membrane micropump", *Proc. MEMS '98*, Heidelberg, Germany, Jan. 25-29, 1998, 432-437

Kim H., Y.-G. Jeong, K. Chun, "Improvement of the linearity of a capacitive pressure sensor using an interdigitated electrode structure," *Sensors and Actuators A* **62** (1997) 586-590

King, L.V. "On the convection of heat from small cylinders in a stream of fluid: Determination of the convection constants of small platinum wires with application to hot wire anemometry", *Proc. Roy. Soc. London* A **90** (1914) 563-570

Kirman R.G., "A vibrating quartz force sensor," *Transducers Tempcon Conf. Papers 1983*, London, June 14-16, 1983.

Kloeck B., N.F. de Rooij, *Proc. Transducers '87*, p. 116

Kloeck B., S.D. Collins, R.L. Smith, N.F. de Rooij, *IEEE Trans. Electron Devices* (1989)

Kloek B., "Piezoresistive sensors", in: *Sensors*, Vol 7, ed. W. Göpel, J. Hesse and J.N. Zemel, VHC Verlagsgesellschaft, Weinheim, Germany, 1994, 145-172

Kniffin M.L., M. Shah, "Packaging for silicon micromachined accelerometers," *Int. J. Microcircuits Electron. Packaging* **19** (1996) 75-86

Ko W.H., J.T. Suminto, G.J. Yeh, "Bonding techniques for microsensors," in *Micromachining and micropackaging of transducers*, C.D. Fung, P.W. Cheung, W.H. Ko, D.G. Fleming (eds.), Elsevier, 1985, 41-61

Ko W.H., M.-H. Bao, Y.-D. Hong, "A high sensitivity integrated-circuit capacitive pressure transducer," *IEEE Trans. Electron Devices* **ED-29** (1982) 48-56

Koen E., F. Pourahmadi, S. Terry, "A multilayer ceramic package for silicon micromachined accelerometers," in *Tech. Dig. 8th Int. Conf. Solid-State Sensors and Actuators (Transducers'95)*, Stockholm, Sweden, June 1995, 273-276

Kohl, F., A. Jachimowicz, J. Steurer, R. Glatz, J. Kuttner, D. Biacovsky, F. Olcaytug, G. Urban: "A micromachined flow sensor for liquid and gaseous fluids", *Sensors and Actuators* A **41** (1994) 293-299

Komiya, K, F. Higuchi, K. Ohtani, "Characteristic of a thermal gas flowmeter", *Rev. Sci. Instrum.* **59** (1988) 477-479

Kovacs G.T.A., *Micromachined transducer sourcebook*, WCB McGraw - Hill, Boston, 1998

Kress H.-J., F. Bantien, J. Marek, M. Willmann, "Silicon pressure sensor with integrated CMOS signal-conditioning circuit and compensation of temperature coefficient," *Sensors and Actuators A* **25-27** (1991) 21-26

Kuttner, H., G. Urban, A. Jachimowicz, F. Kohl, F. Olcaytug, P. Goiser: "Microminiaturized Thermistor Arrays for Temperature Gradient, Flow and Perfusion Measurements", *Sensors and Actuators* A **25-27** (1991) 641 - 645

Lammerink T.S.J., M. Elwenspoek, R.H. van Ouwerkerk, S. Bouwstra, J.H.J. Fluitman, "Performance of thermally excited resonators," *Sensors and Actuators* **A21-A23** (1990) 352-356

Lammerink T.S.J., S.J. Gerritsen, "Fiber-optic sensors based on resonating mechanical structures," *SPIE Vol. 798 Fiber optic Sensors II*, 1987, 67-71

Lammerink T.S.J., W. Wlodarski, "Integrated thermally excited resonant diaphragm pressure sensor," *Proc. 3rd Int. Conf. Solid-State Sensors and Actuators (Transducers '85)*, Philadelphia, 1985, 97-100

Lammerink T.S.J., W. Wlodarski, "Integrated thermally excited resonant diaphragm pressure sensor," *Proc. 3rd Int Conf. Solid-State Sensors and Actuators (Transducers'85), Philadelphia, PA, USA*, 1987, 97-100

Lammerink, T. S. J., M. Elwenspoek, and J. H. J. Fluitman, "Integrated Micro-Liquid Dosing System," Proc. of MEMS'93,1993, 254 - 259

Lammerink, T.S.J., F. Dijkstra, Z. Houkes, J. van Kuijk, "Intelligent gas-mixture flow sensor", *Sensors and Actuators* **46-47** (1995) 380-384

Lammerink, T.S.J., N.R.Tas, M. Elwenspoek, J.H.J. Fluitman, "Micro-liquid flow sensor", *Sensors Actuators* **37-38** (1993) 45-50

Landau and Lifshitz, Lehrbuch der theoretischen Physik, Vol. VII, Akademie Verlag Berlin, GDR 1965

Landau L.D., E.M. Lifschitz, *Lehrbuch der theoretischen Physik-Mechanik*, Akademie-Verlag, Berlin, 3rd edn., 1964, 101-104

Landau, L.D, E.M. Lifshitz, *Hydrodynamik*, Akademie Verlag Berlin 1974, pp.165ff.

Legtenberg R., S. Bouwstra, and J.H.J. Fluitman, "Resonating microbridge mass flow sensor with low-temperature glass bonded cap wafer," *Sensors and Actuators A* **25-27** (1991) 723-727

Legtenberg R., J. Elders and M. Elwenspoek, "Stiction of Surface Micromachined Structures after Rinsing and Drying: Model and Investigation of Adhesion Mechanisms", *Transducers '93*, Yokohama, Japan, June 7-10, 1993, p198-201.

Legtenberg R., A.W. Groeneveld, M.C. Elwenspoek, "Comb-drive actuators for large displacements," *J. of Micromechanics and Microeng.* **6** (1996) 320-329

Lehmann H.W., in *Thin Film Processes II*, Eds. J.L. Vossen, W. Kern, Academic Press Inc., Boston (1991), Chapter 5.

Lemkin M.A., B. Boser, J. Smith, "A 3-axis surface micromachined ΣΔ accelerometer," in *Tech. Digest Int. Solid-State Circuits Conf. (ISSCC'97)*, San Francisco, 1997, 202-203

Lemkin M., B.E. Boser, "A three-axis micromachined accelerometer with a CMOS position-sense interface and digital offset-trim electronics," *IEEE J. Solid-State Circuits* **34** (1999) 456-468

Linder C., N.F. de Rooij, "Investigations on free-standing polysilicon beams in view of their applications as transducers," *Sensors and Actuators* **A21-A23** (1990) 1053-1059

Linnemann, H., P.Woias, C.-D. Senfft, J.A. Ditterich, "A self-priming and bubble-tolerant piezoelectric silicon micropump for liquids and gases", Proc. MEMS '98, Heidelberg, Germany, Jan. 25-29, 1998, 532-537

Liu, C. Y.C. Tai, J.B. Huang, C.M. Ho, "Surface micromachined thermal shear stress sensor", *ASME Application of Microfabrication to fluid mechanics*, Chicago (1994), 9-15

Liu, J.Q. Y.-C. Tai, "In situ monitoring and universal modelling of sacrificial PSG etching using hydrofluid acid", *Proc. MEMS*, Ft. Lauderdale, Fl., USA, Feb. 1993, 71-76

Liu, J.Q. Y.-C. Tai, "Micromachined channel/pressure sensor systems for micro flow studies", *Proc. Transducers*, Yokohama, Japan 1993, 995-997

Löfdahl, L., G. Stemme, B. Johansson, "Silicon based flow sensors used for mean velocity and turbulence measurements", *Experiments in Fluids* **12** (1992) 270-276

Lu C., M.A. Lemkin, B.E. Boser, "A monolithic surface micromachined accelerometer with digital output," *IEEE J. Solid-State Circuits* **30** (1995) 1367-1373

Lutz M., W. Golderer, J. Gerstenmeier, J. Marek, B. Maihofer, S. Mahler, H. Munzel, U. Bischof, "A precision yaw rate sensor in silicon micromachining," in *Tech. Dig. 9th Int. Conf. Solid-State Sensors and Actuators (Transducers'97)*, Chicago, IL, 1997, 847-850

Lyons, C. A. Friedberger, W. Welser, G. Müller, G. Krötz, R. Rasing, "A high-speed mass flow sensor with heated silicon carbide bridges", *Proc. MEMS*, Heidelberg, Germany 1998, 356-360

Madou, M., "*Fundamentals of Microfabrication*", CRC Press, New York, USA, 1997

Mallon J.R., F. Pourahmadi, K. Petersen, P. Barth, T. Vermeulen, J. Bryzek, "Low-pressure sensors employing bossed diaphragms and precision etchstopping", *Sensors & Actuators* **9** (1986) 345-351

Mallon J.R., F. Pourahmadi, K. Petersen, P. Barth, T. Vermeulen, J. Brezek, "Low pressure sensors employing bossed diaphragms and precision etch-stopping," *Sensors and Actuators* **A21-A23** (1990) 89-95

Martin K., "A voltage-controlled switched-capacitor relaxation oscillator," *IEEE J. Solid-State Circuits* **SC-16** (1981) 412-413

Masana, F., M. Domínguez, L. Castañer, V. Jiménez, "Plastic packaging using specially designed lead-frames. Application to a flow sensor", *Proc. MME* '98, Ulvic (Norway) June 3-5, 1998, 276-278

Matsumoto Y., M. Nishimura, M. Matsuura, M. Ishida, "Three-axis SOI capacitive accelerometer with PLL C-V converter," *Sensors and Actuators A* **75** (1999) 77-85

Mayer F., A. Häberli, H. Jacobs, G. Ofner, O. Paul, H. Baltes, "Single-chip CMOS anemometer," *Proc. IEEE Int. Electron Devices Meeting (IEDM)*, 1997, 895-898

Mayer, F. G. Salis, J. Funk, O. Paul, H. Baltes, "Scaling of thermal CMOS gas flow microsensors: experiment and simulation", *Proc. MEMS*, San Diego, CA, USA 1996, 116-121

Meirovitch L., *Analytical methods in vibrations*, Collier-Macmillan Ltd., London, 1967

Meleshenko V.V., "Bending of an elastic rectangular clamped plate: exact versus 'engineering' solutions", *J. Elasticity* **48** (1997) 1-50

Menedez, A.N., Ramaprian, B.R., The use of flush mounted hot film gauges to measure skin friction in unsteady boundary layers", *J. Fluid Mechanics* 161 (1985) 139-159

Menz W., P. Bley, *Mikrosystemtechnik für Ingenieure*, VCH Verlagsgesellschaft mbH, Karlsruhe 1992

Micromachine No. 24, (1998) Micromachine Center, Tokyo, Japan, p. 20

Minami K., H.Tosaka, M.Esahi "Optical in-situ Monitoring of Silicon Diaphragm Thickness During Wet Etching", *Proc. MEMS-94*, 217-222.

Mineta T., S. Kobayashi, Y. Watanabe, S. Kanauchi, I. Nakagawa, E. Suganuma, M. Esashi, "Three axis capacitive accelerometer with uniform axial sensitivities," *J. Micromech. Microeng.* **6** (1996) 431-435

Moldovan C., Institute of Microtechnology, Bucharest, ROMANIA, personal communication (1998)

Morrissey A., G. Kelly, J. Alderman, "Low-stress 3D packaging of a microsystem," *Sensors and Actuators A* **68** (1998) 404-409

Moser, D., R. Lenggenhager, H. Baltes, "Silicon Gas flow sensors using industrial CMOS and bipolar technology", *Sensors and Actuators* A **25** (1991)

Mucha J.A. and D.W Hess, in *Introduction in Microlithography*, Eds. L.F. Thompson, C.G. Willson and M.J. Bowden, ACS Symp. Ser. 219, Washington D.C. (1983).

Mukherjee T., Y. Zhou, G.K. Fedder, "Automated optimal synthesis of microaccelerometers," in *Tech. Dig. Twelfth IEEE Int. Conf. Micro Electro Mechanical Systems (MEMS'99)*, Orlando, FL, 1999, 326-331

Mulhern G.T., D. S. Soane and R.T. Howe, *Proc. 7th Int. Conf. Solid-State Sensors and Actuators (Transducers '93)*, June 1993, 296.

Mullem C.J. van, F.R. Blom, J.H.J. Fluitman, and M. Elwenspoek, "Piezoelectrically driven silicon beam-force sensor," *Sensors and Actuators A* **25-27** (1991) 379-383

Muller, R.S., *Microactuators*, IEEE, A.P. Pisano ISBN: 0780334418, 1999

Muller, R.S., R.T. Howe, S.T. Senturia R.L. Smith and R.M. White, *Microsensors*, IEEE, New York, 1991

Nathanson H.C., R.A. Wickstrom, "A resonant-gate silicon surface transistor with high-Q band-pass properties," *Appl. Physics Letters* **7** (1965) 84-86

Nathanson H.C., W.E. Newell, R.A. Wickstrom, J.R. Davis Jr., "The resonant gate transistor," *IEEE Trans. Electron Devices* **ED-14** (1967) 117-133

Ng, K., Shajii, K., Schmidt, M.A. "A liquid shear stress sensor fabricated using wafer bonding technology", *Proc. Transducers* 1991 (San Francisco, USA, June 24-27, 1991) 931-934

Øhlckers P., R. Holm, H. Jakobsen, T. Kvisteroy, G. Kittilsland, "An integrated resonant accelerometer microsystem for automotive applications," *Int. Conf. Solid-State Sensors and Actuators (Transducers '97)*, Chicago, IL, USA, 1997, 843-846

Olcaytug, F., K. Riedling, W. Fallmann: "A low temperature process for the reactive formation of Si3N4 layers on InSb", *Thin Solid Films* **67** (1980) 321-324

Oosterbroek, R.E., T.S.J. Lammerink, J.W. Berenschot, A. van den Berg, M. Elwenspoek, "Design, realisation and characterisation of a novel capacitive pressure/flow sensor", *Proc. Transducers*, Chicago, June 16-19, 1997, 151-154

Oosterbroek, R.E., T.S.J. Lammerink, J.W. Berenschot, G.J.M. Krijnen, M. Elwenspoek, A. van den Berg, "A micromachined pressure/flow sensor", *Sensors and Actuators* (1999) (in the press)

Padmanabhan, A., H. Goldberg, K.S. Breuer, M.A. Schmidt, "A silicon micromachined floating element shear stress sensor with optical position sensing by photodiodes", *Proc. Transducers* 1995 (Stockholm, Sweden, June 25-29), 436-439

Padmanabhan, A., M. Sheplak, K.S. Breuer, M.A. Schmidt, "Micromachined sensors for static and dynamic shear-stress measurements in aerodynamic flows", *Proc. Transducer*, Chicago, Il, USA 1997, 137-140

Pan J.Y., P. Lin, F. Maseeh, S.D. Senturia, "Verification of FEM analysis of load-deflection methods for measuring mechanical properties of thin films", *Proc. Hilton Head workshop*, June 4-7, 1990, 70-73

Pan, T., Hyman, D. Mehregany, M., Reshotko, E., Willis, B, "Calibration of microfabricated shear stress sensors", *Proc. Transducers* 1995 (Stockholm, Sweden, June 25-29), 443-446

Parameswaran M., H.P. Baltes, Lj. Ristic, A.C. Dhaded, A.M. Robinson, "A new approach for the fabrication of micromechanical structures," *Sensors and Actuators* **19** (1989) 298-307.

Park J.-S., and Y.B. Gianchandani, "A low cost batch-sealed capacitive pressure sensor," *Proc. MEMS '99*, Orlando, USA, 1999, 82-87

Park Y.E., K.D. Wise, "An MOS switched-capacitor readout amplifier for capacitive pressure sensors," *Rec. of the IEEE Custom IC Conf.*, 1983, 380-384

Pedersen M., W. Olthuis, P. Bergveld, "High-performance condenser microphone with fully integrated CMOS amplifier and DC-DC voltage converter," *IEEE J. Micromech Syst.* **7** (1998) 387-394

Peeters E., S. Vergote, B. Puers, W. Sansen, "A highly symmetrical capacitive micro-accelerometer with single degree-of-freedom response," *J. Micromech. Microeng.* **2** (1992) 104-112

Petersen K., C. Kowalski, J. Brown, H. Allen, J. Knutti, "A force sensing chip designed for robotic and manufacturing automation applications," *Rec. of the 3rd Int. Conf. on Solid-State Sensors and Actuators*, 1985, 30-32.

Petersen K., J. Brown, T. Vermeulen, P. Barth, J. Mallon, J. Bryzek, "Ultra-stable, high-temperature pressure sensors using silicon fusion bonding," *Sensors and Actuators* **A21-A23** (1990) 96-101

Petersen K., P. Barth, J. Poydock, J. brown, J. Mallon Jr., J. Bryzek, "Silicon fusion bonding for pressure sensors," *IEEE Solid-State Sensor and Actuator Workshop*, 1988, 144-147

Petersen K.E., "Silicon as a mechanical material", *Proc. IEEE* **70** (1982) 420

Peterson K., F. Pourahmadi, J. Brown, P. Parsons, M. Skinner, and J. Tudor, "Resonant silicon beam pressure sensor fabricated with silicon fusion bonding," *Proc. 6th Int. Conf. Solid-State Sensors and Actuators (Transducers '91), San Fransisco, CA, USA*, 1991, 664-667

Pons P., G. Blasquez, R. Behocaray, "Feasibility of capacitive pressure sensors without compensation circuits," *Sensors and Actuators A* **37-38** (1993) 112-115

Prak A., *Silicon resonant sensors: operation and response*, Ph.D. thesis, University of Twente, Enschede, The Netherlands, 1993.

Puers B., E. Peeters, A. van den Bossche, W. Sansen, "A capacitive pressure sensor with low impedance output and active suppression of parasitic effects," *Sensors and Actuators* **A21-A23** (1990) 108-114

Puers R., S. Reyntjens, "Design and processing experiments of a new miniaturized capacitive triaxial accelerometer," *Sensors and Actuators A* **68** (1998) 324-328

Putty M.W., K. Najafi, "A micromachined vibrating ring gyroscope," in: *Tech. Dig. Solid-State Sensor and Actuator Workshop*, Hilton Head Island, SC, 1994, 213-220

Putty M.W., S.C. Chang, R.T. Howe, A.L. Robinson, and K.D. Wise, "Process integration for active polysilicon resonant microstructures," *Sensors and Actuators* **20** (1989) 143-151

Rehn, L.A., R.W. Tarpley, K.C. Wiemer, K.M. Durham, "Dual-element, solid state fluid flow sensor", *SAE Trans.* **89** (1980) 705-710

Reichl H., "Packaging and interconnection of sensors," *Sensors and Actuators A* **25-27** (1991) 63-71

Reinert W., K. Dittmer, *Quality in packaging*, UETP-MEMS course, FSRM, Switzerland, 1995.

Richter, M., M. Wackerle, P. Woias, B. Hillerich, "A novel flow sensor with high time resolution based on differential pressure sensor principles", *Proc. MEMS*, Orlando, Fl, USA Feb. 1999, 118-123

Ristic, L. (Eds.), *Sensor technology and devices*, Artech House, Boston, 1994

Roesengren L., J. Söderquist and L. Smith, "Micromachined sensor structures with linear capacitive response," *Sensor and Actuators A* **31** (1992) 200-205

Rosengren L., *Silicon microstructures for biomedical sensor systems*, Ph.D. Thesis, Uppsala University, Sweden(1994)

Royer M., J.O. Holmen, M.A. Wurm, O.S. Aadland, M. Glenn, "ZnO on Si integrated acoustic sensor," *Sensors and Actuators* **4** (1983) 357-362

Roylance L.M., J.B. Angell, "A batch fabricated silicon accelerometer," *IEEE Trans. Electron. Devices* **ED-26** (1979) 1911-1917

Rudolf F., A. Jornod, J. Bergqvist, H. Leuthold, "Precision accelerometers with μg resolution," *Sensors and Actuators* **A21-A23** (1990) 297-302

Sanchez S., M. Elwenspoek, C. Gui, M. de Nivelle, R. de Vries, P. de Korte, M. Bruijn, J. Wijnbergen, W. Michalke, E. Steinbeiss, T. Heidenblut, B. Schwierzi, A high T_c superconductor bolometer on a silicon nitride membrane, MEMS 97 (Nagoya, Japan, January 26-30), 506-511

Sander C.S., J.W. Knutti, J.D. Meindl, "A monolithic capacitive pressure transducer with pulse-period output," *IEEE Trans. Electron. Devices* **ED-17** (1980) 927-930

Satchell D.W., and J.C. Greenwood, "A thermally excited silicon accelerometer," *Sensors and Actuators* **17** (1989) 241-245

Scheeper P.R., A.G.H. van der Donk, W. Olthuis, P. Bergveld, "A review of silicon microphones," *Sensors and Actuators A* **44** (1994) 1-11

Scheeper P.R., A.G.H. van der Donk, W. Olthuis, P. Bergveld, "Fabrication of silicon condenser microphones using single wafer technology," *IEEE J. Micromech. Syst.* **1** (1992) 147-154

Scheiter T., H. Kapels, K.-G. Oppermann, M. Steger, C. Hierold, W.M. Werner, H.-J. Timme, "Full integration of a pressure-sensor system into a standard BiCMOS process," *Sensors and Actuators A* **67** (1998) 211-214

Schellin R., G. Hess, "A silicon subminiature microphone based on piezoresistive polysilicon strain gauges," *Sensors and Actuators A* **32** (1992) 555-559

Schmidt M.A., "Wafer-to-wafer bonding for microstructure fabrication," *Proc. IEEE* **86** (1998) 1575-1585.

Schmidt M.A., R.T. Howe, S.D. Senturia, "A micromachined floating-element shear sensor," *Proc. 4th Int. Conf. Solid-State Sensors and Actuators (Transducers '87), Tokyo, Japan, June 2-5*, 1987, 383-386.

Schmidt, H.-J, F. Holtkamp, W. Beneke, "Flow measurement in micromachined orifices", *Proc. MME*, Ulvic, Norway 1998, 156-159

Schmidt, M.A., R.T. Howe, S.D. Senturia, "A micromachined floating element shear sensor", *Proc. Transducers*, Tokyo, Japan, 1987, 383-386

Schmidt, M.A., R.T. Howe, S.D. Senturia, J.H. Haritonidis, "Design and calibration of a micro-fabricated floating element shear stress sensor", *ITTT Trans. Electron Devices* **ED-35** (1988) 750-757

Schneider M., T. Müller, A. Häberli, M. Hornung, H. Baltes, "Integrated micromachined decoupled CMOS chip on chip," *IEEE Workshop Micro Electro Mechanical Systems (MEMS'97)*, Nagoya, Japan, 1997, 512-517

Schröpfer G., S. Ballandras, M. de Labachelerie, P. Blind, Y. Ansel, "Fabrication of a new highly-symmetrical, in-plane accelerometer structure by anisotropic etching of (100) silicon," *J. Micromech. and Microeng.* **7** (1997) 71-78

Seidel H., L. Csepregi, A. Heuberger, and H. Baumgärtel, *J. Electrochem. Soc.* **137** (1990) 3612

Seidel H., R. Reidel, R. Kolbeck, G. Muck, W. Kupke, M. Koniger, "Capacitive silicon accelerometer with highly symmetric design," *Sensors and Actuators* **A21/A23** (1990) 312-315

Selvakumar A., K. Najafi, "High-density comb array microactuators fabricated using a novel bulk/poly-silicon trench refill technology," in: *Tech. Dig. Solid-State Sensor and Actuator Workshop*, Hilton Head Island, SC, 1994, 138-141.

Senturia S.D., R.L. Smith, "Microsensor packaging and system partitioning," *Sensors and Actuators* **15** (1988) 221-234

Seo H., G. Lim, M. Esashi, "Hybrid-type capacitive pressure sensor," *Sensors and Materials* **4** (1993) 277-289

Shames I.H., and C.L. Dym, *Energy methods and finite element methods in structural mechanics*, McGraw-Hill, New York, 1985, chapter 7.

Shirley T.E., *Frequency-pulling effects in microfabricated resonant structures*, M.Sc. thesis, Massachusetts Institute of Technology, 1993

Slikkerveer, P.J., P.H.W. Swinkels and M.H.Zonneveld, *Philips Journal of Research*, Vol 50, No3/4 1996

Smith M.J.S., L. Bowman, J.D. Meindl, "Analysis, design and performance of a capacitive pressure sensor IC," *IEEE Trans. Biomed. Eng.* **BME-33** (1986) 163-174, and "Analysis, design and performance of micropower circuits for a capacitive pressure sensor IC," *IEEE J. Solid-State Circuits* **SC-21** (1986) 1045-1056

Smits J.G., H.A.C. Tilmans, K. Hoen, H. Mulder, J. van Vuuren, and G. Boom, "Resonant diaphragm pressure measurement system with ZnO on Si excitation," *Sensors and Actuators* **4** (1983) 565-571

Smits, G.J., "A piezoelectric micropump with three valves working perestaltically," *Sensors and Actuators* **15** (1990) 203 - 206

Sniegowski J.J., H. Guckel, T.R. Christenson, "Performance characteristics of second generation polysilicon resonating beam force transducers," *IEEE Solid-State Sensors Workshop, Hilton Head Island, SC, USA, June 4-7*, 1990, 9-12

Sparks D.R., S.R. Zarabadi, J.D. Johnson, Q. Jiang, M. Chia, O. Larsen, W. Higdon, P. Castillo-Borelley, "A CMOS integrated surface micromachined angular rate sensor: It's automotive applications," in: *Tech. Dig. 9th Int. Conf. Solid-State Sensors and Actuators (Transducers '97)*, Chicago, IL, 1997, 851-854

Spencer R.R., B.M. Fleischer, P.W. Barth, J.B. Angell, "The voltage controlled duty-cycle oscillator; basis for a new A-to-D conversion technique," *Rec. of the 3rd Int. Conf. On Solid-State Sensors and Actuators*, 1985, 49-52

Spiering V.L, S. Bouwstra and J.H.J. Fluitman, *Sensors & Actuators* A **37-38** (1993) 800

Spiering V.L., S. Bouwstra, J.F. Burger, M. Elwenspoek, "Membranes fabricated with a deep single corrugation for package stress reduction and residual stress relief," *J. Micromech. Microeng.* **3** (1993) 243-246

Stalnaker W.M., L.J. Spangler, G.S. Fehr, G. Fujimoto, "Plastic SMD package technology for accelerometers," in *Proc. 1997 Int. Symp. Microelectronics*, Philadelphia, PA, Oct. 1997, 197-202

Stanton, T.E., Marshall, D., Bryant, C.N., "On the conditions at the boundary of a fluid in turbulent motion", *Proc. Roy. Soc. London A*, **97** (1920) 413-434

Starr J.B., "Squeeze-film damping in solid-state accelerometers," *Proc. IEEE Solid-State Sensors and Actuators Workshop*, Hilton Head Island, June 4-7, 1990, 44-47

Steckenborn, A. Siemens AG, Berlin, personal communication (1998)

Stemme E., G. Stemme, "A balanced resonant pressure sensor," *Sensors and Actuators* **A21-A23** (1990) 336-341

Stemme E., G. Stemme, "A capacitively excited and detected resonant pressure sensor with temperature compensation," *Sensors and Actuators A* **32** (1992) 639-647

Stemme G., "Resonant silicon sensors," *J. Micromech. Microeng.* **1** (1991) 113-125

Stemme, G., " A monolithic gas flow sensors with polyimide as thermal insulator", *IEEE Trans. Electron Devices*, **ED-33** (1986) 1470-1474

Stengl R., T.Tan and U.Gösele, "A model for the silicon wafer bonding process", *Jpn. J. Appl. Phys.* **28** (1989) 1735

Stephan, C.H., M. Zanini, "A micromachined, silicon mass-air-flow sensor for automotive applications", *Proc. Transducers*, San Francisco, USA, 1991, 30-33

Sugiyama S., K. Kawahata, M. Abe, H. Funabashi, I. Igarashi, "High-resolution silicon pressure imager with CMOS processing circuits," *Transducers '87, Rec. of the 4th Int. Conf. on Solid-State Sensors and Actuators*, 1987, 444-447

Sugiyama S., K. Shimaoka, O. Tabata, "Surface-micromachined microdiaphragm pressure sensors," *Sensors and Materials* **4** (1993) 265-275

Sugiyama S., M. Takigawa, I. Igarashi, "Integrated piezoresistive pressure sensor with both voltage and frequency output," *Sensors and Actuators* **4** (1983) 113-120

Sugiyama S., T. Suzuki, K. Kawahata, K. Shimaoka, "Micro-diaphragm pressure sensor," *Rec. IEEE Int. Electron Devices Meeting*, 1986, 184-187

Sugiyama S., T. Suzuki, K. Kawahata, K. Shimaoka, M. Takigawa, I. Igarashi, "Micro-diaphragm pressure sensor," *Proc. IEDM, Los Angeles, California, USA, Dec. 7-10*, 1986, 184-187

Sun, Xi-Qing, Masuzawa and M. Fujino, Micro ultrasonic machining and self aligned multilayer maching/assembly technologies for 3D micromachines, Proc. MEMS'96, San Diego, Ca, USA February 11-15, 1996, 312-317

Suzuki S., S. Tuchitani, K. Sato, S. Ueno, Y. Yokota, M. Sato, M. Esashi, "Semiconductor capacitance-type accelerometer with PWM electrostatic servo technique," *Sensors and Actuators* **A21-A23** (1990) 316-319

Svedin, N., E. Stemme, G. Stemme, "A new bi-directional gas flow sensor based on lift force", *Proc. Transducers* 97, June 16-19, 1997, Chicago, USA, 145-148

Swart, N. A. Nathan, M. Shams, M. Parameswaran, "Numerical optimisation of flow-rate microsensors suing circuit simulation methods", *Proc. Transducers*, San Francisco, USA 1991, 26-29

Sze, S.M. (ed), Semiconductor Sensors, John Wiley & Sons, New York, 1994

Tabata, O., H. Inagaki and I. Igarashi, "Monolithic pressure-flow sensor", *IEEE Trans. Electron Devices* **ED 34** (1987) 2456-2462

Tabib-Azar, M., *Microactuators: electrical, magnetic, thermal, optical, mechanical, chemical & smart structures by 1998*, Kluwer Academic Publishers 1998,

Tang W.C., M.G. Lim, R.T. Howe, "Electrostatic comb drive levitation and control method," *J. Microelectromechanical Systems*, **1**, 1992

Tanigawa H., T. Ishihara, M. Hirata, K. Suzuki, "MOS integrated silicon pressure sensor," *IEEE Trans. Electron Devices* **ED-32** no. 7 (1985) 1191-1195

Tas, N.R. Elwenspoek, M.C. & Legtenberg, R. Side-wall spacers for stiction reduction in surface micromachined mechanisms. Proc. Micromechanics Europe 1996 (Barcelona, Spain, October 21, 1996) 92-95

Thornton K.E.B., D. Uttamchandani, and B. Culshaw, "A sensitive optically excited resonator pressure sensor," *Sensors and Actuators* **A24** (1990) 15-19

Tilmans H.A.C., M. Elwenspoek, "Quasi-monolithic planar load cells using built-in resonant strain gauges," *J. Micromech. Microeng.* **3** (1993) 193-197

Tilmans H.A.C., M. Elwenspoek, and J.H.J. Fluitman, "Micro resonant force gauges", *Sensors and Actuators A* **30** (1992) 35-53

Tilmans H.A.C., *Micromechanical Sensors Using Encapsulated Built-in Resonant Strain Gauges*, Ph.D. Thesis, University of Twente, The Netherlands (1993)

Tilmans H.A.C., S. Bouwstra, "A novel design of a highly sensitive low differential-pressure sensor using built-in resonant strain gauges," *J. Micromech. Microeng.* **3** (1993) 198-202

Tilmans H.A.C., S. Bouwstra, J.H.J. Fluitman, and S.L. Spence, "Design considerations for micromechanical sensors using encapsulated built-in resonant strain gauges," *Sensors and Actuators A* **25-27** (1991) 79-86

Timoshenko S. and S. Woinowsky-Krieger, "Theory of Plates and Shells", McGraw-Hill, New York 1959

Timoshenko S.P., D.H. Young, and W. Weaver, *Vibration problems in engineering*, John Wiley & Sons, 4th ed., 1974, chapter 5.

Tjerkstra R.W., M. de Boer, E. Berenschot, J.G.E. Gardeniers, A. van den Berg, M. Elwenspoek, "Etching technology for microchannels", *proceedings MEMS '97*, (1997) 147-152

Tjerkstra, R.W. M.J de Boer, J.W. Berenschot, J.G.E. Gardeniers, A. van den Berg, M. Elwenspoek, "Etching technology for chromatography microchannels" *Electrochimica Acta* **42** (1997) 3399-3406

Toth F.N., *A design methodology for low-cost, high-performance capacitive sensors*, Ph.D. thesis, Delft: Delft University Press, 1997.

Toth F.N., G.C.M. Meijer, "A low-cost, smart capacitive position sensor," *IEEE Trans. Instrum. Meas.* **41** 1992

Totta P.A., S. Khadpe, N.G. Koopman, T.C. Reiley, M.J. Sheaffer, "Chip-to-package interconnections," in *Microelectronics Packaging Handbook, Part II, Semiconductor Packaging*, R.R. Tummala, E.J. Rymaszewski, A.G. Klopfenstein (eds.), Chapman & Hall, 1997, Chapter 8.

Trautweiler S., *Silicon Hot Film Flow Sensors*, Ph.D. thesis, no. 12185, ETH Zurich, 1997.

Tudor M.J., M.V. Andres, K.W.H. Foulds, J.M. Naden, "Silicon resonator sensors: interrogation techniques and characteristics," *IEEE Proc.* **135** (1988) 364-368

Ueda T., F. Kohsaka, and E. Ogita, "Precision force transducers using mechanical resonators," *Measurement* **3** (1985) 89-94

Urban, G., A. Jachimowicz, F. Kohl, H. Kuttner, F. Olcaytug, P. Goiser, O. Prohaska: "High resolution thin-film temperature sensor arrays for medical applications", *Sensors and Actuators A* **22** (1990) 650 - 654

Van der Wiel, A.J., C. Linder and N.F. de Rooij, "A liquid velocity sensor based on the hot-wire principle", *Sensors and Actuators A* **37-38** (1993) 693-697

Van Heerwaarden, A.P., P.M. Sarro, "Thermal sensors based on the Seebeck effect", *Sensors and Actuators* **10** (1986) 321-346

Van Huffelen W.M., M.J. de Boer, T.M. Klapwijk, Appl. *Phys. Lett.* **58** (1991) 2438

Van Kuijk, J., T.S.J. Lammerink, H.-E. de Bree, M. Elwenspoek, J,H.J. Fluitman, "Multi parameter detection in fluid flows", *Sensors and Actuators A* **46-47** (1995) 369-372

Van Kuijk, J.C.C., "Numerical Modelling of Micro Mechanical Devices", Ph.D. Thesis, University of Twente, 1996

Van Lintel, H.T.C., F.C.M. van de Pol and S. Bouwstra, "A piezoelectric micropump based on micromachining of silicon", *Sensors and Actuators* **15** (1988) 153-167

Van Mullem C.J., F.R. Blom, J.H.J. Fluitman, and M. Elwenspoek "Piezoelectrically driven silicon beam force sensor," *Sensors and Actuators A* **25-27** (1991) 379-383

Van Oudhuisden, B.W., "Integrated flow friction sensor", *Sensors and Actuators* **15** (1988) 135-144

Van Oudhuisden, B.W., A.W. van Herwaarden, "High-sensitivity 2-D flow sensors with an etched thermal isolation structure", *Sensors and Actuators A* **21-23** (1990a) 425-430

Van Oudhuisden, B.W., J.H. Huysing, "An electronic wind meter based on a silicon flow sensor, *Sensors and Actuators A* **21-23** (1990b) 420-424

Van Putten, A.F.P., "An integrated silicon double bridge anemometer", *Sensors and Actuators* **4** (1983) 387-396

Van Putten, A.F.P., S. Middelhoek, "Integrated silicon anemometer", *Electron. Lett.* **10** (1974) 425-426

Van Riet, R.W.M., J.H. Huysing, "Integrated direction sensitive flow meter", *Electron. Lett.* **12** (1976) 647-648

Van Rijn C., M. van der Wekken, W. Nijdam and M. Elwenspoek, "Deflection and maximum load of microfiltration membrane sieves made with silicon micromachining," *Journal of microelectromechanical systems* **6** (1997) 48-54

Vittoz E.A., M.G.R. Degrauwe, S. Bitz, "High-performance crystal oscillator circuits: theory and application," *IEEE J. Solid-State Circuits* **SC-23** (1988) 774-783

Wachutka, G., R. Lenggenhager, D. Moser, H. Baltes, "Analytical 2D-model of CMOS micromachined gas flow sensors", *Proc. Transducers* 1991, 22-25

Wagner H.J., T. Fabula, and A. Prak, *UETP-MEMS Course: Resonant microsensors*, FSRM, Switzerland, 2nd edition, 1995.

Wallis G. and D.I. Pomerantz, "Field assisted glass-metal sealing", *J. Appl. Phys.* **40** (1969) 3946

Wang Y., M. Esashi, “The structures for electrostatic servo capacitive vacuum sensors,” *Sensors and Actuators A* **66** (1998) 213-217

Warkentin D.J., J.H. Haritonidis, M. Mehregany, S.A. Senturia, “A micromachined microphone with optical interference readout,” *Rec. of the 4th Int. Conf. Solid-State Sensors and Actuators (Transducers ’87)*, 1987, 291-294

Wensink H., M.J. de Boer, R.J. Wiegerink, A.F. Zwijze, M.C. elwenspoek, “First micromachined silicon load cell for loads up to 1000 kg,” *Proc. SPIE ’98*, vol. 3514 (Micromachined Devices and Components IV), 1998

White, F.W., *Fluid Mechanics*, McGraw-Hill, New York 1994

Wiegerink, R., R. Zwijze, G. Krijnen, T. Lammerink, M. Elwenspoek, “Quasi-monolithic silicon load cell for loads up to 1000 kg with insensitivity to non-homogeneous load distributions,” *Proc. MEMS ’99*, 1999

Winter, K. "An outline of the techniques available for the measurement of skin friction in turbulent boundary layers", *Progress in aerospace sciences* **18** (1977) 1-55

Yamada K., M. Nishihara, R. Kanzawa, R. Kobayashi, “A piezoresistive integrated pressure sensor,” *Sensors and Actuators* **4** (1983) 63-69

Yamagata ,Y., S. Mihara, N. Nishioki and T. Higuchi, “A new fabrication method for microactuators with piezoelectric thin film using precision cutting technique”, Proc. MEMS’96, San Diego, Ca, USA February 11-15, 1996, 307-311

Yang, C., H. Soeberg: "Monolithic flow sensor for measuring millilitre per minute liquid flow", *Sensors and Actuators* A **33** (1992) 143-153

Yazdi N., A. Salian, K. Najafi, “A high sensitivity capacitive microaccelerometer with a folded-electrode structure,” in *Tech. Dig. Twelfth IEEE Int. Conf. Micro Electro Mechanical Systems (MEMS’99)*, Orlando, FL, 1999, 600-605

Yazdi N., F. Ayazi, K. Najafi, “Micromachined inertial sensors,” *Proc. IEEE* **86** (1998) 1640-1659.

Yoon, E. K. Wise, “An integrated mass flow sensor with on-chip CMOS interface circuitry”, *IEEE Trans. Electron Devices* **ED-39** (1992) 1376-1386

Zook J.D., D.W. Burns, H. Guckel, J.J. Sniegowski, R.L. Engelstad, Z. Feng, “Resonant microbeam strain transducers,” *Proc. 6th Int. Conf. Solid-State Sensors and Actuators (Transducers ’91), San Fransisco, CA, USA*, 1991, 529-532

Index